RECENT ADVANCES IN THE PSYCHOLOGY OF LANGUAGE

Formal and Experimental Approaches

NATO CONFERENCE SERIES

I Ecology
II Systems Science
III Human Factors
IV Marine Sciences
V Air—Sea Interactions

III HUMAN FACTORS

RECENT ADVANCES IN THE PSYCHOLOGY OF LANGUAGE

Formal and Experimental Approaches

Edited by
Robin N. Campbell
and Philip T. Smith
University of Stirling, Scotland

Published in coordination with NATO Scientific Affairs Division

PLENUM PRESS · NEW YORK AND LONDON

Library of Congress Cataloging in Publication Data

Stirling Psychology of Language Conference, University of Stirling, 1976.
  Recent advances in the psychology of language.

  (NATO conference series: III, Human factors; v. 4a & b)
  "Published in coordination with NATO Scientific Affairs Division."
  "Proceedings of the . . . Stirling Psychology of Language Conference held at the
University of Stirling, Scotland, June 21–26, 1976, sponsored by the NATO Special
Program Panel on Human Factors."
  Includes index.
  CONTENTS: A. Language development and mother–child interaction. B. Formal
and Experimental Approaches.
  1. Psycholinguistics–Congresses. I. Campbell, Robin N. II. Smith, Philip T. III. North
Atlantic Treaty Organization. Division of Scientific Affairs. IV. Nato Special Program
Panel on Human Factors. V. Title. VI. Series.
BF455.S72   1976   401'.9   77-28734
ISBN 0-306-32884-4 (v. 4a)
ISBN 0-306-32885-2 (v. 4b)

Proceedings of the second half of the Stirling Psychology of Language
Conference held at the University of Stirling, Scotland, June 21–26,
1976, sponsored by the NATO Special Program Panel on Human Factors

© 1978 Plenum Press, New York
A Division of Plenum Publishing Corporation
227 West 17th Street, New York, N.Y. 10011

Printed in the United States of America

PREFACE

The Stirling Psychology of Language Conference was held in the
University of Stirling, 21-26 June 1976. 250 people attended the
conference and 70 papers were presented. The two volumes of Proceedings present a selection of papers from the conference reflecting as far as possible the range of topics that were discussed.

Volume 1 is concerned exclusively with language acquisition. In
recent years the 'centre of gravity' of acquisition research has
shifted from syntactic and phonological description to the amorphous domains of semantics and pragmatics. This shift is reflected
in the two large sections (II and III) devoted to these aspects of
language development. In addition the volume contains three smaller
sections dealing with general problems of acquisition theory, syntax
and the development of comprehension, and applied developmental
psycholinguistics.

Volume 2 contains a substantial section of papers which stress the
formal aspects of psycholinguistics: these include papers in which
artificial intelligence figures prominently, papers which apply recent developments in syntax and semantics to psycholinguistic problems, and papers that are broadly critical of the use psychologists
have made of linguistic theories. Volume 2 also contains a section
dealing with the experimental study of sentence comprehension and
production, and there is a final section concerned with phonology
and its development.

We wish to thank the NATO Scientific Affairs Division for substantial financial support; Jerry Bruner, John Lyons, Sinclair Rogers
and Colin Stork for help with the planning of the conference; John
Riddy and the University of Stirling for provision of the conference
facilities; Robert Coales, Barry Conlin, Liz Russ and Derry Yates
for help with administration, and Marion Capitani and Philippa
Guptara who typed the manuscript.

                                        Robin N. Campbell
                                        Philip Smith

                                        April 1977

CONTENTS OF VOLUME 4B

SECTION I: FORMAL APPROACHES

CONTENTS OF VOLUME 4A

LEARNING THE SYNTAX OF QUESTIONS

WILLIAM and TERESA LABOV

University of Pennsylvania

This paper is a first report on a longitudinal study of
the acquisition of a syntactic rule: the inversion of
WH- questions.[1] The general aim of this presentation
is to show how certain modifications in the theory and
practice of linguistics can contribute to the study of
acquisition, and to the psychology of language generally.
These new developments spring from an approach to linguistic analysis that is based on the observation of
speech in every-day contexts, and experiments carried
out in every-day situations. In this approach, intro-
spection may be taken as a handy guide to the issues,
but not as evidence for the developing theory.

These methods necessarily involve new techniques and a
new theoretical basis for approaching the variation
that is characteristic of language as it is used. Until
recently, variation theory has had little application
to acquisition, or indeed to the study of language generally, which has absorbed almost completely the set-
theoretical approach that we refer to as the "categorical view" (see section 4 below).[2] Yet the study of
acquisition is necessarily a study of linguistic change
and variation. It is hardly conceivable that one set of
rules or categories simply changes into another, and
it follows in both theory and practice that the view
of language as a discrete, invariant set of categories
cannot deal with change in any rational way. We will
argue, following Weinreich, that a model that presents
a child's language as a homogeneous object is needlessly

1

unrealistic and represents a backward step from structural theories that can accommodate the facts of orderly
heterogeneity (Weinreich, Labov and Herzog 1968:100).

Studies of acquisition so far have been based on two
kinds of data: sampling by recording at intervals,
typically one-half to two hours every two weeks; or
diaries in which a parent took care to note the first
occurrence of particular forms, gave illustrative anec-
dotes, and more rarely noted disappearances. The first
kind of data gives a clear view of the distribution of
forms during short periods, and could be used for quantitative studies. This is seldom done, since the analysts are primarily interested in describing a stage of
development within a homogeneous view of language, and
discount variations as remnants of previous stages or
anticipations of new ones. Transitions from stage to
stage are reconstructed by the same type of formal
arguments that are used in grammars based on the intro-
spections of the theorist. It is possible to use more
accountable procedures to analyse variation in these
recorded samples, but there is seldom enough data in
two hours of speech to display the development of syntactic rules. The diarists give us precise information
on transition points, but their qualitative records do
not follow the basic principle of accountability that
is needed to study variation: to evaluate the use of
each form against the envelope of all possible alternatives that are in use during that period.

This report is based on a more extensive type of data:
a near-total record of all the linguistic forms in the
speech of one child - in this case all the WH- questions
- from the very first occurrence to the acquisition of
the adult system. For the past four and a half years,
we have been engaged in a systematic study of the acquisition of language by our youngest child, Jessie.[3]
Given the sociologist's interest in the social context
of acquisition, we have kept detailed notes on that
context. Given the linguist's interest in quantitative
analysis, we have accumulated a unique record of everything that Jessie said - within hearing - up to 30
months, and a total record of selected features after
that.[4]

We have applied recent modifications of linguistic
theory to study both the acquisition of both syntax and
semantics, in order to describe and explain the way in
which Jessie's language has developed. We have already
had considerable interaction with psychologists interested

in category formation: our study of Jessie's first two
words, "The Grammar of Cat and Mama" (Labov & Labov
1974) utilised the approach to variability in denotation
that was first reported in W. Labov (1973). This involves serious modifications of the conception of "distinctive" and "redundant" features that has dominated semantic analysis, and fits in with recent empirical results
in psychology that also questions this framework.[5]
The present report will confront one of the most chal-
lenging and sharpest questions in the acquisition of
syntax: how are transformational rules learned?

1.   Some inconsistencies in the inversion of WH-questions

The following twenty questions were among those asked
by Jessie on July 16, 1975, at 3:10 - when she was
exactly 3 years,10 months old.

(1)   Where's the chickens?
(2)   Where's Philadelphia?
(3)   Where this comes from?

(4)   What's that?
(5)   What is this?
(6)   What do you do when you want to be rich?
(7)   What does sun do to snow?
(8)   What that means?
(9)   What NOT starts with?
(10)   What the address is?

(11)   How do babies get inside the mommies?
(12)   But how them buy their tents?

(13)   Why are we going down?
(14)   Why you said to Daddy you might be kidding?
(15)   Why we can't wear sandals for walking in the wood?
(16)   Why when we don't know what it is we have to press
       it?
(17)   Why when a child grows up there's no daddy?

(18)   Does you have it or you?
(19)   Is peaches bigger than apricots?
(20)   Is there some on top of the cars?

It is evident that Jessie showed variation in the degree
of acquisition of the adult rules for WH- questions,
which are based on two transformations upon declarative
forms:

(T1) WH- FRONTING

$$X \quad - \quad WH\text{-}NP \quad - \quad Y$$
$$1 \qquad\qquad 2 \qquad\qquad 3 \rightarrow 2 \quad 1 \quad 3$$

(T2) INVERSION

$$\left\{ \begin{matrix} Q \\ WH\text{-}NP \end{matrix} \right\} - NP - Tense - \left\{ \begin{matrix} \emptyset \\ Modal \\ have \\ be \end{matrix} \right\} - (NEG) - (Verb) - X$$
$$\quad\; 1 \qquad\; 2 \qquad 3 \qquad\quad 4 \qquad\qquad 5 \qquad\; 6 \qquad\;\; 7$$
$$\rightarrow 1 \quad 3 \quad 4 \quad 5 \quad 2 \quad 6 \quad 7$$

The formalism used in generative grammar should not be
allowed to obscure the close connections of these rules
to the process of questioning and the relationship of
sentences used in every-day life. X and Y are dummy
symbols for unanalysed portions of the string that can
be used to include any number of words or be null. The
analysis indicated in (T1) simply locates an item to be
questioned, with the admittedly abstract notion of WH-
attached to a noun phrase that ultimately appears in
forms that register certain semantic features of the
item questioned. Thus for (T1), the question (2) is
analysed

(21) Philadelphia is - [WH-locative noun phrase]?
            X   WH-NP
            1   2

into a first and second part as shown (3 is null here)
and, as the arrow directs, the second part is moved to
the front of the sentence:

(22) [WH-locative noun phrase] - Philadelphia is?
            WH-NP   X
              2   1

We can re-state the rule in ordinary language as: put
the word that is questioned at the front of the question.

The output of (T1) is then subject to the more complex
(T2), which states that when a question word stands at
the beginning of the sentence, the tense marker is
attracted to that word; in doing so, it must move around
the subject noun phrase.[6] The concept of "tense marker"
is of course quite abstract; it may appear as the difference between is and was, or as a zero form in should.
But the speaker must locate something equivalent to

that tense marker in order to know what element to bring
to the front of the sentence and how to carry out inversion in a way that matches the adult grammar. If there
is no other member of the auxiliary, the tense marker
will appear as the difference between present do and
past did; if it is the former, with number agreement as
in (6) and (7). If the next word is a modal (can,
should), or the auxiliary be, or the verb be, this
also will be attracted to the WH- word, along with the
tense marker and (optionally) a negative if it is present. Thus (T2) requires the following more abstract
analysis of (22):

(23) $[$WH-loc. NP$]$ - $[$Philadelphia$]_{NP}$ - Tense - be
            1   2   3   4

This is then transformed into

(24) $[$WH-loc. NP$]$ - Tense - be - $[$Philadelphia$]_{NP}$
            1   3   4   2

Other rules of grammar realise WH-locative NP as where,
combine the tense marker with the following be to yield
is, and contract is and where to give (2) Where's
Philadelphia?

Within the framework of generative grammar, (T1) and
(T2) are projected as formal statements of the minimally
complex set of relationships that must be grasped in
order to produce the adult pattern. Without joining
in a search for other formalisms, our task is to decide
when and if Jessie's speech during this two and a half
year period shows evidence of syntactic relations that
can be registered in statements of this generality.

After a first glance at Jessie's twenty questions, it
seems as if something equivalent to (T1) has been learned completely, and that (T2) has been learned for yes-
no questions, but not for WH- questions. So far, this
fits in with regularities observed in other studies of
acquisition (Brown, 1973; Ravem 1974; Labov, Cohen,
Robins and Lewis, 1968: 291-300). First there is a general implicational relation that inversion is learned
earlier in yes-no questions than in WH- questions. It
does not seem hard to advance functional arguments on
why this should be so: in yes-no questions, inversion
is the only signal of a question, while in WH- questions
it is redundant. There is no doubt that What does that
mean and What that means are two alternative ways of
"saying the same thing." Yet there is still something
odd here.

In the first place, we find that Jessie's WH- questions are among the first sentences that she used, and appear - with inversion - before yes-no questions. At 2:3:28 she asked

(25) What's that?

and produced five other examples before her first yes-no question, nineteen days later

(26) Are you work?

Secondly, there seems to be no evidence of failure to apply (T1), either in Jessie's data or in previous studies of WH- questions (Ravem 1974). We don't find early sentences of the form

(27) He can't do it why?

This is not so easy to understand, and some feel uncertain about transformational analyses as a consequence (Ravem, 1974). Yet sentences such as (3), (8), (9) or (10) cannot be understood very easily without reference to (T1). The child does not hear such sentences from adults,[8] and may be producing them by perceiving a relationship between (9) What NOT starts with? and

(9') NOT starts with the letter N.

We must then account for two puzzling questions; if Jessie learned to use inversion in WH- questions first, why did she show such inconsistency 18 months later, long after she acquired the consistent use of the adult forms for yes-no questions? And why did she, like other children, show perfect control over (T1) but variable use of (T2)?

As a first step toward answering these questions, we will look at the overall frequency for WH- questions. This proved a valuable strategy in the study of Jessie's first words, where we found that a burst of high frequency in the use of a form preceded advances in semantic and syntactic structures.

2. The frequency of WH- Questions and Inversions of
    them.

For about fourteen months after Jessie's first WH-questions, she showed only sporadic uses of this syntactic form, less than one a day. In the first three months of 1975 (3:4-3:8), there was a small but significant increase to two or three a day. Then there was a sudden increase to an average of 30 questions a day in

April, May and June, and another sudden jump to an
average of 79 questions a day in July, with a peak of
115 questions on July 16th, the day from which we drew
our sample of twenty questions.

Figure 1 shows this progression of the average number
of WH- questions per day for each month. After the peak
in mid-July, the average frequency fell off slowly over
the next two months (to 4:0), then fell more sharply
through October and December to a stable plateau of 14-
18 per day for the next seven months.

The data come close to a total record of Jessie's WH-
questions. Each question was recorded in writing within minutes after Jessie said it, usually by Teresa
Labov.[9] Of course they do not include some questions
Jessie asked herself or other people.[10] Figure 1 represents a sample of one sixth of the approximately
25,000 questions, and all analyses to follow are based
on this sample of 3,368.[11]

The notion that children learn syntactic regularities
with great ease and rapidity does not find support in
these data. We are dealing with a period of two and a
half years, and an enormous volume of questions. The
problem then is not to ask, how does the child learn
this syntactic rule so quickly, but rather: why does
it take so much time and so many trials?

Figure 1 also shows the percentage of inversion in WH-
questions (the dotted line) for the second half of the
two-and-a-half year period. It begins at a low 9%, and
rises steadily with some fluctuations to 76% at the end
of this period. For the first half, the parallel between the percent inversion and frequency is startling.
The sharp increase of frequency in July is accompanied
by a jump of twenty percent in inversion, and the peak
frequency occurs just before inversion passes fifty
percent; then while inversion continues to climb, frequency declines.

The over-all pattern of Figure 1 is strikingly parallel
to the relation between frequency and semantic development of Jessie's first words. We suspect that this pattern will become a familiar one as additional detailed
longitudinal studies are carried out. There is an intimate interrelation of cause and effect here: the use of
more questions is accompanied by a wider range of syntactic forms, and allows more practice in the operation

of the rules. One cause of high frequency may very
well be interest in the syntactic process itself - the
formation of sentences. As the rule becomes stabilised,
the practice effect appears to diminish. However, it
must also be born in mind that the frequency pattern is
the product of many other factors.[12]

However we view the regularities of Figure 1, they fall
outside of the traditional linguistic theory that has
been referred to as the categorical view. Within that
framework, the rule system that Jessie is trying to
learn is said to have rules that are either

(a) obligatory or (b) optional.

The adult inversion rule would then have to be characterised as optional, or as an obligatory rule with
varying numbers of "performance errors". Neither of
these formulations helps to answer questions that must
be considered fundamental to the theory of acquisition.
How does an optional rule become obligatory? What was
the state of Jessie's knowledge of the inversion rule
during that period?

A first approach to these questions is analytic. It is
possible that the variation in Jessie's use of inversion
was the result of a mixture of invariant components:
that she always applied the rule in some environments
and never in others. To explore this possibility we
will want to examine the major environments in which
inversion takes place. The first and most obvious
variable is the item that is questioned, depending on
whether Jessie wanted to know about a person, an object,
a place, a location, a time, a manner or a reason,
reflected in the WH- forms, who, what, when, where, why
and how.

3. Differentiation of the WH- words.

The examples of (1)-(17) make it seem as if Jessie
showed the same degree of variation in the inversion of
all her WH- forms, but even a casual inspection of the
data will show that this is not the case. There are
sizeable differences among the WH- words. Figure 2
shows the monthly percentages of inversion for the four
most common: how, where, what and why, for the same
span as Figure 1, 3:3 to 4:9. An inspection of this
figure suggests a division into four periods.

Period (a):   January to March 1975. During this initial
low-frequency period there was no significant difference
between the WH- forms except for how, which is generally
separated from the others by a higher frequency of inversion.

Period (b):   April to September 1975. From 3:6 to 4:0,
Jessie showed a regular stratification of the four forms
in the order: how, where, what,why, but with why decli-
ning to a minimum in June while the others rose.

Period (c):   September 1975 to February 1976. In the
next six months 4:0 to 4:6, Jessie seemed to have formed
a stable rule of inverting after how, where and what
but not after why, which remained stable at about 10-
15% inversion.

Period (d):   March to June 1976. In this period, Jessie
showed a rapid increase in inversion after why, approaching the adult rule of obligatory inversion.

The third period is the most remarkable section of
Figure 2 in that here it seems clear that Jessie has
formed a special rule of her own:

(28) Invert after all WH- words except why.

There is no doubt that for such a rule, and it is perhaps the most dramatic illustration that we have seen
of the language acquisition device as a rule-forming
mechanism. Not only did Jessie form this rule over
the course of period (b), but maintained it consistently
over the six months of period (c).

The maintenance of rule (28) is all the more striking
when we consider that Jessie did know about and could
imitate the adult rule. On July 14th, near the peak
frequency of WH- questions when inversion after what
had reached 50%, we asked Jessie, "Which is a more
grown-up way to say it, What is this? or What this is?"
She thought for a moment and said, "What is this?"
We were careful not to repeat this experiment, to avoid
teaching the very phenomenon we were studying; but five
days later, we initiated an experiment - the question
game - that gave us more data on Jessie's ability to
control the adult form of inversion.

In the question game, one parent first asked a question,
usually a why form. It was then Jessie's turn to ask
a question. She always followed suit with the same WH-

form, but showed independent control of inversion as in
her first four turns:

Parent form   Jessie question

(29) Why do. . .   Why does, why you wearing hair?

(30) Why does. .   Why do you have sneakers?

(31) Why do. . .   Why do you always wear watch?

(32) Why is. . .   Why are you wearing sunglasses?

Except for the first question, Jessie's turns show
appropriate do-support, tense and number agreement that
could not have been the result of direct imitation.
Her over-all performance in the question game throughout
July and August showed 22 out of 26 inversions after
why: her success outside of the game amounted to only
13 out of 292 or 4%.

It should not be surprising that Jessie had two different styles of speech. Shatz and Gelman have shown that
four-year-olds have different styles for talking with
grown-ups, with two-year-olds, and with their peers
(1973). There are also other special conditions on
inversion connected with meta-linguistic situations.
Questions of the form (8) What that means?occurred
without inversion more often than other what questions.
In this same July-August period, only 4 out of 26 what
questions on meaning were inverted, as opposed to 66%
generally. We have therefore excluded all meta-linguistic questions - on orthography, naming and meaning -
from the percentages of Figure 2, since they alter the
proportions of each form, raising why and lowering what,
in ways that obscure the operation of internal constraints.

The disfavouring of inversion by why was noted by Ravem
(1974) in his study of the acquisition of English by
his children, native speakers of Norwegian and in Brown's
data. Ravem suggested a semantic explanation: that the
concept why is difficult to acquire and that this fact
somehow affects the acquisition of inversion. We will
see below that this is an unprofitable, and unnecessar-
ily speculative solution, at least as far as Jessie's
evidence is concerned.

It will appear below that Jessie's underlying rule system is even more regular and systematic than is suggested by the simple rule (28): invert except after why.
But to capture even this first and most obvious general-

isation about Jessie's system within the traditional
categorical view, it would be necessary to throw away
a considerable amount of data. It would be easy enough
to say that throughout period (c) the small percentages
of non-inverted how, where and what were performance
errors, but it would be less palatable to say that the
5-10% of inversions after why were also performance
errors. Without this strategy, the traditional analysis
is forced to return to the first unenlightening statement that (T2) was optional, with inverted and non-
inverted forms in free variation, for two and a half
years.

To see why this is so, it will be necessary to say a
few more things about the traditional "categorical"
view, and then see what can be done to construct a more
direct approach to the study of variation in Jessie's
WH- questions.

4. The systematic study of variation.

The categorical view. At a number of points in this
discussion, we have indicated that the point of view
which has dominated linguistic thinking until recently
was not well suited to the study of variable properties
of speech production. The conception of language structure projected was a set of discrete, qualitatively
distinct, invariant categories. Data that showed variation was set aside as irrelevant to linguistic theory
("data flux"), contaminated by the influence of competing
systems ("dialect mixture"), or the result of imperfec-
tions in perception and memory ("performance errors").
The proper object of linguistic study was said to be an
ideal construction of a homogeneous speech community
in which all speakers learned the language perfectly
and instantly (Chomsky 1965:3).

This rejection of quantitative studies may appear strange
to workers in other disciplines who are intimately
involved with the correlation of variables and the analysis of variance. But it should be noted that the categorical view was a good first approximation to linguistic structure, which can be conceived as a way of locating invariance in the continuous substrata of phonetic
and semantic substance.

This search for invariance begins with the fundamental
postulate of linguistics enunciated by Bloomfield in
1926: that some utterances are the same (1926: 26-7).
But the fundamental fact of phonetics, attested in

almost every introductory text, is that no two utterances are exactly alike. This contradiction is resolved
by what we may call the fundamental corollary of linguistics: that some differences don't make a difference.
In other words, certain differences between alternative
ways of saying the same thing are in free variation.
The existence of free variation is the necessary correlate of the fundamental linguistic postulate of invariant structure.

Operating strictly within this view, nothing linguistically significant can be said about whether one or the
other of two free variants occurs: that variation can
not be constrained in any way by linguistic rules. It
follows that no linguistic rule can express relations
of more or less.[13] But the notion that variation can
not be constrained mistakenly assumes that linguistic
structure is limited by the notion of referential
"same." For historical, social and psychological
reasons, there are many structural relations between
alternative ways of saying "the same thing" that do not
convey referential information, yet are preserved over
many generations as stable aspects of linguistic structure. Though many such constraints are stable and uniform throughout the speech community, some vary in
strength in ways that depend on the grammatical knowledge of the speaker. The deletion of the final /t/ or
/d/ in past, old, missed or fined is heavily constrained
if that consonant is a separate past tense signal, and
this condition regularly becomes stronger with age
(W. Labov, 1972; Summerlin, 1972).

Linguistic variables and variable rules. A systematic
approach to variation requires that we recognise and
define the existence of variants that are alternative
ways of "saying the same thing".[14] These variants are
then grouped in mutually exclusive sets of possible
alternatives. We are then able to follow a principle
of accountability: instead of merely reporting the
occurrence of a variant, we report the number of times
it occurred along with the total number of occurrences
of the environment in which this variant can appear.

In the course of the investigation, the description of
the environment that defines the variable is continually
refined. Invariant conditions are recognised and set
aside, along with environments where the variable can
not be identified clearly. In the study of inversion,
for example, one must set aside cases of zero copula,

as in <u>Why he gonna go?</u>, since it is impossible to tell
if inversion had occurred before deletion. As the process continues, one may have to redefine at what level
the variation occurs: what seemed like phonetic variation may turn out to be the result of variation at a
higher level, or vice versa. For example, in an
earlier study, the deletion of the copula was originally
conceived as a grammatical variable, and later turned
out to be a phonological process of deleting a single
consonant (W. Labov, 1969). Variation within a rule
may be resolved into variation between competing rules
or competing derivations.

The study of variation is primarily concerned with the
analysis of variable constraints, features of the environment that favour or disfavour the occurrence of one
or the other variant. Within the generative framework
that is currently utilised for this purpose, variation
appears as a directional <u>variable rule</u>; in the case of
phonological variation, as a restricted re-write rule
of the form ^{15}

(33) $A \rightarrow \langle B \rangle \ / \ C \ \_\_ \ \langle D \rangle$

Here the angled brackets around the output of the rule
indicate that the process is variable, and angled brackets around an element of the environment sign "/"
indicate variable constraints that favour the rule. We
may restate (33) as: "A is variably realised as B when
it is preceded by C (an invariant condition) and more
often when it is followed by D than when it is not (a
variable constraint)."
For syntactic processes like Jessie's WH- inversion,
the variable constraints are entered in the structural
analysis of a transformation. Thus we can re-write (T2)
to include the data of Figure 3 during period (c):

$$
(T2') \quad
\left\{ \begin{matrix} Q \\ \text{WH-NP} \end{matrix} \right\}
\left\langle \begin{matrix} \text{+Manner} \\ \text{+Locative} \\ \text{+Concrete} \\ \text{+Temporal} \\ \text{+Reason} \end{matrix} \right\rangle
- NP - Tense -
\left\{ \begin{matrix} \emptyset \\ \text{Modal} \\ \text{have} \\ \text{be} \end{matrix} \right\}
- (NEG) - (Verb) - X
$$

$$
\begin{matrix} 1 & 2 & 3 & 4 & 5 & 6 & 7 \end{matrix}
$$

$$
\rightarrow 1 \ \langle 3 \ 4 \ 5 \rangle \ 2 \ 6 \ 7
$$

Here parentheses continue to indicate optionality, and
the angled brackets in the output indicate the variable

character of the transformations: that sometimes it
applies and sometimes it does not. Angled brackets in
the structural analysis indicate that five syntactic
features of WH-NP are variable constraints in the
environment of the rule[16] and the vertical ordering
indicates their relative strength, so that the rule
applies most often if the WH- form is how, next most
often if it is where, and so on.

A variable rule is therefore a rule of grammar with a
variable output and a statement that at least one factor
in the environment influences that output. By incorporating variable rules into the current framework of
generative grammar, we recognise the existence of variation as a stable and reproducible aspect of linguistic
structure. Equivalent modifications can be made within
other formal frameworks. The fundamental change is an
alteration in the relation between the theory and the
data, so that the theory can utilise as evidence the
language that is used in the speech community and the
speech production of children, our primary data for the
acquisition of grammar.

By constraining free variation, we also constrict the
number of possible relations that can exist within optional rules, and so define more closely the concept of
the grammatical system. A further consequence is the
strengthening of the evidence that can be used to prove
or disprove a given hypothesis about the grammar. Within
the traditional view, the data was restricted so that the
major theoretical activity was conceived as a selection
among alternative formalisms on the basis of an internal "simplicity metric": on the whole, these solutions
have not proved stable or convincing. By contrast, the
comparison of variable constraints offers us an unlimited
number of possibilities for investigating rule systems
to see if they are the same or different on the basis
of the data itself, and for tracing change and development within an ordered series of rule systems. We will
follow this strategy in approaching the questions about
Jessie's inversion rules that were raised in section 2.

The further exploration of variable constraints quickly
encounters a multi-variate situation where the arithmetic methods of Figure 3 are inadequate to analyse the
many intersecting relations. At this point, one can
turn to the methods developed by H. Cedergren and D.
Sankoff in 1972, who provided a program for calculating
the contributions of each variable constraint to the

over-all probability of a rule operating (Cedergren and
Sankoff, 1974). Other methods of multi-variate analysis
were not suitable for internal linguistic relations
because of two special characteristics of linguistic
rules:

(1)   the inevitability of small and empty cells in the
      matrix: for example, the negative occurs primarily
      with <u>why</u> questions in Jessie's data, not with <u>how</u>,
      <u>when</u>, <u>what</u> or <u>where</u>.

(2)   the linguistic constraints are typically indepen-
      dent; whereas we expect interaction with social
      variables, internal linguistic constraints typi-
      cally show no interaction.

The Cedergren-Sankoff program takes these characteristics into account by operating directly upon irregularly
filled cells, preserving the information on the quantity
in each and giving larger cells more weight than smaller
ones. Secondly, it assumes independence of the factors
and immediately proceeds to test that assumption. The
following example from the inversion data will illustrate this process.

Let us consider two groups of mutually exclusive variable
constraints on inversion: three WH- forms will make up
one group of factors, and the presence or absence of
auxiliary contraction the second. The data is submitted
to the program as the number of sentences with inversion and the total number of cases for each combination
of factors represented by one or more tokens. In this
case, all cells are full, but some are much larger than
others.

No. of Inversions	Total	Group 1 WH-form	Group 2 Contraction	Percent Inversion
14	38	where	no	36
16	20	where	yes	80
50	242	what	no	20
28	36	what	yes	77
13	209	why	no	6
1	26	why	yes	3
——	——			——
122	571			21

The program produces a probability p_i for each factor
in each group G_i plus an input probability p_0. After
15 iterations of the maximum likelihood program we obtain:

p_0		p_1		p_2	
Input	.66	Where	.72	Contraction	.67
		What	.50	No contraction	.33
		Why	.28		

In this simple case, the program has given us the relations that we might have inferred from the original percentages above, in a somewhat more exact form: where
favours inversion, what has no effect, and why disfavours
it; it is also favoured by contraction. The assumption
of independence is reflected in the fact that each factor in p_1 is associated with a single probability which
applies no matter what value is selected for p_2. We can
now proceed to test that assumption by combining these
independent probabilities to get an expected value for
each combination, and then comparing this with our
original observations. In the model of the variable
rule program we are using here, the over-all probability
P is related to individual probabilities by (34):

$$(34) \quad \frac{P}{1-P} = \frac{(p_0)\,(p_1)\,(p_2)\,\ldots\,(p_n)}{(1-p_0)(1-p_1)(1-p_2)\ldots(1-p_n)}$$

A factor with a probability of .5 contributes nothing
to the application of the rule, since $p_i/(1-p_i)$ =
.5/(1-.5) = 1. Entering our values into (34) and solving for P, we get the following expected values for
applications of the rule, compared below with the observed values:

WH-Form	Contraction	Observed Inversions	Expected Inversions	x^2
where	no	14	13.91	0.00
where	yes	16	16.14	0.00
what	no	50	54.01	0.39
what	yes	28	20.04	7.13
why	no	13	9.15	1.68
why	yes	1	·8.69	10.22
				19.41

The fit is quite good for where, but contracted what
shows a sizeable gap between expectation and observation,
and why is very much off, contributing the most to the
total chi-square. This confirms what we might have
found from inspection of the percentages in this simple
case: there is an interaction between contraction

and WH-form, since contraction in <u>where</u> and <u>what</u> heavily
favours inversion, but disfavours it in <u>why</u>.

The program has thus demonstrated that (T2') will not
do as a representation of inversion for the data on hand.
Since <u>why</u> contributes most to X^2 the next step is to
segregate <u>why</u> from <u>what</u> and <u>where</u>. When we do so, and
re-run the program, we find the following probabilities:

	p_0		p_1			p_2
Input	.66	Where	.60	Contraction		.70
		What	.40	No contraction		.30

and the following fit of expectation and observation:

WH-Form	Contraction	Observed Inversions	Expected Inversions	X^2
where	no	14	13.10	0.10
where	yes	16	18.08	2.48
what	no	50	52.64	0.17
what	yes	28	24.02	1.98
				4.72

Clearly the fit is much better: the original X^2 of 7.13
for contracted <u>what</u> has dropped to 1.98, and the ratio
of X^2 per cell from 3.23 to 1.18. We can provisionally
conclude that <u>where</u> and <u>what</u> are here governed by the
same rule. A separate rule for <u>why</u> is needed: what
kind of rule it would be remains to be seen.

The chi-square figure is not used here as a confidence
measure but as a comparative measure of fit, to show
at various stages of development how components of a
rule can be fitted together with greater or lesser
success, and to locate the specific areas where lack of
independence interferes with such a fit. In the example
just given, the fit is still not particularly good,
since there are obviously other prominent sources of
variation that are not being taken into account. As we
include other factors in the analysis, chi-square per
cell should drop well below 1.00.

The variable rule program is not a pre-theoretical
analysis that converts raw data into factors. It can
only operate on an initial analysis that sets up factors
and factor groups. This analysis must in turn proceed
from a theory of language structure that would predict
what the most likely constraints on such a rule would

be. This is of course applies equally to the first
steps in arithmetic analysis as well as to the more
refined testing of hypotheses that we can carry out with
the variable rule program.

We have already seen that the WH- forms have a powerful
influence on inversion: there is no prior theoretical
expectation for such an effect, and the reasons for it
will be explored in following sections. Beyond this,
we must consider the nature of movement transformations,
The form of (T2) and (T2') reflects the view that
syntactic relations involve two distinct processes:
constituent analysis and constituent movement. Our
approach to locating variable constraints shows the
same conceptual approach. We would look for:

(1) Factors influencing the size and complexity of the
constituents that change position. This is the most
obvious strategy: we would naturally examine the realisations of items 3,4, and 5 in rule (T2) to see what
has to be moved: a tense marker alone, a modal, or in
addition, a negative. One might expect that a lone
tense marker would be more difficult to work with since
it requires an additional operation of do support. At
the other extreme, combining a negative with a modal
and a tense marker requires more complex adjustments.
In this area, we would also be on the alert for complex
noun phrases in subject position, since this complicates
the task of locating the tense marker.

(2) Factors influencing the analytical separation of
the items to be moved. Contraction is one of the
major factors that interfere with the child's analysis:
the tendency of the auxiliary to fuse with the subject
as a fixed form would make (T2) more difficult to carry
out. Since we already know that contraction occurs
more often with pronouns than with other noun phrase
subjects, and especially with those ending in vowels
(Brown 1973:339; W. Labov, 1969), we would have to
distinguish the various pronoun forms as well as the
forms of the verb or auxiliary for contractability.
Larger problems of rule ordering fall into this category, such as those created by the foregrounding of
whole sentences as in (16) and (17).

(3) Stylistic factors that influerce the relation of
the speaker's grammar to the grammars of higher and
lower status people: typically parents on the one hand
and babies on the other. We have already mentioned a
number of such factors in section 3.

These are the considerations that led us to select 41
factors that might account for variation in Jessie's
inversion. These factors are not entered into the rule
along the lines just given, since their syntactic locations and their possible effects on inversion are quite
distinct. In order to set up mutually exclusive, exhaustive sets we must organise them into eight groups
according to their substitutability in the surface
structure. The eight groups are listed below, following
the order of their occurrence in the string analysed by
(T2).

Group 1: the eight WH-forms: <u>who</u>, <u>whose</u>, <u>what</u>, <u>which</u>,
          <u>where</u>, <u>when</u>, <u>how</u> and <u>why</u>.

Group 2: the various types of subjects: different pro-
          nouns[17], other singular and plural noun phra-
          ses.

Group 3: the auxiliary and verb forms: modals, <u>can</u>,
          <u>will</u>, and <u>should</u> (the only ones used); <u>is</u> and
          <u>are</u> as forms of auxiliary and main verb <u>be</u>[18];
          main verbs without lexical auxiliary.

Group 4: the complement: a noun phrase alone, as in
          (1), (2), (4), (5); sentential complements,
          as in (6), (14); foreground clauses, as in
          (16).

Group 5: whether or not the verb or auxiliary is con-
          tracted with the preceding element.

Group 6: tense and aspect: the choice of past, progres-
          sive, <u>gonna</u>, past progressive, or the residual
          case of the general present.

Group 7: the presence or absence of negation.

Group 8: meta-linguistic functions: the question game,
          questions on meaning, orthography and naming
          (specifically, "what is this called").

5.   The variable analysis of WH- inversions.

The 41 factors of our full analysis are more than we
can cope with or explain in this first report: they are
the resources from which we can draw explanations. The
3,368 questions of our first sample also represent a
resource that we can draw upon and sub-divide in various
ways. If we analyse all of these questions with all of
the factors, we will not expect to obtain a particularly
good fit, since we have already recognised the presence
of change and interaction among the WH- forms. We can

use such an over-all analysis to sum up constraints that
we have found to operate consistently throughout each
period.

Table 1 shows the variable rule output for this complete
analysis.[19] The fit is no better than we would expect:
an average chi-square of 2.17 per cell. But with this
volume of data the program converged quite swiftly (9
iterations), and we can be confident that those relations that are consistently represented within sub-
sections of the data[20] are stable and reliable.

Meta-linguistic effects (Group 8). The strongest factor
in favour of inversion in any group is the question
game at .95 (primarily a constraint operating on why).
Almost equally powerful in the opposite direction are
questions on meaning ("What that means?") at .14. Questions on spelling have a moderate effect in disfavouring
inversion at .35. We can classify all of these as special routines that became fixed in their original form.

Negation (Group 7). There are 194 negative questions,
and only 11% show inversion. These are heavily concentrated in the why category (all but 8) but the negative is still lower with these than with the bulk of
why questions (15%). The program accordingly shows a
moderately negative effect of the negative at .42.

Tense and aspect (Group 6). For reasons we do not yet
understand, the past tense consistently favours inversion (.69). It may be that the use of the past was
developed at about the same time as the inversion rule,
and so was coupled with it. On the other hand, there
are two clear reasons why the progressive (.40) and
gonna (.36) should disfavour the rule. First, because
the redundant auxiliary is often omitted with the -ing
suffix (and even more often with gonna); and second,
because an additional act of analysis is required in
selecting and moving the auxiliary. The same reasoning
applies to the modal should (.33).

Foreground clauses (Group 4). Sentences such as (17)
Why when a child grows up there's no daddy? strongly
disfavour inversion (.20). The complications here are
manifold and involve almost all of the problems that
arise under category (2) above.

The WH- forms. We can now return to the elements that
we know are associated with change and development in
Jessie's WH- forms. The pattern of the more common

Table 1

	N	% inversion	pi
Group 1: WH- form			
whose	17	94	.81
how	248	89	.78
which	59	88	.67
who	16	87	.55
where	364	78	.53
what	1294	66	.41
when	57	56	.38
why	921	15	.05
Group 2: subject			
it	226	60	.63
that	287	70	.62
singular NP (-pro)	724	63	.57
you	595	48	.56
we	407	42	.48
plural NP (-pro)	175	60	.48
they	126	44	.44
this, these, those	181	59	.44
he, she	146	37	.43
I	109	33	.34
Group 3: verb form			
will	18	72	.82
are	340	65	.57
can	150	35	.47
main verb (-COP,-Modal)	1172	48	.40
is, am	1259	59	.36
should	37	40	.33
Group 4: complement			
sentence complement	161	60	.72
NP only	728	76	.58
foregrounded sentence	20	10	.20
other	2055	46	.48
Group 5: contraction			
auxiliary contracted	393	64	.57
no contraction	2583	52	.43
Group 6: tense and aspect			
past	246	59	.69
past progressive	29	51	.55
general present	2312	55	.50
present progressive	254	41	.40
gonna	135	44	.36
Group 7: negation			
negative	194	11	.42
positive	2782	57	.58
Group 8: meta-linguistic			
question game	40	77	.95
naming	113	66	.45
spelling	94	67	.36
meaning	73	30	.14

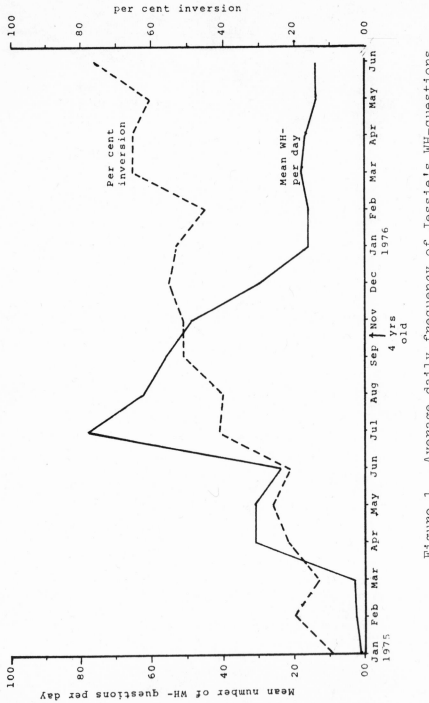

<u>Figure 1.</u> Average daily frequency of Jessie's WH-questions and percent inversion by month.

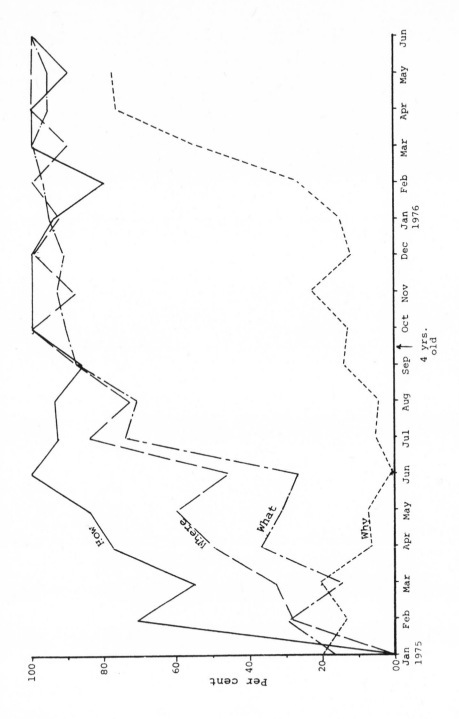

Figure 2.   Development of inversion in Jessie's WH-ques-
            tions: percent by month.

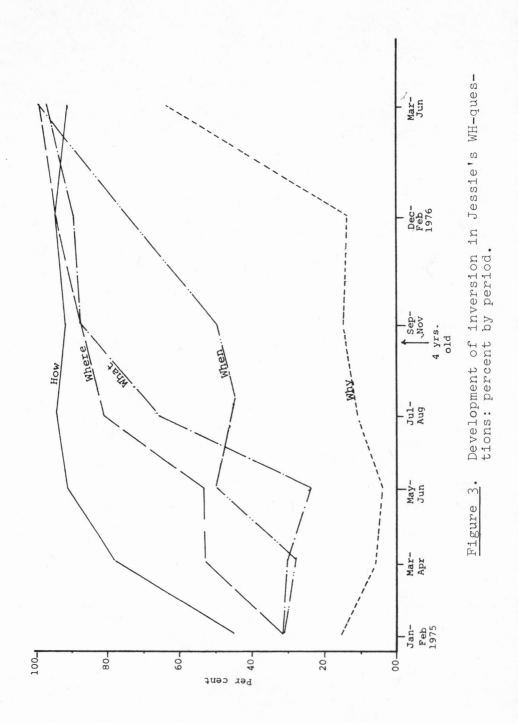

Figure 3.   Development of inversion in Jessie's WH-questions: percent by period.

WH- forms in Table 1 reflects the wave effect of period
(b) in Figure 2:

    how → where → what → why

Our major purpose is now to use the variable rule program to trace more precisely the development of the
inversion of the major WH- forms, compensating for any
differences in the distribution of other variable constraints among them. It is not possible at this stage
to analyse each month separately: if N is reduced to a
hundred or so, too many of the smaller variables will
become invariant and will then have to be eliminated
from the program.[21] It is then necessary to group the
data by two or three month periods: figure 3 shows the
raw percentages for this re-organisation. Here we have
been able to add data for when questions, though the
numbers for who, whose, and which are too sparse to permit accurate quantitative study in these sub-divisions.

When complicates the picture a bit, and begins to suggest more stratification in period (c) than the simple
statement of Jessie's rule (28). We can now take the
data for each of these periods and submit it to the
variable rule program, approaching as closely as possible the full analysis of Table 1.[22] Figure 4 shows
the values for the five WH- forms of Figure 3.

One can observe a number of differences between the raw
percentages of Figure 3 and the variable rule analysis
of Figure 4. In Figure 4, how does not rise at the
outset as it does in Figure 3, but is consistently
high throughout except for a rapid fall at the end.
This fall is an artifact of the very high percentages
of inversion among the three leading factors in period
(d): when Jessie failed to invert only one how question
its value dropped below what and where, which are more
frequent than how and are not as easily affected by
single tokens. Otherwise, we see that how is regularly
the highest factor, where is next, then what, when and
finally why. This regular stratification holds from
the peak in July to February 1976, over a period of
eight months. The equivalence of the WH- forms at the
top of Figure 3 has disappeared.

The reason for the difference between the two figures
is that part of the high values for where and what questions - as far as raw percentages are concerned - is
their co-occurrence with other favouring constraints
such as contraction, favoured subject forms, and so on,
while the high values for how must be attributed

primarily to the effect of that form itself.[23]

By incorporating WH- constraints into a single rule
(T2'), we indicate that for Jessie, inversion is a
single process. In the light of other recent studies
of acquisition, [24] it seems quite plausible that Jessie
could learn a stable set of variable constraints, and
maintain them for six or eight months. But such a
claim cannot be made without specific evidence related
to WH- inversion. How do we know that (T2') operates
as a single rule, and at what point did it begin to do
so? Though periods (a) and (b) do not show the regular
stratification of WH- factors that we find in (c), that
does not mean that (T2) was not operating at that time
with a different ordering of factors in Group 1.[25]

The switch from a fluctuating relationship of WH- forms
to a stable one does suggest some kind of internal re-
organisation, but it might also represent a stable relationship among separate and competing rules. We will
need to draw upon two kinds of evidence to resolve
these questions: (1) data prior to periods (b) and (c)
that might explain how and when the ordering of the WH-
factors arose, and (2) data internal to (c) that would
show whether the various factor groups indeed contribute
independently to the application of the rule.

6. The origins of the WH- constraints.

Jessie's first WH- questions throw considerable light
on the problems just raised. Beginning in November
1973 (2:2) we find the following over a period of five
weeks:

(35) Nov 24   Where the boy? (Walking around the living
                  room)
(36) Nov 26   How 'bout that, Mama? (T. dressing J.)
(37) Nov 27   How 'bout that, daddy? (Walking into W.'s
                  study.)
(38) Nov 28   How 'bout the wash? (to T.getting fork and
                  spoon from the dishwasher).
(39)   "   How 'bout these? (getting toy animals from
                  from sister Jo).
(40)   "   Where Daddy? (After T. gives her her bath)
(41) Nov 30   How 'bout a baby? (J. and T. are looking
                  at a book with pictures of a baby in a
                  chair; J. puts up her shirt and "nurses"
                  the baby)
(42)   "   How 'bout a boy? (J. and T. looking at a
                  picture of a family in a book)

(43) Nov 30   How'bout a girl? (same situation)
(44) Dec 7   How 'bout these? (to self, taking off shoes)
(45) Dec 13   How 'bout that, yeah? (holding cat Toby,
              who has just subsided struggles to be
              free)
(46) Dec 29   How 'bout the face? (T. washing J.'s hands;
              she has just said she would wash J.'s
              face)
(47) Dec 30   How 'bout these, mommy? (bringing clothes
              to T., as T. tries to get J. dressed in
              them)
(48)   "   How [də] Nam? (twice, as T. starts to fix
              breakfast and J. sees dog Sam and starts
              to fix his)

The sudden burst of activity at the end of November (14
WH- questions) was succeeded by a sparser use in December (8), but was followed by a sudden upturn in January
1974 (70) and February (117). It is clear that the dominant form for questions was "How about (NP)", and that
the use of a WH- form how is associated with other WH-
question where... The use of "How about (NP)" begins
as a vague kind of identification, but develops into
clear requests for action by the end of December. The
first verb occurs with How about on January 18th,
repeating the message of (48) three weeks earlier:

(49) How 'bout [də] eat? (to T. as T. puts food in Sam's
                         dish)

This is not entirely clear in its syntactic structure,
since [də] is structurally ambiguous, but on the next
day we get a form that could not come closer to the
adult suggestion:

(50) How 'bout get ketchup? (eating French fries, to T.)

This use of how 'bout continued sporadically throughout
1974 and the 1975-6 period covered by Figures 2-4. At
3:4 the following forms illustrate the definiteness of
its use as well as the general advance in linguistic
maturity:

(51) Jan 25 '75   How 'bout you move so we both can have
                  some space for lie down?

(52)   "   How 'bout I move your hair out of the
                  way: it's really hard for brush it.

It is noteworthy that there is no clear use of how to
question adverbs of manner until well into the end of
the period (a). On the day following we note:

(53) Jan 26 How we get tired?

but we do not observe a clear questioning manner until
forty days later:

(54) Mar 7 How did Simon and Joanna get back from
             there? (twice, in reference to their trip
             to Sears Roebuck the night before)

Nevertheless, Figure 4 shows that <u>how</u> questions strongly
favour inversion from well before this period. We suggest that this is the result of a formal phrase structure
rule that crosses semantic lines:

(S1) Q_{HOW} $\rightarrow$ How + V_1 + NP   (VP)

where V_1 is a class of function words that originally
was confined to <u>about</u>, but developed to include <u>come</u>
and <u>do</u>. It was natural enough for this <u>do</u> to develop
the alternations of tense and number characteristic of
the tense marker, since the optional VP which follows
as in (49) - (52) never contains a tense marker.[26]
'Bout and <u>come</u> could then be seen as parallel to other
items that are not overtly marked for tense, such as
<u>should</u>.

We can therefore argue that the high initial inversion
rates for <u>how</u> do not represent inversion according to
(T2') at all, but rather a heavy favouring of (S1).
At the same time, the occasional occurrence of uninver-
ted forms such as

(55) Mar 7 '75 How it goes?

implies a different phrase structure, closer to the
adult role:

(SØ) S $\rightarrow$ NP + Aux + VP

A later rule would develop in VP an adverb of manner
as (Prep + WH-NP) which in turn requires (T1) to produce (55). It follows that at some time yet to be determined, the inverted forms were (suddenly or gradually)
analysed by Jessie as the result of (SØ), (T1), and
(T2) rather than (S1). It will remain for us to show
how that point can be located.

<u>What and where</u>. It also follows naturally that the
initial occurrences of <u>where</u> were "inversions" produced
directly by a phrase structure rule. There is no
reason to derive (35) or (40) from an underlying phrase
structure based on (SØ):

(40') Daddy - Tense - is - [WH-locative NP]

but rather from a simple rule

(S2) $Q_{WHERE} \rightarrow$ where(s)   +   NP

The first <u>what</u> questions follow a similar rule:

(56) Jan 26 '74   What's that? (J. bringing T. a book
                               with double pages that have become
                               unglued)
(57) Jan 28 '74   What's that? (imitating T.'s question
                               while looking at a horse on a tele-
                               vision show)
(58) Jan 30 '74   What's this? (to herself, seeing a merry-
                               go-round in a book)

These sentences seem to be produced by a phrase structure
parallel to (S2):

(S3) $Q_{WHAT} \rightarrow$ what(s)   +   NP

(S2) and (S3) may indeed be a single rule, but we will
not attempt to resolve that question in this report.
Our principal argument is that sentences such as (8)
and (9) cannot be produced by (S3), and that at some
undetermined point in Figure 4, inverted sentences like
(4-6) and (56-58) will be reanalysed in the same way as
the result of (SØ) followed by (T1) and then (T2).

<u>Why</u>. It is certainly not the case that <u>why</u> questions
develop late in Jessie's speech. A major aspect of the
earlier period, preceding Figures 2-3 (2:4-3:4), is the
dominant use of <u>why</u> questions. The first examples occur
in January 1974 shortly after the development shown in
(35-48) and are all of the simple form <u>Why</u>? These one-
word questions quickly rise in frequency to four a day
(Jan 31), ten per day (Feb 2), and twenty a day (Feb
31). This high frequency continues to accelerate, to 56
per day (April 26), when <u>why</u> first occurs with a full
sentence following:

(59)   Why Mommy put up curtains?

The following day Jessie again produced 56 isolated
<u>why</u>'s but also the following fuller forms:

(60)   Why water on these? (T. watering flowers)
(61)   Why over our picnic, why? (T. has said picnic is
           over)
(62)   Why pieces stops? (T. has stopped putting pieces
           in a stack)
(63)   Why move fence back? (T. and J. moving fence back
           to protect a seeded area)
(64)   Why you pick macaroni? (T. says macaroni is ready)

(65) Why you drink milk? (T. drinking milk; she says
         she likes it)
(66) Why that bowl?

The sudden jump from one to seven full <u>why</u> questions
indicates that Jessie has a ready means for producing
<u>why</u> questions which is simply:

(S4) $Q_{WHY} \rightarrow$ why + NP (S)

Our confidence in (S4) is reinforced by the fact that
(62) and (63) were produced in response to T.'s request
for an amplification of <u>Why</u>?

(63')   J.:   Why?
          T.:   Why what?
          J.:   Why move fence back?

Rule (S4) is different from (S1-3) in one important respect: it indicates that <u>why</u> is not integrated into the
structure of the sentences that it questions. This is
readily understandable when we consider the relative
degrees of integration in the adult grammar of the
various noun phrases that might be questioned. A
number of syntactic arguments indicate that the direct
object is the most tightly integrated into the verb
phrase: next comes the locative, then adverb of manner,
temporal and final reason adverbial. It is evident
that the surface ordering reflects this degree of integration, as in <u>He took the train downtown in a hurry</u>
<u>last Wednesday for a very good reason</u>.

The difference in syntactic integration can be seen by
the privileges of inserting other syntactic units before
the constituent in question, the ease of foregrounding,
and operation of reflexive transformations and other
syntactic rules that have scopes limited to "same sentence". Very often we find that reason adverbials are
presented by separate sentences that are very loosely
tied to the main clause. Thus the independent (S) of
rule (S4) is easily identified, especially in response
to "Why what?". All of these syntactic arguments show
<u>why</u> at one extreme of the continuum of integration, then
<u>when</u>, <u>where</u> and <u>what</u>, and this is reflected in the order-
ing of inversion probabilities in Jessie's questions.
The major exception is the place of <u>how</u>, which we might
have expected to find between <u>when</u> and <u>where</u> if it had
not been produced by an entirely different mechanism at
the outset. With <u>how</u> removed from the series, and
<u>when</u> operating at very low frequencies, we obtain a
sharp opposition between the behaviour of <u>what</u> and <u>where</u>

on the one hand, and <u>why</u> on the other. The fact that
this opposition is most marked in June, 1975, at the
end of period (b), would make us suspect that this is
a turning point in syntactic re-organisation, but this
is only a suspicion, and more material arguments are
required.

7. <u>Contraction</u>

The simplified example of a variable rule analysis in
Section 4 was not made up for the occasion: it represents the actual data for the period March-June 1975.
We recall that contraction of the auxiliary had oppo-
sing effects, favouring inversion for <u>what</u> and <u>where</u>,
but disfavouring it for <u>why</u>. We concluded that a separate rule could be written for <u>what</u> and <u>where</u> on the
one hand, and <u>why</u> on the other, but we did not attempt
to do so.

There is no obvious way that contraction could be
shown as a variable constraint on (T2'). In the adult
grammar, contraction is a morpho-phonemic process that
operates at a lower level than syntactic transformations.[27] In this particular case, the only possible
way we could derive (2) Where's Philadelphia? from
(SØ) is

```
(SØ)   S —→ NP  +  Aux  +  VP
         . . .→ Philadelphia is (WH-Loc.NP)
            —→ Philadelphia is where
(T1)        —→ Where Philadelphia is
(T2)        —→ Where is Philadelphia
(Contr)     —→ Where's Philadelphia?
```

Though our variable rule analysis was carried out as if
all questions were produced by inversion, the inherent
logic of the contraction constraint points to a different derivation. The only way that contraction could
favour inversion in <u>where</u> and <u>what</u> questions is by producing through the phrase structure rules (S2) and
(S3). As these rules are now written, contraction is
obligatory, but they could easily be modified to produce <u>Where is Philadelphia</u> or <u>What is this?</u> Since (S2)
could not produce <u>Where Philadelphia is</u> we must have
some use of (SØ) and (T1), and given those rules,
some use of (T2) to follow. Schematically:

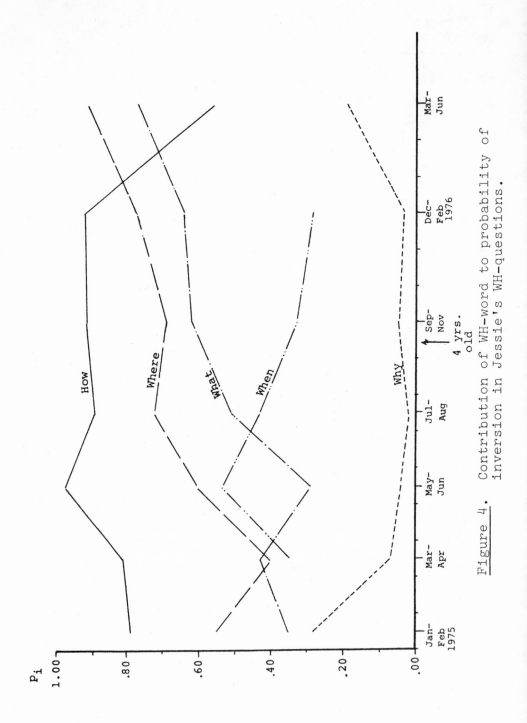

<u>Figure 4.</u>   Contribution of WH-word to probability of
inversion in Jessie's WH-questions.

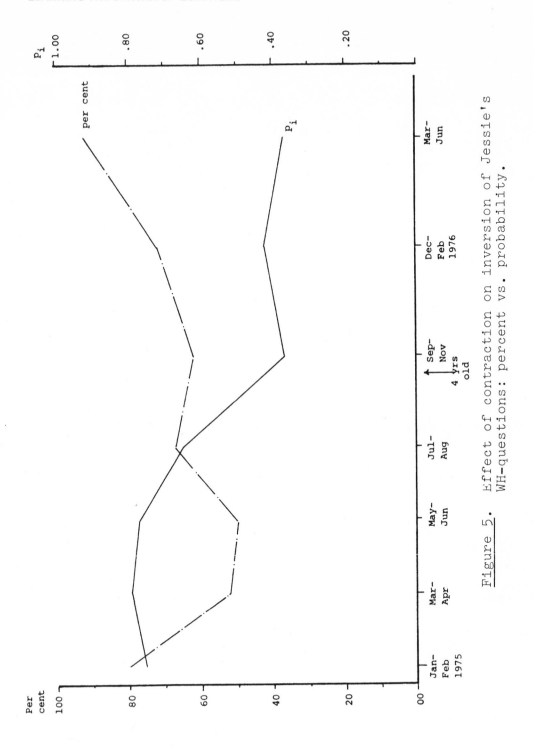

Figure 5. Effect of contraction on inversion of Jessie's WH-questions: percent vs. probability.

	(S2)	(SØ) →	(T1) →	(T2)
Where's Philadelphia?	yes	maybe	maybe	maybe
Where is Philadelphia?	maybe	maybe	maybe	maybe
Where Philadelphia is?	no	yes	yes	no

To the extent that contraction favours inversion, we
can argue that (S2) is definitely required for some
cases, though not for all, since otherwise the relationship would be categorical. An analysis of where's into
where + is under (S2) is possible, but there is also no
reason that it cannot be fitted into (SØ).

The development of the variable constraint on contraction will give us a more precise view of these possibilities. Figure 5 (dotted line) shows the raw percentages for contraction in WH- questions for the
periods of figures 3-4: it falls at first, then rises
steadily from March 1975 onward. The solid line shows
the output of the Cedergren-Sankoff program for the contraction constraint. It has two distinct phrases: a
high plateau around .75 through June, and a low plateau
below .40 from September 1975 on.

This reversal of the effect of contraction yields a
first answer to the question: when does (S2) give way
to (SØ)? It seems that this occurs directly after the
high burst of frequency in July, as we enter the stable
period (c).

It may be helpful to examine the behaviour of contraction
for individual WH- forms. A variable rule analysis for
what and why was run for the two six-month periods of
Mar-Aug 1975 and Sep 75-Feb 76: that is, on either side
of the major transition point of Figure 5. The contraction probabilities are:

	Mar-Aug 75	Sep 75-Feb 76
what	.78	.65
why	.20	.17

The over-all decline in p_i for contraction is therefore
due to a decline in the favouring of inversion with
what (and where). Why remains unchanged in this respect during the entire period: only 1 inversion out of
55 contracted forms in the first six months, and only 1
out of 40 in the second.

Why should contraction have such a powerful reverse

effect on why? It may indeed be the case that contraction with subject pronouns is not a productive process
at this stage (Brown 1973:392). In sentences like

(17) Why when a child grows up there's no daddy?

there's is as much a fixed form as Where's in (1-2),
and this situation may continue throughout period (c).

On the other hand, the favouring effect of contraction
on what and where inversion declines as the differences
between contracted and uncontracted forms diminish in
the period when both approach 100%.

	Mar-Aug 75	Sep 75-Feb 76
contracted what	91%	97%
uncontracted what	53%	89%

Thus strongly suggests the growth of (T2), though as we
noted above, it is not impossible that uncontracted
inversions could be produced directly by phrase structure. Such a development is not motivated by any
other evidence, however, and there is no reason why such
a change in the primitive phrase structure rule (S3)
would be motivated by an increase in the use of inversion or that (T1) should now begin to decline after its
initial rise.

Stronger evidence can be provided by using the fundamental logic of the variable rule program developed in
Section 4: if the inversion rule (T2) is in fact developing in Jessie's system, there should be a better fit
between expectation and observation.

8. Fitting the rule.

The next step in our analysis is quite straightforward:
to examine the chi-square figures for the successive
periods of Figures 3-5. To do so, we must take into
account the fact that the chi-square statistic is not
well suited to small cells, particularly cells with
only one member.[28] Considering chi-square for cells of
two or more members we observe the following progression:[29]

	No. of Cells	No. of Tokens	X^2 per cell
Jan-Feb 75	32	94	0.61
Mar-Apr 75	144	427	1.15
May-Jun 75	95	216	0.73
Jul-Aug 75	241	886	1.32
Sep-Nov 75	143	400	1.12
Dec 75-Feb 76	144	378	0.50

The figures for the last period are the lowest chi-square per cell that we have recorded for any data group. Furthermore, the other chi-squares per cell below 1.00 are for relatively small sub-samples, the first and the third: the decline in this statistic for the last period cannot be attributed to the magnitude of the sample.

We are therefore led to the conclusion that in the second three months of period (c) - December 1975 to February 1976 - the expected fusion of elements did occur, producing a variable rule of the form (T2'); that up to that time, the grammar was characterised by an alternation of the original phrase structure rules (S1-3) with the sequence of (SØ), (T1), and the variable inversion rule (T2'). The decline of contraction as a favouring factor was a clear signal of the decline of the early phrase structure rules: the complete triumph of the inversion rule necessarily marks the end of contraction as a feature favouring or disfavouring any WH- form.

9. Further perspectives.

At this point, we seem to have a view of the historical origins of the variable constraints in Jessie's WH-questions, their relation to possible rule formations, and the dynamics that operate upon them. We expect that our further studies will yield more precise information on the mechanism by which disparate elements are integrated into a single system. It will probably not be necessary to enter all 25,000 questions into the variable rule program to do so: but we have the resources to look more closely at each of the transition points as the argument proceeds.

This analysis has been presented against the background of a prior perspective which holds that the primary purpose of linguistic theory is to choose among rapidly proliferating analyses as the data run out. That will only be true if we make it true, artificially limiting the data by preconceived notions of what it can be.

We find that the world in which language is used is a
very large one, and those who analyse language are very
few: it is also obvious that there are great many children, and they talk a great deal. We believe that we
have available some of the tools we need to find the
system in that talk, to see how systems grow and are
transformed. It should also be clear that the formalisms
we use are pliable, and need not be our masters. There
are many powerful tools of mathematics available that
we can use to analyse our data once we recognise and
set aside the limiting assumptions of the categorical
view.

The coupling of theoretical advance with empirical investigation that is projected here is quite different from
the model that linguistics has presented to psychology
in the past. What is the extent of the shift in the
theory of grammar proposed here? Five points seem to
us to be significant.

(1)   The primary data of linguistics is no longer seen
       as the intuitions of the theorist, but rather obser-
       vations of language in every-day use and experiments
       on behaviour that illuminate that use.

(2)   Constraints on alternate ways of saying the same
       thing may be seen as a part of linguistic structure
       whenever we discover stable evidence of such con-
       straints operating in the every-day use of lan-
       guage.

(3)   These constraints will in turn constrain the theory
       in two ways:

       (a)   They allow us to rule out most of the pos-
              sible relations between variants and pre-
              dict a tighter set of linguistic relations
              than are now conceived;

       (b)   they allow us to show when the data justi-
              fies the writing of one rule operating in
              a variety of environments, assembling par-
              ticular rules into a single rule schema.

(4)   We no longer require that there be only one way of
       producing a given form-meaning construction like
       Where's Philadelphia? As children receive new lin-
       guistic data and form new rules, they do not immedi-
       ately abandon their older rules: that is, there is
       no reason to believe that the grammatical slate is
       a tabula rasa, re-written from scratch with each
       addition to the input data.

(5) Change from one rule system to another is therefore
     seen as two continuous changes in probabilities:
     input probabilities for competing rules and relative
     changes in the weight of the variable constraints
     within a rule.

We do not believe that these modifications affect any of
the positive and insightful approaches to language in
twentieth-century structural theory. The work of the
past two decades has brought home the remarkable intri-
cacy of syntactic relations, focussing our attention on
the extraordinary achievement of the child as language
learner. The study we have presented here continues in
that tradition, and adds a new appreciation for the
industry and ingenuity of that child. We have not presented a picture of sudden and dramatic maturation of
linguistic structure, but rather a remarkable development in the capacity of the language learner.

It is not difficult to see the analogy between the child
as language learner and the researchers we have projected. Both pay close attention to the speech of those
around them, though they are not so foolish as t , abandon their own system every time they hear another one.
But in the course of time, they re-draft their theory
until it is closer and closer to the model that the
every-day world provides, since they find that that
world is richer, more rewarding, and more intricately
put together than anything they could have imagined.
Those who have been fortunate enough to bring their
analyses into close contact with the every-day world
will agree that this is indeed the case.

Footnotes

1. A shorter version of this paper was given by W.
    Labov at the Psychology of Language Conference,
    University of Stirling, Scotland, on 26th June
    1976, as "Recent modifications of linguistic theory
    applied to the study of language acquisition". It
    was submitted to the conference as a joint report
    by W. and T. Labov, and is here presented in its
    full form by the two authors.

2. The study of acquisition has been broadened in
    recent years by consideration of semantic and social
    context (e.g., Bloom 1971; Shatz and Gelman, 1973).
    But to this point there has been little contact
    between variation theory and acquisition of early
    syntactic rules, since the indirect contribution of

W. Labov to the work of Bloom was confined to the
development of the generative framework. It should
be noted, however, that Bloom's research has independently moved forward to confront problems of
variation (see footnote 14).

3. Jessie was born on September 16, 1971. She is our
    fifth child, but is separated from the next youn-
    gest by a gap of 11 years. She has three sisters
    (born in 1954, 1956 and 1960) and one brother (born
    1958).

4. The grat majority of the data was recorded in writ-
    ing by T. Labov (see footnotes 9, 10 for methods);
    the analysis of variation was carried out in writing
    by W. Labov, and the interpretation of the results
    is the responsibility of both authors. We are much
    indebted to Donald Hindle for adapting the Cedergren-
    Sankoff program for the analysis of variation to
    our PDP 11/10 system at the University of Pennsyl-
    vania, which made it possible for us to carry out
    more flexible and varied analyses of the data.

5. Some of the interactions of linguistic and psycho-
    logical research were first developed at the work-
    shop on the Nature and Structure of Categories
    sponsored by the Social Science Research Council
    at the University of California, Berkeley, on
    May 17-19 1974.

6. In addition to the WH-NP constituent, a "Q" marker
    is also shown as an option in constituent 1: this
    is the abstract representation of the question sig-
    nal of yes-no questions, which also triggers inver-
    sion. We have not shown the range of negative and
    affective elements that can produce inversion
    optionally in various dialects (as in the literary
    form Seldom was he seen or the Black English Verna-
    cular Ain't nobody see it), since these are not
    relevant to the immediate problem on hand.

7. When be is the main verb, it is subject to a rule
    that moves it into auxiliary position and is thus
    affected by (T2). The same process optionally
    moves have, though not in colloquial American dia-
    lects. The main Verb constituent is therefore shown
    as optional, in parentheses, since it may be moved
    into the position of constituent 4 by such a rule.

8. Some children hear more than others: if there are
    other young children in the house, or if there are
    speakers of dialects who do not always invert after
    WH- forms (Labov, Cohen, Robins and Lewis 1968).
    It seems inevitable that Jessie would have heard
    a certain number of uninverted WH- sentences from
    children her own age.

9. Questions were written down within minutes after
    they were said on 3x5 slips. The early period
    covered all of Jessie's utterances but during the
    period of Figures 1-5 only selected structures were
    recorded completely. It is of course possible that
    Jessie's production of WH- questions was stimulated
    by T.'s interest, and that the overall production
    was higher than it might have been without obser-
    vation; but this would not affect the relative pro-
    portion of inversion and other variables which were
    the focus of our study.

10. We have recorded some questions that Jessie asked
    herself (see (44) and (58)). In September 1975
    Jessie began to attend nursery school during the
    day, and we therefore have no record of questions
    that she may have asked others there except for the
    days when W. or T. worked at the school. However,
    from our observations there we concluded that she
    asked relatively few questions in that situation,
    and this did not seem to effect the total number of
    questions she asked in any serious way. For example,
    we find that for November 1975 the average number
    of WH- questions was 25 for weekdays, and 31 for
    weekends and holidays. We have further reduced any
    bias by beginning our sampling of the data with all
    Sundays (see footnote 11).

11. This first sample of our data includes all days for
    Jan-Mar 1975, and all Sundays thereafter with the
    addition of Apr 7-18 1975, Jul 17-20 1975, and the
    first halves of January and May 1976.

12. One such factor is exposure to new places and situa-
    tions. In July and August 1975 we took Jessie,
    already an experienced traveller, to Marseille,
    Berlin, Florence and Corsica. The high burst of
    WH- questions coincides with the beginning of this
    trip; however, the decline to a low plateau ocurred
    in October and November, when Jessie was in a stable
    social situation in Philadelphia.

13.   Indeed, some linguists have argued, in defense of
       the traditional view, that it is impossible for
       children to learn such regularities. (Bickerton
       1971; Butters 1971). This position is of course
       difficult to reconcile with the large body of
       observations of variable constraints quite uniform
       throughout the community (Labov, Cohen, Robins and
       Lewis, 1968; Summerlin 1972, Guy 1975) and the
       research of psychologists which shows great sensi-
       tivity of subjects to lexical frequencies (Solomon
       and Howes 1951; Solomon and Postman 1952). Recent
       studies in our laboratory by Sally Boyd show that
       children as young as five have acquired the Phila-
       delphia pattern of -t,d deletion, deleting more
       often before a consonant and less often with gram-
       matical clusters.

14.   Not all variation studies are concerned with variants
       that have equivalent truth values. The development
       of phrase structure rules inevitably involves varia-
       tion in the choice of meaningful options, and two
       research groups are beginning to explore this area:
       the group headed by Bloom at Teachers College,
       Columbia, and the research project on Pidgin
       Deutsch at Heidelberg headed by W. Klein and N.
       Dittmar.

15.   These are not unrestricted re-write rules, but are
       limited to the replacement or deletion of single
       elements. The restricted character of phonologi-
       cal rules may be compared to the wider scope of
       transformations, which have not yet been success-
       fully limited in their operation.

16.   The feature [+concrete] is used here as an econo-
       mical way to refer to what as Jessie uses it. To
       distinguish these noun phrases from those questioned
       by who we would have to add [+animate].

17.   The pronoun categories considered here are I; you;
       he or she; it; we; this; these or those; and that.

18.   Since there is no variation in the assignment of
       is to singular noun phrases and am to first person,
       it is not necessary to set up a separate category
       for am. On the other hand, there is some variation
       in the occurrence of is and are with singular and
       plural noun phrases. The absence of a copula can-
       not be recorded as a separate category, since it
       is invariantly associated with lack of inversion;
       in some analyses, this was included with the
       most likely verb form.

19. The total N for this analysis is not equal to the 3,368 of the full sample since there are a number of invariant sub-categories that must be excluded from the version of the Cedergren-Sankoff program used here.

20. The sub-sections examined include all of the divisions of Figures 3 and 4, as well as larger groupings of them.

21. Earlier versions of the Cedergren-Sankoff program did not require this elimination of invariant categories, although it is a realistic feature which maximises the effect of those variables that are active.

22. Again, it should be noted that comparability of sub-sections is limited by the differences in the numbers of variant cells. As more data is drawn from our record, this will become less of a problem.

23. Looking ahead to the origins of the WH- questions in Jessie's speech, one can see that how would be independent of factors that affect where and what. How is not contracted, and occurs with a wider range of subject noun phrases. The inversion of how is based on an original identification of how questions with how about and how come forms.

24. The studies of the acquisition of -t,d deletion are noted in footnote 13.

25. The re-ordering of variable constraints is a basic mechanism of linguistic change, first observed in Labov 1963; for change correlated with age, see Labov 1972.

26. When How'bout occurs with a full sentence and subject noun phrases, those subjects are normally first and second persons; and since the tense is present, we do not get the third singular -s which is the only overt sign of the tense marker. As (50-52) show, suggestions are normally made for I, you and we.

27. The use of contraction could only favour inversion by the use of derivational constraints and "global rules" (Lakoff, 1970), by which the grammar looks ahead and compares possible derivations. It does not seem a probable solution in this case.

28. The chi-square statistic used here is $(E-O)2/E + (E-O)2/(N-E)$ where E is the expected value, O the

observed, and N the number of tokens. If the expected value for a cell happens to be quite low, say, .04, and there is only one member in the cell, one may still occasionally observe an inverted form, to yield a chi-square value of 24.

29. Progressions parallel to this can be observed for cells of all sizes.

References

Bickerton, D. (1971). Inherent variability and variable rules. Foundations of Language, 7, 457-492.

Bloom, L. (1970). Language Development: Form and Function in Emerging Grammars. Cambridge, Mass.: MIT Press.

Bloomfield, L. (1926). A set of postulates for the science of language. Language, 2, 153-164.

Brown, R. (1973). A first language: the early stages. Cambridge, Mass.: Harvard University Press.

Butters, R.R. (1971). On the notion 'rule of grammar' in dialectology. Papers from the Seventh Regional Meeting of the Chicago Linguistic Society. Chicago: Department of Linguistics, University of Chicago.

Cedergren, H., and Sankoff, D. (1974). Variable rules: performance as a statistical reflection of competence. Language, 50, 333-355.

Chomsky, N. (1965). Aspects of the Theory of Syntax. Cambridge, Mass.: MIT Press.

Guy, G. (1975). Variation in the group and the individual: the case of final stop deletion. Pennsylvania Working Papers on Linguistic Change and Variation, 1:4. Philadelphia: U.S. Regional Survey.

Labov, W. (1963). The social motivation of a sound change. Word, 19, 273-309.

Labov, W. (1969). Contraction, deletion, and inherent variability of the English copula. Language, 45, 715-762.

Labov, W. (1972). Where do grammars stop? In R. Shuy (ed.) Georgetown Monograph on Language and Linguistics, 25, 43-88.

Labov, W.,Cohen, P., Robins, C., and Lewis, J. (1968). A study of the non-standard English of Negro and Puerto Rican speakers in New York City. Cooperative Research Report 3288. Vols. I and II. New York: Columbia University.

Labov, W., and Labov, T. (1974). The grammar of cat and mama. Paper given before the Linguistic Society of America, New York City, December 1974.

Lakoff, G. (1970). Global rules. Language, 46, 627-639.

Ravem, R. (1974). The learning of WH- questions. In Richards, J.C. (ed.) Error Analysis: Perspectives on Second Language Acquisition. London: Longmans.

Shatz, M., and Gelman, R. (1973). The development of communication skills: modifications in the speech of young children as a function of listener. Monographs of the Society for Research in Child Development, 38:5.

Solomon, R.L., and Howes, D.H. (1951). Word frequency, personal values and visual duration thresholds. Psychological Review, 58, 256-270.

Solomon, R.L., and Postman, L. (1952). Frequency of usage as a determinant of recognition threshholds for words. Journal of Experimental Psychology, 42, 195-201.

Summerlin, N.C. (1972). A dialect study: affective parameters in the deletion and substitution of consonants in the deep South. Unpublished Florida State University dissertation.

Weinreich, U., Labov, W., and Herzog, M. (1968). Empirical foundations for a theory of language change. In Lehmann, W., and Malkiel, Y. (eds.) Directions for Historical Linguistics. Austin, Tex.: University of Texas Press.

SEMANTICS AND LANGUAGE ACQUISITION:

SOME THEORETICAL CONSIDERATIONS

DAVID S. PALERMO

The Pennsylvania State University

Now that developmental psycholinguists have been gradually
easing themselves out of the syntactic fixation with which they
have been afflicted for some time, they have come face to face
with the problem of semantics which is, of course, the heart of
any account of language and language acquisition. My plan today
is to begin by presenting some reasons for rejecting one theory
of semantic development - namely, the Semantic Feature Hypothesis -
which has stimulated a great deal of research on semantic development including some of my own. Second, I will sketch an alternative theoretical model which may be more satisfactory. The latter
theory draws from aspects of my own research, the developmental
work of Donaldson and Wales (e.g., 1970) Macnamara (1972), Rosch
(e.g., 1973) and Nelson (e.g. 1974), and research reported by
Bransford, Franks, Posner and others (e.g. Bransford and McCarrell,
1975; Franks and Bransford, 1971; Posner and Keele, 1968).

Looking at the problem of semantics broadly, I have had three
major concerns about feature theories which currently appear in
the literature. First, language involves the communication of
ideas from the mind of a speaker to the mind of the listener.
While most discussions of semantics may pay tribute to this rather
obvious fact, most seem to ignore it in a concern with the details
of a componential analysis of words and the composition rules
relating those words in sentences without regard to the psychological process of the communication framework which constrains the
kind of componential analysis which could apply.

Once having recognized the communication issue, the second major
issue becomes immediately apparent. Every act of communication

is contextually determined. Thus, exactly the same words
embedded in exactly the same syntactic frame have different meanings
in different contexts. Furthermore, some of our work with children
suggests that the context may be more important to communication
than the words themselves (Wilcox and Palermo, 1975). While it
remains to be demonstrated that context is as important for adult
communication, there is little question that context is essential
for giving meaning to language (e.g. Bransford and Johnson, 1972).
If this is so, the strongest version of the argument would hold
that there is no abstract linguistic meaning of words or sentences
in the traditional sense and an interpretive semantics based upon
that assumption is doomed in principle.

Third, most discussions of semantics rule out the metaphoric use
of words as within the domain of semantic theory. Yet it is obvious
that metaphor is a common, if not pervasive, yet, contextually
meaningful use of language. The elimination of phrases such as
"the mouth of the river", "the configuration of ideas", "the
eye of a needle", "crooked people", and so on is to rob language
of its communicative purpose. Furthermore, it is not clear, even
within semantic feature theory, where one is to draw the line
between metaphoric and nonmetaphoric use of language synchroni-
cally (the issue is further complicated by diachronic considerations). For example, when using the word 'mouth' it would be
difficult to determine in the appropriate context when that word
is used in an idiomatic, anomalous, or meaningless way in phrases
such as, "the mouth of the man", "the mouth of the amoeba", "the
mouth of the river", "the mouth of the cave", "the mouth of the
mountain", "the mouth of the church", "the mouth of the mind",
and so on indefinitely. The first of these phrases must be meaningful on any account of semantics but what of the others? It is
obvious that in the appropriate context native speakers of the
language would have no hesitancy in accepting them as semantically
appropriate. Assigning metaphor to the categories of idiom,
anomaly, or meaninglessness is to ignore not only the use of
language but the conceptual base underlying language.

Thus, any theory of semantics, especially if it is to have
psychological appeal, must take into account communication, context,
and metaphor. Feature theory seems to neglect all three or, at
least, proponents of this type of theory have directed little
attention to the problems inherent to these issues. There are three
aspects of features which make a consideration of these problems
difficult. First, an analysis within the feature framework is
atomistic in nature. Each word is broken down into a set of
abstract features which may or may not be hierarchically organized.
The features are considered elements of the meaning but no
account is usually attempted of the source of the elements and,
perhaps more importantly, no account is given of the manner in
which the whole is constructed from the parts. To say that the

meaning of a word consists of the features x, y, and z is to give
no hint of how those features are weighted or integrated in the
single meaning of the idea a speaker is attempting to convey in
some particular context. The hierarchical organization of features
which Clark (1973) discusses does not solve this problem and is
advanced more as an account of developmental processes than as an
account of the structure of meaning. Bierwisch (1970a) has considered some features as natural perceptual characteristics of
the organism and both Bierwisch (1970b) and Katz (1972) have
attempted by somewhat different means to avoid the difficulty of
having an unstructured amorphous set of features for each word.
Both theoretical efforts are admitted by their authors to be far
from complete solutions to this problem and neither seem satisfactory to solve the other problems about feature theory raised
here.

Second, features imply discrete categories with specifiable
boundaries. Such a system seems to have been successful in advanc-
ing our understanding of phonology but the analysis of the abstract
system underlying the realization of speech sounds is a delimited
problem at least relative to the semantic system. The identification of the phonological features can be specified within the
limits of the mechanical system available to realize the actual-
ization of speech. Speech sounds are categorically perceived
while words do not have the same characteristics when viewed
within a semantic framework.   The words 'pill' and 'bill' can
be characterized as differing phonologically on the basis of one
feature in a manner which does not seem analogous to different-
iating, say, 'girl' and 'woman' in terms of one semantic feature.
Further, the concepts of table and liberty, for example, are just
that - concepts which can range over many exemplars depending
upon the scenario in which they occur. There are no specifiable
boundaries which can be defined by a set of features.

The inadequacy of feature theory becomes most obvious in the case
of metaphor. As soon as one allows metaphoric use of words,
features become trivial because one would have to multiply the
features indefinitely for each word to take account of every
conceivable metaphoric use which the mind might generate for a
word in the language.

These concerns have led me to join others (e.g. Nelson, 1974;
Fodor, Fodor and Garrett, 1975) in rejecting a feature theory of
semantics. The task, then, becomes one of developing an
alternative theory of semantics which accounts for the manner in
which the language is used to signify the meanings or ideas
which a person has in mind such that those meanings may be
conveyed to the mind of another person.

It is assumed that the meanings or ideas consist of prototypic concepts, relations among concepts which are also prototypic in nature, and the representation of those concepts and relations in time and space. Thus, semantics is built upon a cognitive base of meaning and/or knowledge. Language is a means of representing that cognitive base of meaning. Language may express some but not necessarily all of the cognitive base at all developmental levels, i.e., not all meaning may be expressed semantically in language. The latter point is most clear prior to the appearance of language when the child obviously has concepts but no language to communicate them. (It is not limited to children, however, since the meaning of many internal states can not be completely expressed in the language mode.) The developmental problem, therefore, is twofold in the sense that one needs a theory of the development of concepts in the base and, second, a theory to account for the manner in which language comes to reflect the meanings of those concepts which the child may wish to express.

Prototypes are assumed to be of two types, those which are the abstract base underlying what we usually refer to as nouns in the language and those which are the abstract base underlying what we usually refer to as verbs in the language. Thus, there are conceptual and relational prototypes, the latter serving the function of relating the former. The concepts of time and space are limited initially, from a developmental point of view, to the here and now and, as a result, probably play no significant role in the early construction of meaning. Of course, as space and time become a part of the conceptual system these concepts will influence others.

Prototypes are assumed to consist of a central core, or abstract best exemplar, meaning and extend from that core meaning to undefinable boundaries which may encompass exemplars with varying degrees of similarity to the prototypic core meaning and to each other. The concept of family resemblance recently taken by Rosch and Mervis (1975) from Wittgenstein captures the relations among attributes which define the structure of a prototype. Thus, almost any exemplar may come under the rubric of a particular prototypic concept but the exemplariness of the instance will vary widely depending upon its similarity to the core and the context. For example, there are many exemplars of table which, within a feature theory, might be identified as having a flat surface supported by four legs. But note that within a prototypic view a cloud may fall into the domain of table when a bevy of angels surround it consuming their manna. In contrast to feature theory, the word 'table' is conceived here as having a prototypic meaning which allows it to be extended to cloud in such a context. A cloud, however, even in this context, would not be considered an exemplar which is close to the core meaning of the prototype for table. It

is the emphasis on the functional similarities of the cloud which
allow the meaningful extension of the concept to this instance.
Similarly, legless and nonflat surfaces may serve as exemplars of
table in other contexts and on the basis of other dimensions of
similarity.

It is clear from the examples above that the investigation of
metaphor may be a particularly fruitful entry to the nature of the
prototypic concept underlying any particular word. It is assumed
that the metaphoric use of a word is successful in communication
because some aspect of the core meaning of the concept is transferred to the metaphoric context. One would not be likely to rely
on a peripheral aspect of a concept's meaning in metaphoric use.
Thus, the investigation of metaphoric construction and comprehension may provide a technique for discovering the structure of
prototypic concepts.

It should be made explicit here that prototypes are not conceived
as being singularly perceptual in nature. Rather, they are
assumed to include emotive and function components as well.
Developmentally the latter may be more important initially as
Nelson (1974) has suggested. The prototype is a configuration of
abstract relations among abstract perceptual, functional and
affective factors. There is, of course, some problem in discussing the latter factors. The perceptual aspects are the easiest
to identify in quantitative terms but the lack of terminology
should not obscure the definite functional and emotional aspects
of meaning. Consider the strong functional component of the common
word 'gasoline' for those who know little about it except that it
is required to make one's car operate or the affective component
of 'death' for many persons. Incidentally, one cannot say that
the person who knows only the above mentioned characteristics of
gasoline does not know the meaning of gasoline. That IS that
person's meaning for gasoline, but that prototypic concept may be
changed given the appropriate experiences. On this view there
is no arbitrary single meaning of a word. The development of
acquired concepts will be expanded below. In any case, it should
be clear that the three components of the prototype identified
here are isolated only conceptually, for any particular prototype
they are intimately interwoven as a single unity which is that
prototype.

Further, it is assumed that the focal point or core of any prototype may be shifted, or transformed by contextual factors.
Linguistically this function is executed in its simplest form by
adjectives, in the case of nouns, and adverbs, in the case of
verbs. For example, the core meaning of the concept bird will be
different from that for 'large bird', or 'cremated bird' and so on.
In much the same way, the relational concept walk will have a

different core meaning when the walking is done 'rapidly',
'drunkenly', or 'haughtily'. The adjective and adverbial modifiers
referred to here are those Nelson (1976) refers to as attributives
with the function of classification. The subcategorization, or
shift in the prototypic core meaning, through the use of adjectives
has been shown by Nelson to occur somewhat later developmentally than
the use of adjectives to describe a specific referent. In any case,
such examples of linguistic contextual transformations can, of
course, be elaborated in most complex ways. The transformations
may also be produced by nonlinguistic contexts, however, as in the
case of water in the context of desert or flood, the transformation
of sword in the context of war or museum, or the transformation of
automobile in the context of the 1920's and the 1970's.

It should be pointed out in connection with contextual transformations that aspects of the prototype which are close to the core will
not ordinarily be made explicit in language. Unless the speaker
understands that the listener does not know the concept or feels
that the listener has a different central prototype, as when an adult
speaks to a child, he will not make explicit attributes inherent to
the prototypic core meaning. For example, in discussing dogs, one
would not ordinarily mention the fact that they have four legs
because that is presumably a part of the core meaning of dog. On
the other hand, attributes of the core are likely to be made explicit
in language when they involve transformations of the core, i.e.,
'collie dogs', or when a differentiation of instances which are
encompassed by the core is being made, i.e., 'the border collie'.
The latter is a point emphasized by Olson (1970).

Prototypes are assumed to be both natural and acquired. As Rosch's
data suggest, for example, some colours and geometric forms appear
to have natural built-in core meanings. There are less well
documented relational prototypes which are natural but certain
aspects of movement are probably naturally conceptualized relations
while it would seem that static relations are less likely to
prove natural. Macnamara (1972) has argued that this should be the
case and Nelson (1973, 1975) has provided evidence that children's
early meanings are tied closely to function and to changing
aspects of referents as opposed to static attributes of those
referents. One might also expect that process verbs would appear
before state verbs in the acquisition process. In addition to
natural concepts as such, there are certainly many natural dimensions which are used as a basis of other classifications or
concepts shading from natural to acquired. Surely figure-ground
relationships are natural, shape constancy, division on the basis
of contour and many more new dimensions currently being explored
in research with infants (cf. Cohen and Salaptatek, 1975). Nelson
(1975) suggests that size is an important dimension during early
stages of language acquisition. There must be many as yet unknown

natural dimensions forming the basis of concept formation which are common to most humans. Certainly there are individual differences but as Macnamara points out, the child seldom forms bizarre concepts indicating that there must be some constraints on what is grouped together. One could probably argue that many of the case relations which form the basis of Fillmore's case grammar (1968) are naturally conceptualized. Acquired concepts presumably derive from or are induced from objects and events classified in part on the basis of natural dimensions as well as acquired distinctions with which the child is forced to deal in his environment. It seems likely that the prototypic meanings of the latter may well change with experience while natural prototypes are stable over time and development.

It should be noted that the sense of words referring to natural concepts is known as a function of the nature of the organism and therefore commonality of meaning of these words is assured by the commonality of the organisms. In the case of the sense of words referring to acquired concepts, however, the commonality of meaning is a function of the commonality of the abstracting characteristics of the organism and the particular experiences themselves. There is likely to be, therefore, greater variability in the sense of the latter than the former words. As has been demonstrated, the child's prototype of a colour will be the same as that of an adult (Mervis, Catlin and Rosch, 1975) but the concept of dog, for example, may initially be a function of the dogs with which the child comes in contact in terms of particular dogs, pictures of dogs, and verbal input from others about dogs. Both a willingness by the child to overextend the concept beyond those exemplars which are included in the concept by adults and a failure to extend the concept to exemplars included by adults should be expected at early developmental points during acquisition. Both over- and underextension should be anticipated because it is unlikely that the child's concept of dog would be the same as an adult's concept. Not that the child's concept is necessarily less complete but it is likely to be different at least with respect to the boundaries. Some preliminary research in our laboratory has found little evidence, in the case of the concept of dog, for either over- or underextension of the concept in three year old children. The particular concept involved may be too well conceptualized in a manner similar to that of adults at three years of age. Less familiar concepts will be explored in future research.

One last point with respect to natural concepts pertains to the concepts of time and space. It is assumed that the here and now are natural concepts but that the child must acquire the concepts which extend time into the past and the future and space into three dinensional distance. In some sense these concepts are of especial interest because they combine aspects of both naturalness and experience in ways which may be different from those of other concepts.

A prototypic concept is assumed to have meaning only insofar as it
is related to other concepts, i.e. concepts related to other objects,
events, actions and at least in the developmentally early stages,
the self. Thus, there cannot be a single prototypic concept but
only a concept in relation to at least one other concept. Simply,
the concept of 'tree', for example, necessitates a concept of
'not tree'. While Nelson (1974) has argued that actions may be
object free, on the present view they may not be free of all other
concepts, i.e., other relational or conceptual prototypes. Inherent
in the prototype, however, are orderings of transformations which
the prototype may undergo and the relations into which it may enter.
Since the child knows the concepts prior to acquiring the language,
he knows the relations into which those concepts may enter. It is
assumed that language acquisition is a matter of determining the
words and syntactic rules for representing those concepts. Since
the sense is known by the child, he must use the referents of
the language used by others to infer the sense which others are
communicating so that he may do the same in attempting to
communicate his ideas. Early in development the limit on
transformations and relations often appears to be quite severe.
Macnamara (1972) has cited the example of children appropriately
identifying instances of their own dolls and their own toys which
included the dolls while at the same time denying that one of the
dolls is a toy. Asch and Nerlove (1960) have demonstrated the
developmental course of adjectival transformations such as 'sweet'
and 'crooked' which are applicable to objects but only later
applicable to people. It seems clear that certain transformations
and relations are developmentally much later than others and should
be investigated as clues to conceptual development. Metaphoric
use of words is likely to be a later developmental phenomenon.

By way of summary, it is being argued here that the child acquires
his language through the process of determining from his language
environment how the concepts he already has available to him may
be expressed. Those concepts are prototypic in nature and thus
the meanings of words used to communicate those concepts have a
prototypic base. In the case of acquired concepts, the child's
prototypes may differ somewhat from that of the adult particularly
in the peripheral regions. It is assumed that with enough
experience the child abstracts the central tendency of the
distribution of functional, perceptual, and affective characteristics of concepts to allow the child's meanings to overlap
with those of others to an extent which allows communication and
the similar classification by both of new instances. Natural
concepts provide a common base for communication even when acquired
concepts may sometimes   fail. It is conceivable, however, that
most of the concepts, both natural and acquired, of the child and
the adult are the same and it is only the conceptualization of the
complexity of the relationships into which they enter which may

differ. If that is the case, the syntax of the child and adult
will only differ in terms of the complexity of expressed relations
and not the relations themselves. The metaphoric use of language
may represent one of the most complex levels of such expression.
Language acquisition then becomes a matter of acquiring ways to
express concepts in more and more complex interrelations. Such a
process would be limited only by the cognitive capacities of the
developing child. It is then those cognitive capacities about
which we need to know more in order to understand the language
acquisition process. Thus, this exercise in formulating a
theoretical sketch has made it clear that our study of language
acquisition has taken us from syntax to semantics to cognition
from which vantage point we may gain some purchase on the meaning
of child language.

References

Bierwisch, M. (1970a). Semantics. In J. Lyons (Ed.),
    New horizons in linguistics. Baltimore, Md.: Penguin Books.

Bierwisch, M. (1970). On classifying semantic features.
    In M. Bierwisch and K.E. Heidolph (Eds.), Progress
    in linguistics. The Hague: Mouton.

Bransford, J.D., & McCarrell, N.S. (1975). Some thoughts
    about understanding what it means to comprehend. In
    W.B. Weimer and D.S. Palermo (Eds.), Cognition and
    the Symbolic Processes. New York: Lawrence Erlbaum
    Associates.

Bransford, J.D., & Johnson, M.K. (1972). Contextual pre-
    requisites for understanding: Some investigations of
    comprehension and recall. J. Verb. Learn. Verb.
    Behav., 11, 717-726.

Clark, E. (1973). What's in a word? On the child's acquisition
    of semantics in his first language. In T.E. Moore (Ed.),
    Cognitive Development and the Acquisition of Language.
    New York: Academic Press.

Cohen, L.B., & S alapatek, P. (1975). Infant perception:
    From sensation to cognition. Vol. I & II- New York:
    Academic Press.

Donaldson, M., & Wales, R.J. (1970). On the acquisition
    of some relational terms. In J.R. Hayes (Ed.) Cognition
    and the development of language. New York: Wiley.

Fillmore, C. (1968). The case for case. In E. Bach and
    R.T. Harms (Eds.), Universals in linguistic theory.
    New York: Holt, Rinehart & Winston.

Fodor, J.D., Fodor, J.A., & Garrett, M.F. (1975). The psychological unreality of semantics representations. Linguistic Inquiry, 6, 515-531.

Franks, J.J., & Bransford, J.D. (1971). Abstraction of visual patterns. J. Exp. Psychol., 90, 65-74.

Katz, J.J. (1972). Semantic theory. New York: Harper & Row.

Macnamara, J. (1972). Cognitive basis of language learning in infants. Psychol. Rev., 79, 1-13.

Mervis, C.C. Catlin, J., & Rosch, E. (1975). Development of the structure of color categories. Developmental Psychology, 11, 54-60.

Nelson, K. (1974). Concept, word, and sentence: Interrelations in acquisition and development. Psychol. Rev., 81, 267-285.

Nelson, K. (1975). The nominal shift in semantic-syntactic development. Cog. Psychol., 7, 461-479.

Nelson, K. (1976). Some attributes of adjectives used by very young children. Cognition, 4, 13-30.

Olson, D.R. (1970). Language and thought: Aspects of a cognitive theory of semantics. Psychol. Rev., 77, 257-273.

Posner, M.J. & Keele, S.W. (1968). On the genesis of abstract ideas. J. Exp. Psychol., 77, 353-364.

Rosch, E.H. (1973). On the internal structure of perceptual and semantic categories. In T.E. Moore (Ed.), Cognitive development and the acquisition of language. New York: Academic Press.

Rosch, E. & Mervis, C.B. (1975). Family resemblances: Studies in the internal structure of categories. Cog. Psychol., 7, 573-605.

Wilcox, S. & Palermo, D.S. (1975). 'In', 'on' and 'under' revisited. Int. J. Cog. Psychol., 3, 245-254.

EMPIRICAL QUESTIONS ABOUT DEVELOPMENTAL PSYCHOLINGUISTICS RAISED BY A THEORY OF LANGUAGE ACQUISITION[1]

KENNETH WEXLER

University of California at Irvine

Formal Constraints and Language Acquisition

I take it that the chief concern of developmental psycholinguistics is the question of how the child learns
language. Following Chomsky (1965, Chapter 1, and many
other places) I take it that this problem is also the
major concern of linguistic theory. Yet it is my
impression that there is a tension between the two disciplines, as if the two disciplines gave two different
kinds of answers to the same question. It seems to me
rather that the situation would be more productive for
science as a whole if the relationship between the two
disciplines were more like a division of labour than
one of competing viewpoints. In my opinion the tension
exists at least partly because of a conceptual confusion.
In this paper I would first like to explicate that confusion and then to sketch out, with examples from my
own work, ways of making the relationship between the
two disciplines more productive.

Specifically, it seems to me that the major substantive
difference between theories of language acquisition
seated in psychology versus those which emerge from
linguistics[2] involves the question of formal universals
of grammar. First, there might be disagreement about
whether they exist. But I will ignore this question
here, in favour of one on which there is probably more
widespread and violent disagreement, namely the question,
given that formal universals exist, could any of them
be innate?

In order to put the questions into perspective, we have
to remember two overwhelming facts, the first true of
the world, the second of theories. First, language is
learned, that is, every normal child can learn any
natural language as a first language, under the appropriate conditions. Second, no theory can explain this
fact. That is, we have no model which can show how
language is learned, under the appropriate circumstances. The first fact won't go away. It is our task to
make the second statement false at some future time.

The linguistic method has been to study adult language
in an attempt to characterise it formally. As more
and more linguistic data have been described, it has
been argued that certain highly abstract conditions are
necessary to describe the grammars of natural languages.
Some of these conditions, it is argued, are universal.
It is further argued that these conditions are so
abstract, and the data presented to children so unclear,
and so tangentially related to the conditions, that the
conditions must be innate.

Along with this argument, Chomsky (1965, 1975) has presented an "instantaneous" model of language acquisition,
which goes roughly like this. A child can be conceived
of as a learning device, L, which contains a number of
statements about the nature of language together with
an "evaluation measure". The device, L, is presented
with a body of "primary data" drawn from one language.
L first decides which possible grammars are "compatible"
with the set of data and then using the evaluation
measure chooses the most highly valued of the compatible
grammars. Chomsky of course observes that this model
contains false assumptions, but offers reasons
why it is useful to adopt it for the purpose of linguistic theory. To repeat, the model is not intended to
represent the time-course of linguistic development,
but is supposed to be used as a frame-work for the development of linguistic theory.

Now, the conceptual confusion that I think exists is
the following: psychologists can view the "instantaneous
model" as a model of the time course of language acquisition and can present factual arguments against it.
For, example, it is clear that language is not learned
at one fell swoop. Rather, pieces of knowledge about
language are added by the child over a number of years.
However, it seems to me that the argument against the
"instantaneous model" has mistakenly been taken to lead
to the conclusion that there is an argument against

innate principles of language. Rather, I think, exactly
the opposite is the case.

Psychology can put empirically-founded cognitive constraints on the language learner in a number of ways.
For example, experiments can be devised which show that
the short-term memory of the child is strictly limited.
One can then extrapolate and say that the short-term
memory of the child when idealised as the language learner is also strictly limited. But what is to be concluded from this result? That the instantaneous model
is wrong as a point-by-point model of language acquisition. But there is no reason to argue that this result
is evidence that there are no innate formal constraints.
In fact, I would argue, the opposite is true.

We can distinguish two parts of the language learning
mechanism L. The first part, we can call "formal constraints" that the language learner uses. The second
part is what has been traditionally called a "learning
mechanism", say, something like a hypothesis-formation
system. Of course, the two parts of the system have to
be intimately inter-related, since the learning mechanism must use the formal constraints while it is constructing its hypothesis. But for the conceptual purposes
we can keep these two systems apart.[3]

The usual kinds of cognitive constraints that are discussed in psychology are constraints on the "learning
mechanism". It seems to me that many students of child
language think that by discovering cognitive constraints
on language acquisition they can show that the theory of
language acquisition will not need formal constraints;
in other words, that all the work can be done by the
learning mechanism. The argument is that these cognitive constraints will make the learning problem more
manageable.

But it is precisely here where the confusion lies. It very much depends on the kind of cognitive constraint that
is being discussed. If the cognitive constraints are
actually limitations on the learning mechanism, such
as memory and computational limitations, then for a
fixed attained (adult) system, the stronger the cognitive
constraints,then the more will formal constraints be
needed. The situation can be seen in the following
diagram of the learning theory.

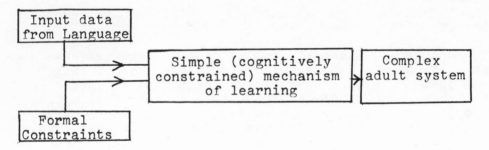

I make two assumptions.

1) Developmental psychology can show that the child
   has a large number of cognitive constraints, for
   example, limitations on processing while learning
   is going on.

2) Linguistics can show that the attained language is
   very complicated.

The conclusion from 1 and 2, of course, is that there
must be formal constraints on language in order to a
account for the learning of a complicated system (assumption 2)by a simple learning procedure (assumption 1).

A number of cognitive constraints that have been proposed may be treated in just this fashion, that is, as
if they add to the difficulty of the language learning
problem by simplifying the abilities of the learning
mechanism. Thus the cognitive constraints require
stronger formal constraints. These include,

a) all kinds of memory limitations, both short-term
   and long-term.

b) any kind of computational limitation, for example,
   inability to deal with embedded sentences of a particular kind.

c) time-bound aspects of the learning procedure, that
   is, of course, language is learned gradually over
   time, and not instantaneously.

At first glance it might appear as if this gradual nature
of the learning process can be seen as making the language learning problem easier and thus requiring fewer
formal constraints than under the instantaneous model.
But, of course, this is not true. The "instantaneous"
model could actually allow a large amount of time for
language learning to take place - there is no compulsion

that it be done in an "instant". Rather, the instantaneous model actually simplifies the problem for the
learner, since it allows all the primary data to be
available to the learner at one time. Thus the learner
may, for example, compare large numbers of sentences.
The realistic course of acquisition, which unfolds in
time, makes it more difficult for the learner - for
example, only small amounts of data are available at a
time[4]. This particular aspect of the learning procedure
has been noted by Braine (1971) who writes:

> The human discovery procedure obviously differs
> in many respects from the kinds of procedures
> envisaged by Harris (1951), and others ... A
> more interesting and particularly noteworthy dif-
> ference, it seems to me, is that the procedure
> must be able to accept a corpus utterance by
> utterance, processing and forgetting each utter-
> ance before the next is accepted, i.e., two
> utterances of the corpus should rarely, if
> ever, be directly compared with each other.
> Unlike the linguist, the child cannot survey
> all his corpus at once. Note that this restric-
> tion does not mean that two sentences are never
> compared with each other; it means rather, that
> if two sentences are compared, one of them is
> self-generated from those rules that have already
> been acquired.

The conclusion that we come to is that those kinds of
cognitive constraints which put limitations on the language learning mechanism must increase the need for formal constraints on grammars.

There may, however, be other kinds of empirical results
attainable by developmental psycholinguistics, results
which actually make the language learning problem simpler and thus reduce the need for formal constraints.
An example is in the question of whether or not children
hear many ungrammatical sentences. One of the arguments
for the necessity for formal constraints in the language
learning process is that the speech that children hear
is complex and often ungrammatical so that in acquiring
a grammar for their language they have to discard, or
reformulate, much of the data available to them. Thus,
if the class of possible grammars were not heavily constrained, the child could never know which sentences to
ignore.

Here, empirical studies have provided evidence which

makes the language learning problem simpler. Many
papers at this conference, of course, are devoted to
an analysis of the speech that the child hears. I only
want to mention a review of some evidence by Ervin-
Tripp(1971, pp. 191-194).

The evidence is based on analysis of observations of
family interactions. The data support the claims that,

1) "The syntax of speech addressed to children is
    simple, containing short sentences with relative-
    ly few passives, conjoined phrases, or sub-
    ordinate clauses."

2) "Speech to children is markedly lacking in
    hesitations, false starts and errors."

3) ". . . Speech to children is highly repetitive."

In light of this and other evidence, it makes sense, in
my opinion, to adopt as a working hypothesis, that the
input to the language learner is free from error, and
is relatively simple.[5] That the input to the learner
is free from error, if true, is an empirical result
developed in psychology which makes the language learning problem simpler. On the other hand, the simplicity
of the input can be seen as making the language learning problem more difficult (and thus requiring more
formal constraints) because, though the input is simple,
the output (attained grammar) is complex. The instantaneous model can allow data as complex as the learner
needs. A more realistic model must simplify the data.
Thus more formal constraints on grammars will be
necessary under the realistic model which simplified
the input data.

A Theory of Syntax Acquisition

I would now like to describe a formal theory of syntax
acquisition that I and my colleagues have been developing over the last few years. For an over-all view of

the theory, see Wexler, Culicover and Hamburger (1976
in press). For other discussions and mathematical
proofs see Hamburger and Wexler (1973, 1975) and
Wexler and Hamburger (1973). For the syntactic evidence
see Culicover and Wexler (1976, in press).

In this theory the kinds of issues that I have been discussing are made explicit. The theory makes particular
assumptions about the nature of adult competence, the
nature of the linguistic environment of the child, the
linguistic capacities that the child brings to bear on
the learning problem, and the nature of the child's
learning procedure (hypothesis generating and changing
mechanism). We have proved that this model of the
child's learning procedure can learn grammars of the
kind posited, which are close to the kinds of grammars
that many linguists claim represent adult competence
(transformational grammars of a certain form).

A more detailed account of these assumptions follows.
The class of grammars that we assume that the learner
has to have the capacity to learn is the class of grammars specified by a finite number of transformations
on a universal base grammar. This last assumption is
clearly falsifiable, since base grammars are not universal (e.g., there are SVO and SOV languages), but we
have also proposed ways of weakening this restriction
(see Wexler and Culicover 1974 and Culicover and Wexler
1974). Now the assumption that would have made it
easiest to prove that this class of languages could be
learned is that the learner can receive negative information, that is, that he can be corrected for speaking
ungrammatical sentences. But the empirical result of
Brown and Hanlon (1970) is that children are **not** corrected for speaking ungrammatical sentences. Thus the
language learning problem is made more difficult. But
if we assume that only grammatical sentences are presented to the language learner, then we cannot prove that
any transformational grammar is learnable.

The next move, then, is to figure out a way of enriching
the information available to the language learner. We
do this using the following rationale. Sentences are
not presented to the child in isolation. Rather, he
hears them in context. Perhaps, during the period when
he is learning language, the sentences are simple enough,
so that he can figure out their meaning. That this
assumption may be roughly correct empirically receives
support from Slobin (1975, p. 30) who writes:

Most studies of child language comprehension
put the child into a situation where there are
no contextual cues to the meaning of utterances,
but, in real life, there is little reason for a
preschool child to rely heavily on syntactic factors to determine the basic propositional and
referential meaning of sentences which he hears.
Judith Johnson and I have gone through transcripts
of adult speech to children between the ages of
two and five, in Turkish and in English, looking
for sentences which could be open to misinterpre-
tation if the child lacked basic syntactic knowledge, such as the roles of word order of an
utterance which could possibly be misinterpreted. That is, the overwhelming majority of
utterances were clearly interpretable in context, requiring only knowledge of word meanings
and the normal relations between actors, actions,
and objects in the world.

Thus empirical studies of the context of language acquisition may once again aid us in simplifying the learning
problem, by allowing us to assume that the child can
construct an interpretation of the sentence from one
context. The theory makes one other assumption - that
the learner has the cognitive capacity to convert the
semantic interpretation into a syntactic deep structure
(thus the theory assumes the Katz-Postal Hypothesis,
which says that semantics is defined on deep structure)
To a certain extent, of course, this capacity must be
learned, since there is no universal relation between
an (ordered) deep structure and semantic interpretation.
We assume that we are studying the learner at a point
at which he has already learned the relation between
semantic interpretation and deep structure. For a
suggestion toward a theory of the learning of this
relation, see Wexler and Culicover (1974).

Thus we can formalise the input data to the child as a
sequence of (b,s) pairs, where b is a base phrasemarker and s is a surface sentence, with b and s related
by the transformational component. The learner receives
this sequence of (b,s) pairs from a particular language.
He constructs a transformational component by applying
a learning procedure to these data. The learning procedure, as in our earlier diagram, contains two parts.
First, it contains a mechanism for adding and discarding
hypotheses. Second, it contains a set of formal constraints on the kinds of possible grammars. We have

proved (Hamburger and Wexler 1975) that the procedure
will learn with probability 1 any of the allowed grammars, when presented with data of the appropriate form.

The learning procedure can be shown to work only if
a) semantic information of the kind described is presented to the learner and

b) particular formal constraints are placed on the
    class of grammars.

A number of these constraints are not the usual constraints described in grammatical theory. We have found
quite a bit of linguistic evidence that the constraints
are correct (Culicover and Wexler, 1976, in press).
Thus we have two kinds of evidence that the constraints
should be part of linguistic and psychological theory.
First, they are necessary for language acquisition and
second, there is linguistic evidence in their favour.

The learning procedure hypothesises or discards at most
one transformation each time a datum is presented. Also,
the procedure has no memory for past data. Thus it
meets the condition laid down in the quote above from
Braine. In line with our previous arguments, it should
be noted that these two conditions, which modify the
instantaneous model, do not make it easier for us to
prove language learnability, but harder. They are
empirical psychological conditions which we want our
theory to meet. Because of them, we need to add further
formal constraints to the theory, which we then hypothesise are part of the mental capabilities of the
child.

Questions for Developmental Psycholinguistics

1) The theory, as we have developed it, is only a first
step. Obviously, to be a general theory of language
acquisition, it would have to include a theory of the
learning of phonology, semantics and pragmatics (to the
extent that those are learned). But even with respect
to syntax the theory must be greatly developed. We
have not considered other kinds of syntactic devices,
for example rule features or more complicated kinds of
transformations. We have assumed only obligatory transformations, because of the particularly difficult learnability problems of optional transformations. This is
the kind of formal problem where empirical psychological
knowledge could be quite helpful, so it is perhaps worth
discussing in somewhat more detail.

Recall that obligatory transformations are ones which
must apply when their structural descriptions are met,
whereas optional transformations may apply or not when
their structural descriptions are met. To see the problem, suppose that a learner over-generalises, that is,
he hypothesises a too general optional transformation
T_C (C for "child") whereas the correct transformation
T_A (A for "adult") is less general. Now suppose a
datum is presented which the adult's transformation T_A
does not fit, but which the learner's transformation
T_C does fit. Assuming that all the other transformations in the adult and the learner components are identical, if the learner applies T_C he will make an error,
since in the adult (correct) component the datum is
untransformed by T_A. But if the learner does not apply
T_C he will be correct on the datum. Thus he takes the
latter course and is correct, and thus does not modify
his transformational component.

We could assume that some of the time when he is correct
(i.e., with a non-zero probability) the learner rejected
or added a transformation. Then this particular problem
would be solved, but then obviously the learning procedure, if it ever converged to a correct grammar, would
not stay there with probability 1. Thus the general
learning problem would not be solved.

Thus, ignoring this last suggestion, if the learner has
an incorrect component with a too general optional
transformation, he will never change it. In other
words, the learner will be able to derive every (b,s)
pair that is presented to him. But he will also be
able to generate (b,s) pairs that are not presented to
him. If there is no correction by the adult, (i.e.,
no negative information) then the error will remain.

What are possible solutions to the problem of optional
transformations? First, perhaps there is some correction. I earlier indicated why I think that negative
information should not be assumed (the Brown and Hanlon
results), but there may be a more limited way in which
it can play a role. This is the kind of question on
which we need detailed information from developmental
psycholinguistics. It may be that certain kinds of
errors are corrected. For example, if the child over-
generalised the Passive transformation, so that it
applied to the verb weigh, deriving, "160 pounds are
weighed by Daddy" from "Daddy weighed 160 pounds," his
parent might correct him, or not understand him. Or it

might be that, while parents don't correct and are
understanding, even of ungrammatical sentences, other
children aren't so sanguine, and don't understand such
sentences. Second, perhaps a more powerful learning
procedure could be found, which remembered to which
analyses transformations actually apply. Although we
do not know whether this can be done, and suspect that
it cannot, this procedure might yield a solution.
Third, perhaps the procedure might be modified so that
it selects the least general (i.e., the most specific)
analysis first, and only discards this analysis when
evidence against it is presented. We suspect that,
once again, this procedure might not solve the problem
in general and, furthermore, it will not allow for the
most striking observation of developmental psycholinguistics: overgeneralisation. A fourth possibility
is that there may be some restrictions possible on the
statement of what transformations can do, and that this
restriction will allow overgeneralisation of a particular kind only, which can be corrected.

Alternatively, perhaps the solution is that <u>no</u> "solution" is needed, at least for some cases. Perhaps, in
fact, such overgeneralisations of optional transformations remain in the adult grammar of the language learner and are a mechanism of linguistic change. In essence the situation is that the adult grammar can generate the base-surface pair (b,s) where the optional transformation T_C does not apply to b. The child's grammar
does apply T_C to b, which, since the transformation is
optional, yields (b,s) and (b,s'). Thus we have the
following situation.

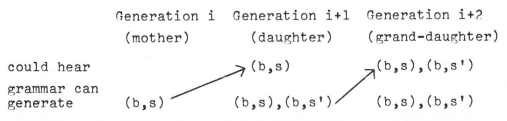

	Generation i (mother)	Generation i+1 (daughter)	Generation i+2 (grand-daughter)
could hear		(b,s)	(b,s),(b,s')
grammar can generate	(b,s)	(b,s),(b,s')	(b,s),(b,s')

The child's grammar (generation i+1) possibly contains
an extra derivation (b,s') which we assume is not corrected. Thus (b,s') becomes a part of the input to the
child (generation i+2) and thus will definitely be generated by that child's (generation i+2) grammar. Thus a
possible change in one generation definitely remains in
succeeding generations. (The transformation, of course,
is always optional).

Change of this kind, in which the domain to which a
transformation applies has been broadened, is called
"simplification" in traditional studies of linguistic
change (usually phonological) because when a domain is
broadened, the statement of the domain's structure is
usually simplified.

If this analysis is correct, then we can make the following prediction (about syntax at any rate; the conditions of input to the child, etc., might be considerably different for phonology):

2) Historical simplification (generalization) will affect
optional transformations, but not obligatory ones. This
statement, of course, assumes that no other processes
are at work. The reason that obligatory transformations
will not simplify (generalize) is that if they do, there
will be contradictory input to the child (i.e., evidence
against them).

    Our original learnability proof (Hamburger and
Wexler 1975) derived a finite bound on the complexity
of the data that a learner would need in learning a
grammar.   But this bound is huge. I once calculated it
for a too simple artificial grammar, and the result was
40,000.   (Complexity is measured as the depth of
Sentence embedding in the base phrasemarker.) However,
in our current work, we are attempting to lower this
bound drastically. We suspect, in fact, that we can
reduce the bound to 3, although this takes new assumptions about constraints on grammars. In other words,
if we can prove this result, we can show that even
though a language learner will eventually learn a
grammar which can deal with complex sentences, still,
in the process of learning, he need only deal with
sentences which contain embedded sentences which contain
sentences which contain sentences, and no sentences
more complex than these. The results of Ervin-Tripp
that we quoted earlier indicated that children did not
hear many complex sentences. However, at some later
stage they will have to hear embedded sentences in order
to receive the data on which they will build transformations which only apply to complex sentences. It
would be useful to have an account of the kinds of complex sentences that children hear. In particular, how
complex can the sentences they hear (and understand)
get? Is a depth of 3 too large? Or, in fact, can
children deal with more complex embeddings than these?

3) We need to know much more about the kinds of information that a child receives about sentences. Can expansions ever play the role of negative information? What
kinds of expansions are there, and what kind of information do they give the child?

4) Is it true, as Slobin has suggested, that the meanings
of sentences directed toward young children are clear
from the words and the context? If only partially true,
what kinds of sentences is it true for?

5) Do children exhibit memory for sentences that they
have heard in the past, in particular for sentences that
would not fit into their grammatical frame-work, so
that these sentences may be used for grammatical construction? I doubt that they do, but a positive answer
even of a limited kind would prove helpful.

It is questions like these whose answers are important
to the theory of language acquisition. Some answers
will make the problem harder, so that further formal
constraints on grammars will be necessary. Others will
make the problem easier. We can hope that linguistic
and psychological efforts together will eventually
converge on a satisfactory theory of language acquisition.

                              Footnotes

1 This work was supported by the National Science
   Foundation under Grant NSF SOC 74-23469.

2 Actually, most linguists ignore the question of
   language acquisition, and many of those who discuss
   the problem actually take the position which I am
   characterizing as the "psychological" position.
   Nevertheless, it seems to me that there are two
   positions here, with histories which can be
   traced back to linguistics and psychology.

3 This distinction reflects traditional terminology,
   but Chomsky (1975) uses the term "learning theory"
   to include the specification of the formal constraints. Chomsky offers arguments as to why, in
   fact, these constraints are the most interesting
   part of "learning theory".

4 Of course, it may be true that the "information
   processing" capacities of the child are more suited
   for small amounts of data at a time. But we are

simply arguing here that the instantaneous model
will require the fewest formal constraints in
order to allow the learning mechanism to learn the
language. Since the instantaneous model does not
have these information-processing limitations, it
will not be affected by all the data being available at once.

5   Noam Chomsky (personal communication) points out
that these results, even if true, are relevant only
if it is the case that children can learn language
only from speech addessed to them and that data
that is not addressed to them cannot play a role in
language learning. If it is possible for children
to learn language by being immersed in an essentially
adult environment, then our theory will have to
allow for this fact. I know of no empirical work
which bears on this point, which should be of major
interest for developmental psycholinguistics. Note,
though, that the complexity of adult speech can
help to provide <u>more</u> data for the language learner
and thus will make our problem easier. Errors,
however, work in the opposite direction. If there
are any errors in the data which children use, and
if these errors are not marked in some way by extralinguistic considerations (e.g. long-pauses), then
this will create a problem for learnability theory
which will have to be dealt with.

References

Braine, M. D. S. (1971). On two types of models of the
    internalization of grammar. In D. Slobin (Ed.),
    <u>The Ontogenesis of Grammar</u>. Academic Press, New
    York.
Brown, R. and Hanlon, C. (1970). Derivational complex-
    ity and order of acquisition of child speech. In
    J. R. Hayes (Ed.), <u>Cognition and the Development
    of Language</u>. Wiley, New York.
Chomsky, N. (1965). <u>Aspects of the Theory of Syntax</u>.
    MIT Press, Cambridge, Ma.
Chomsky, N. (1975). <u>Reflections on Language</u>. Athen-
    aeum Press, New York.
Culicover, P. and Wexler, K. (1974). <u>The Invariance
    Principle and universals of grammar</u>. Social
    Sciences Working Paper No. 55, University of
    California, Irvine.
Culicover, P. W. and Wexler, K. (1976). Some syntactic
    implications of a theory of language learnability.

In P. W. Culicover, T. Wascow, and A. Akmajian (Eds.), Studies in Formal Syntax. Academic Press, New York (in press).

Ervin-Tripp, W. (1971). An overview of theories of grammatical development. In D. Slobin (Ed.), The Ontongenesis of Grammar, Academic Press, New York.

Hamburger, H. and Wexler, K. (1973a). Identifiabllity of a class of transformational grammars. In K. J. J. Hintikka, J. M. E. Moravcsik and P. Suppes (Eds.), Approaches to Natural Language. D. Reidel, Dordrecht, Holland.

Hamburger, H. and Wexler, K. (1973b). A mathematical theory of learning transformational grammar. Social Sciences Working Paper No. 47, University of California, Irvine; also Journal of Mathematical Psychology, 12, 2, May 1975.

Slobin, D. I. (1975). Language change in childhood and in history. Working Paper No. 41, Language Behavior Research Laboratory, University of California, Berkeley.

Wexler, K. and Culicover, P. (1974). The semantic basis for language acquisition: The Invariance Principle as a replacement for the Universal Base Hypothesis. Social Science Working Paper No. 50, University of California, Irvine.

Wexler, K., Culicover, P. and Hamburger, H. (1975). Learning-theoretic foundations of linguistic universals. Theoretical Linguistics, 2, 3, 215-253.

Wexler, K. and Hamburger, H. (1973). On the insufficiency of surface data for the learning of transformational languages. In K. J. J. Hintikka, J. M. E. Moravcsik and P. Suppes (Eds.), Approaches to Natural Language. D. Reidel, Dordrecht, Holland.

THE INTERPLAY OF SEMANTIC AND SURFACE STRUCTURE ACQUISITION

LARRY H. REEKER

University of Arizona

This paper is designed to report on a theory of language
acquisition embodied in a computational model, which is rather
unique among theories of language acquisition in the degree to
which it assumes an interplay between surface syntax and considerations of meaning. Since the theory is explicit as to where this
interplay takes place, it will be possible to discuss the necessity
of both semantic and surface structure in concrete terms. The
theory and its computational model are both called PST - the Problem
Solving Theory. The creation and improvement of PST has been
taking place, with some interruptions, since 1967. The latest
version of the computational model is called Instance 3.

MODELLING SCIENTIFIC THEORIES ON THE COMPUTER

Language acquisition studies at the present time appear to be
concentrating on data collection with little overt concern or
agreement as to underlying acquisitional mechanisms. While it is
easy to point to inadequacies in earlier theories of child language
development, it is much more difficult to come up with new theories.
It is clear that there is a tendency to consign behaviour less
quickly to innateness than was the case a few years ago, but there
is also a reluctance to suggest in any detail how language might be
learned. This is regrettable, since theory formation has histor-
ically been an integral part of science for very good reasons.
Theories have guided experimentation and provided a framework in
which to arrange the facts, enabling generalizations which predict
new phenomena. Even theories that have turned out to be incorrect
(as most have, in any field of science) have served these purposes,
often for long periods of time. Granting the importance of further

data gathering in the field of language acquisition, then, it is
also important to formulate plausible theories that account for
the known data.

The difficulties of theory formulation and testing in the behavioural sciences are well known. In order to cope with these difficulties, an increasing number of behavioural scientists have turned
to one of science's most flexible tools - the electronic digital
computer. A scientific theory expressed in terms of a computer
programme is as precise as any mathematical theory, yet the
medium is sufficiently flexible that the theory need not be
forced into any particular mould.

Computational modelling consists of writing programmes for a
digital computer which, when executed on the computer, produce
behaviour similar or identical to that produced by a naturally-
occurring system (when executed, the programmes are said to
'simulate' the behaviour, so the term 'simulation' is sometimes
used instead of 'modelling'). The mere resemblance by the computer
output of the behaviour in question is no assurance that the
model works in the same manner as the natural system. Many computer simulations do not pretend to go any deeper than the most
superficial behaviour. If, on the other hand, the components of
the model emobdy everything that is known concerning components
of the natural system, if they are 'reasonable' on the basis of
the usual requirements on scientific theories, and if the behaviour
coincides with the known behaviour of the modelled system one
would be hard-put to find any differences between the computational
model and any other sort of scientific theory, medium excepted.

In the absence of a generally-accepted model of mechanisms responsible for some sort of behaviour (such as language learning) computer
modelling is an experimental approach. Plausible mechanisms of
some complexity can readily be tested on a computer to determine
whether or not they generate plausible behaviour.   Models of
very complex systems are likely to be so complex that simulation
provides the only feasible method for determining their behaviour.
With such models, intuitions are far too imprecise for the task of
evaluating the behaviour, and presently available mathematical
techniques are not adequate.

There are a number of dangers inherent in computer modelling. One
danger lies in becoming overly infatuated with a model on the
basis of a restricted range of outputs, when the model's overall
behaviour may be at variance with the empirical data, or the model's
internal structure is not consonant with well-founded theories of
related phenomena. Basically, this is a danger in any sort of
theorizing, but it is somehow very easy to get carried away when
the execution of a complex computer model produces 'interesting'

behaviour.

Another danger lies in the fact that the digital computer, while
ultimately general in the range of procedures that it can execute,
is itself organized in a rather arbitrary fashion, and there is no
reason to assume that its organization in any way reflects the
organization of any naturally-occurring system (e.g. the human mind).
The same can be said for the programming languages used to write
computer models of behaviour. Thus at some level, the model will
tend to reflect the computer organization or the organization of the
programming language, and one must be careful to separate legitimate
theoretical constructs from the constructs required by the modelling
medium. The same precaution applies, of course, to mathematical
modelling in general.

The computational modelling process forces a degree of explicitness
that is often absent in discursive expositions of theories, and
this can be beneficial if one keeps in mind the cautions enunciated
above. In addition, it is often possible to approximate unknown
components with 'rule-of-thumb' or statistical components within
the overall model, and to experiment with variations on other components. Later, the components in question can themselves be modelled
explicitly. The technique of working from the larger system down
to subsystems is called 'top-down modelling', and it is the strategy
that has been employed in the PST system, discussed below.

MODELLING LANGUAGE ACQUISITION

Linguistic systems are sufficiently complex that much controversy
remains concerning what an adequate model will look like. The
assumptions made in this paper are quite basic. Excluding consideration of phonology, the three components of a grammar that must be
acquired by a child are

(1) Some representation for the surface syntactic structures
of the possible utterances in the language.

(2) A representation for the meanings expressible by
utterances in the language.

(3) A mapping (which, since it represents the relation between
utterances and their meanings, we will call the semantic
mapping) between the surface structures and their meaning
representations, and vice-versa.

In this paper, we will make a common simplifying assumption and
confine utterances to sentences.

Contemporary linguistic theories can shed some light on the space

of possible grammatical theories, since some models have been shown
inadequate to handle the complexity of natural languages. It is
not necessarily true, however, that linguistic theories have any
direct correspondence to the systems that are acquired by the human
being. We can refer to the grammar that is acquired by a human as
the <u>mental grammar,</u> and since it is the mental grammar that will be
of primary importance in our discussion, the term 'grammar' used by
itself will denote that construct, when there appears to be no
danger of confusion.

For the surface structure description, the usual array of alternatives is available, including context free and context sensitive
grammars. Transformational grammars may be used to effect economy
in surface structure descriptions; in fact, they were originally
introduced for precisely that purpose.

For the meaning representation, a number of psychological and
linguistic theories have been proposed, including that of Anderson
and Bower (1973). The Katz-Fodor semantic theory began with a
rather simple meaning representation, consisting of semantic
features (Katz and Fodor (1963)) and has moved toward increasing
complexity in more recent publications (Katz (1972)). In some
cases, it almost seems as if something like a visual image provides
the most satisfactory meaning representation. Within computational
theories, the most highly developed meaning representation is that
of Schank (1969, 1973). The representation used in PST and
illustrated in Fig. 1, called semantic dependency, resembles that
of Schank (Reeker (1969, 1975)).

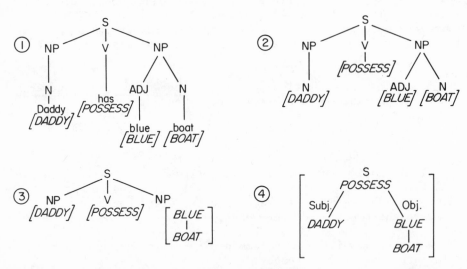

Figure 1. Semantic mapping of phrase marker to meaning representation.

A similarly large variety of alternatives has been suggested for
the semantic mapping, though we will disregard some of the more
ad hoc suggestions. Generally semanticists mix syntactic description and semantic mapping, using transformations for each.   In
addition to the other virtues enumerated by its proponents, this
approach captures the intuition that syntax has developed merely to
facilitate and clarify the semantic mapping. Transformations map
phrase markers to phrase markers, so the meaning representation in
generative semantics is ultimately in the form of a phrase marker
also. A more general (though not necessarily more powerful)
technique for defining mappings on trees is an extension of the
method used to define 'truth' in formal logic.   Knuth (1968) has
discussed this very general method with an eye to the definition
of programming languages.   In Knuth's terms, the method of Katz
and Fodor is an attribute grammar with synthesized attributes
(i.e. attributes are defined in terms of attributes of lower nodes
within the phrase marker), though much of the power of the Katz-
Fodor method comes from the fact that a portion of the mapping
has already been accomplished by a set of transformations. The
mapping used in the example later in this paper is an attribute
grammar with attributes that may themselves be complex trees (that
is, semantic dependency trees). Another alternative that merits
exploration is the augmented transition network (Woods (1970)).

In modelling grammatical acquisition, one may choose to start with
a system for learning any of the components of the grammar; yet
it is not possible to ignore the others. It is highly doubtful,
for instance, if the learning of the surface syntactic structure
can take place in the absence of both a meaning representation and
a semantic mapping. Consider, for example, a child who has never
heard language except from a radio, and thus is provided only
surface input. It is not to be expected that he would learn the
meanings of sentences in the language, of course. It is an open
empirical question as to whether he could determine the system
inherent in the surface structure, given enough time, but it does
not seem credible to this author. Yet researchers have sometimes
proposed models of syntax acquisition based upon surface input
alone.

Once one decides that all three components of the grammar must be
included in the model, another problem arises: how is it possible
for the child to obtain the meanings of sentences for which he has
not yet acquired a grammar? Here it seems necessary to assume that
meaning initially is learned separately from the sentences which
the child hears. There may later be elements of meaning derived
from portions of adult sentences which the child cannot fully
understand. In building a model of the process, it is not unreason-
able to provide as input both the adult sentences and their meanings.

The next decision that needs to be made is what one is going to
assume that the child has learned at any point in his acquisitional
history. In some modelling, it has been assumed that the child
acquires subsets of the adult language, yet the phenomena do not
support that view. Not only are children's own utterances of a
'telegraphic' sort, but their constructions are consistent enough
to make it appear that the phenomena are best described as a mental
grammar. A system that models the learning of full adult sentences,
forming a larger and larger subset until the full adult grammar
is acquired will be called a <u>sentence-acquiring</u> system. The PST
system, described below, is not a sentence-acquiring system, but a
successive approximation system. That is, the sentences generated
by the child grammar become more and more 'adult-like' as time
goes on.

DIVIDING UP THE LEARNING TASK

Despite the interaction that must take place among the various
linguistic systems, it will be the position of this paper that
language learning is not a monolithic task involving a single type
of learning, but rather, that there are several identifiable components, each of which should be termed 'learning'. We shall now
consider what these components might be.

If the grammar of the adult language is to be induced from utterances heard by the child, there are two logically separable
processes that must take place. The first is some determination of
the structure of the language. As a simple example of this,
consider the sentences 'He walked' and 'He talked'. Both of these
sentences are of the form 'He X', where 'X' represents either
'walked' or 'talked'; and this fact can be reflected in the surface
grammar

$$S \longrightarrow he\ X$$
$$X \longrightarrow walked$$
$$X \longrightarrow talked$$

Notice, however, that such a grammar cannot be written until there
has been some structured analysis of the sentences. For consider
the sentence 'He walked and talked'. This might be expressed as
'He X' also, but in the absence of any evidence to the contrary,
the language learner might just as well analyze it as 'He walked
and X', or even 'He walked X', which would lead to a bad generalization about the language, or to no generalization at all.

<u>Generalization</u> is the second learning process that must take
place, and it will interact with the first process, which we shall
call <u>structural learning,</u> or structural induction, to form a surface
grammar of the language. That there are real problems concerning
generalization mechanisms, even when the structural data are

available, is one of the first lessons to be learned in computer
simulation. The sentences 'They eat fish', 'They may fish' and
'They and I fish' can be generated by the following very
economical surface grammar:

$$S \longrightarrow \text{They X fish}$$
$$X \longrightarrow \text{eat}$$
$$X \longrightarrow \text{may}$$
$$X \longrightarrow \text{ and I}$$

Clearly, any economy gained in such a surface specification is no
compensation for the difficulty that will be encountered in trying
to specify the semantics of these sentences, if that semantic
specification must be based on the surface grammar. Generalization,
then, must be based on more than raw surface data. Furthermore,
when the 'obvious' sorts of heuristics are incorporated within a
computer programme and supplied with a reasonably large, 'typical'
sort of input, the amount of overgeneralization produced is so
massive as to be ludicrous.

Since more is being learned than just surface structure, additional
learning processes must be involved. One of these is what we might
call 'lexical semantics learning' - the learning of the meanings
of individual lexical items. What is acquired is a semantic mapping
from an individual terminal symbol in the surface grammar to a
meaning representation for that symbol. Although this sort of
learning concerns individual lexical items, it does not follow that
it takes place in isolation from full utterances. In fact,
linguistic, as well as situational, context must be of increasing
importance in acquiring lexical semantics as language learning
progresses. Somehow, the meaning representation itself must be
learned (although a predisposition to certain representations may
be innate), but we will not deal with this aspect of learning, since
it moves beyond language to the realm of representation of
general knowledge.

Given the knowledge of the meanings of individual lexical items,
how is it possible to acquire the more complex semantic mapping
for utterances? For a number of reasons, it appears that the
learning of the semantic mapping for sentences must take place
simultanesouly with the acquisition of surface grammar, and be a
basic part of the structural learning process. This is the
assumption made in the Problem Solving Theory, where surface
acquisition is aided by the learning of the semantic mapping, and
vice-versa.

A last form of learning has to do with making behaviour automatic,
or mechanical. This is clearly necessary with language, which does
not ordinarily require conscious thought beyond that enunciated in

the behaviour itself, either for perception or for production. It
is the system which has been abstracted from instances of adult
language and generalized that will be 'mechanized' rather than the
individual instances, and this fact has made the mechanization of
language difficult to explain on the basis of the best-understood
types of mechanization, such as conditioning and mediation. PST
does not try to deal with the mechanization problems, either, though
some remarks have been made concerning it in defending some of
PST's mechanisms (Reeker (1974)). Mechanization is an interesting
problem, but it is certainly not peculiar to language.

We have now discussed some of the problems that must be dealt with
in setting up a computational model, and the solutions that have
been incorporated in one such model, PST. In the next section, PST
will be examined in a little more detail. If the preceding
discussion sounded very much like a discussion of formulating
theories of language acquisition, that is because computational
modelling is not really any different from theory formulation in
general.

THE PROBLEM SOLVING THEORY

One computational model of a language acquisition process, the
Problem Solving Theory, deals with what we might call 'grammatical
learning', that is, the acquisition of a surface structure grammar
and a syntactic mapping. It assumes that lexical semantics learning
is taking place separately. PST is an ongoing project, of which a
completely new version, Instance 3, has recently been written. We
shall sketch very briefly what is included in Instance 3 and what
other problems need to be faced.

PST was developed to model the changes that take place in a child's
language from the beginning of one 'time period' to the next. Since
the theory says that the change takes place as a result of adult
language input, the time in which an input is received and causes
a change in the child's grammar is the time period used. The
presumption is that many utterances have no effect at all, or act
only as an aid in mechanization, so the time periods are not
uniform in duration. One then envisages a succession of child
grammars, which produce a succession of child languages.

The Problem Solving Theory says that language is learned as a
by-product of the child's solution of problems that arise naturally
in his attempts to reproduce adult sentences, or what he perceives
of adult sentences, using a child grammar. The particular form of
problem solving behaviour that is used involves detecting a
difference and making a simple change designed to remedy that
difference. For instance, if the adult sentence is 'Here's
Johnny's bear' and the child perceives this as 'Here Johnny bear',

but the child's own grammar generates only 'Johnny bear', the
difference to be remedied requires the addition of 'here'.
If 'here' is already present in the child's grammar, then it
belongs to some class H. Further, 'Johnny bear' will belong to
some class J. The new grammar will be able to generate HJ.
Furthermore, the semantic mapping of HJ will be added to the
grammar, based upon the mapping of H, J, and the perceived meaning
of the adult sentence, 'Here's Johnny's bear'. The overall
scheme of this structural induction portion of PST, with certain
of the interaction not detailed here, is shown in Fig. 2.

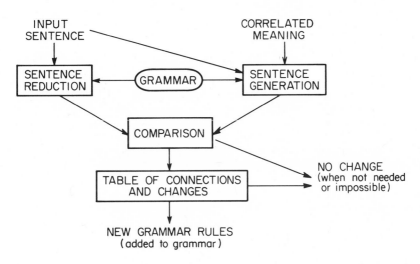

Figure 2. Structural induction portion of PST.

Problem solving in PST consists of remedying the type of difference
noted in the example above, in accordance with the table below
(called for historical reasons, partially to acknowledge a structural
similarity to Newell, Shaw and Simon's GPS Programme (Newell and
Ernst, 1969), the Table of Connections and Changes, or CC-table.
A portion of the table has been omitted).

Reduction (RS) / Child Form (ChS)	Sentence Change	Grammar Change
ZX / X	Prefix Z	$C1(X) \rightarrow C1(Z)C1(X)$
XZ / X	Suffix Z	$C1(X) \rightarrow C1(X)C1(Z)$
XYZ / XZ	Infix Y in X–Z	$C1(XZ) \rightarrow C1(X)C1(Y)C1(Z)$
XQZ / XYZ	Replace Y by Q in X–Z	$C1(XYZ) \rightarrow C1(X)C1(Q)C1(Z)$

Notice that this processes 'preserves' generalizations implicit in
the child's grammar, since anything that has previously been
included in the classes H and J can participate in the construction
HJ. However, the process itself has no provision for making the
generalizations in the first place; PST was originally designed
primarily to model structural learning, and experiments with it
have always provided a grammar with some generalizations already
made. It is clear from these experiments that the generalization
mechanism must be very conservative.

There are two respects in which PST may appear to be 'circular',
when in fact, it is merely 'recursive'. One of these is in the
reduction mechanisms, and the other is in the mechanism which
generates the child sentence. In both of these mechanisms, a
grammar is used, and if the child is just acquiring a grammar, it
seems paradoxical that he could be using it in the process. It
must be remembered, however, that the grammar being used at any
time is the child grammar acquired up to that time, not a full
adult grammar. All that is needed to complete a recursive specification is a description of what the child grammar consists of at
'time zero'. PST assumes that individual words and their meanings
were known at that time, and that the child could then perceive
strings of individual words, up to the limits imposed by a limited
short-term memory.

More thorough expositions of PST, with examples, can be found in

other publications (Reeker (1971, 1974, 1976), but it might be
well to discuss generalization in more detail, since it is a major
concern of Instance 3. As remarked earlier, the most obvious
schemes for generalization always seem to lead to massive overgeneralization - far more than is observed in the available data on
language acquisition. The structural learning mechanisms of PST
will magnify the influence of this overgeneralization by extending
it to larger and larger portions of the language. It is as
necessary, therefore, to be cautious with generalization as it is
to include it in the language acquisition model.

The generalization mechanisms of Instance 3 results in a certain
amount of overgeneralization - probably more than is realistic
in terms of observed child behaviour. It may be necessary to
discover and add additional constraints to keep things from getting
out of hand, and if so, some study will be necessary to determine
what sorts of constraints are reasonable. The generalization
mechanisms are already constrained by a number of principles. The
first of these is that meaning, as well as syntax, must provide a
minimal contrast. If the grammar has produced the rules

$$X ---> C_1 Z$$
$$X ---> C_2 Z$$

then a generalization is possible which merges C_1 and C_2. As
explained in Reeker (1974), however, there is a principle of
semantic consistency. In this case, this principle states that in
order for generalization to take place, Sem(X ---> C_1 Z) and
Sem(X ---> C_2 Z) must be exactly the same, except that Sem (C1)
in the first is replaced by Sem(C2) in the second. ('Sem (A --->B)'
can be read as 'the semantics of A when the rule A ---> B has
been applied in the generation'; Sem(A) is just the semantic
mapping of A.)

It is more difficult to deal structurally with another seemingly
reasonable constraint on generalization - that of semantic
commonality. One would like this constraint to require that C_1
and C_2 have 'something in common' semantically. Since PST
does not really come to grips with semantic decomposition of
lexical items (by, for instance, dealing with semantic primitives,
such as features) at the present time, it is not possible to get
at semantic commonality directly. On the other hand, it is not
clear that children do, either. Two ways of enforcing a measure
of semantic commonality are (1) requiring that C_1 and C_2 appear
in two (or some arbitrary larger number) common contexts, and (2)
requiring that Sem(C_1) and Sem(C_2) be structurally the same if they
are not single items. The second of these is included in
Instance 3, while the first is really not in the spirit of PST,
since it requires that information about ungeneralized previous
instances be stored for an unspecified length of time.

Sometimes it may be necessary to group items together in order
to make generalizations. For example,

$$X \longrightarrow C_1 Z$$
$$X \longrightarrow C_2 C_3 Z$$

might be generalized to

$$X \longrightarrow CZ$$
$$C \longrightarrow C_1$$
$$C \longrightarrow C_2 C_3$$

In that case, it seems that the meaning of $C_2 C_3$ must consist of
a single entity. This provision is necessary to assure that a
semantic mapping of the type used in PST will work, but it also
seems to be empirically valid and psychologically reasonable. To
illustrate this principle of semantic coherence it is perhaps
easiest to indicate a case in which it is not satisfied. Consider
the above rules with the semantic mapping

$$\text{Sem}(X \longrightarrow C_2 C_3 Z = \text{Sem}(Z)$$
$$\text{Sem}(C_2) \quad \text{Sem}(C_3)$$

PST would not allow this to be produced.

THE MODEL AND THE PHENOMENA

In principle, one wants to judge a theory of language acquisition
on how well its performance predicts the performance of a child.
There is one thing immediately wrong with that proposal in this
case, however: PST predicts that the learning of the child and
the order of acquisition are both highly dependent on the linguistic input. Because of the relative complexity of various grammatical constructions, and because of the simplicity of the elementary changes that PST allows, certain features of language will
have to be learned before others if a language is to be learned at
all, but that is predicted by most theories of acquisition. It
is not presently feasible to test a model such as this by giving
it exactly the same input as a child over a period of several years.
Nevertheless, it is the case that PST is not only the most
explicit extant model, but the most reasonable in terms of its
overall predictions of child performance. We will now examine
briefly the reasons for making such a statement. (No references
are included in this section. The material is available in the
literature, and much of it will be familiar to this audience. A
detailed examination of PST in light of the phenomena, including
data collected by the author, is included in (Reeker (1974)),
and though that material was written about an earlier version of
PST, and primarily in light of data available through 1970
an evaluation today would be similar).

As already pointed out, PST is a successive approximation theory, rather than a sentence-acquiring theory. That in itself sets it apart from most other models of language acquisition (see Reeker (1976) for a survey of other computational models). Not only the overall form of the language being learned, but the rate of learning as well, tend to favour PST. PST does not learn explosively, nor massively overgeneralize, even after adding the capability for affixing larger segments, so as to be able to deal with clauses. The learning rate is slow at first, when very few input sentences will trigger any learning at all, then increases, based both on increasing short-term memory and on the growing grammar being formed by the programme, then decreases again, as few sentences will trigger changes because most are understood without difficulty.

The initially small short-term memory is an important part of PST, since it enables the formulation of a rudimentary grammar that can later be expanded. If the short-term memory were very large at the beginning (as it does not seem to be in children), that would handicap learning in a PST-like model, providing a possible explanation for the 'critical age', after which language acquisition appears to be much more difficult. (The author hastens to add that the 'critical age' may not be so very critical as has often been thought. There have just not been any very good tests, since non-linguistic children have tended to be handicapped mentally - perhaps as a result of language deprivation, or of other forms of neglect, and second language learners suffer from the interference of other symbolic systems).

It should be remarked that the short-term memory model included in Instance 3 not only incorporates the commonly-held assumptions about short-term storage being in terms of 'hunks' (in this case, partially-parsed pieces of the adult sentence), but it produces sentence reductions remarkably like those produced by children.

The original formulation of PST was influenced by child monologue phenomena, such as the presleep monologues recorded by Ruth Weir, since the author observed in those monologues what could be interpreted as a vocalization of the type of comparison process now incorporated in PST. Similar phenomena are noted in various diary-based studies, such as those of Grégoire and Leopold. In seeking a theory of language acquisition, the author's conviction was that one ought to look first at general mechanisms, rather than specialized linguistic mechanisms, though the general mechanisms have undoubtedly been adapted in various ways to such specialized domains as language. The mechanisms of PST are general ones. In the structural induction portion, in particular, the reduction phase constitutes assimilation, the changes in the child's sentence and in the grammar constitute short-term and

long-term accommodation, in Piaget's terms. The comparison and
change mechanisms are like those used by Newell, Simon and
co-workers as part of a model of general problem-solving behaviour.
(These and other resemblances to other psychological theories are
detailed in (Reeker 1974)).

Although extensive experiments with PST involving later syntactic
acquisition have yet to be performed (and in fact, the implementation of the model is being rewritten – without changing the mechanisms postulated – to overcome limitations in the amount of
material that it can handle in a reasonable time), there do not
appear to be any difficulties in principle with a broad range of
advanced phenomena. The treatment of relative clauses, for
instance, is sketched below. Experiments at the early syntactic
level show it to be satisfactory in producing the usual 'pivot-
open' languages, without assuming salience of position, but merely
as a natural consequence of the inputs.

Adult speech to children does appear to facilitate learning – as
PST would predict – by providing a variety of utterances that are
'one stage ahead' of the child, or can be reduced to something of
that sort. In response to a child's utterance('Want book') a
parent will repeat back expansions to the child, often as part
of a question ('You want the book?' or 'You want the book, huh?'
or something similar). These expansions tend to provide the
sort of input that PST would use for structural induction. (Such
expansions are not essential, but speed PST-type learning,
provided they are accurate expansions, semantically). Although
more recent studies show that adult speech to children is actually
quite good, there do inevitably occur the false starts, hesitations, etc. that have been cited as evidence of the degeneracy
of the child's data. Most such degenerate utterances will not be
handled by PST at all, and thus will not affect the language being
learned, according to PST.

The philosophy of the PST project has been to start as simply as
possible, and gradually to refine and extend the system, without
changing its theoretical underpinnings. This process is continuing,
and will continue in the future. A module for mechanization
learning and provision for learning semantics of lexical items
through verbal context (that learned through nonverbal context
being beyond the scope of PST) will be added soon.

Although the preceding exposition of the background, purpose,
assumptions, and structure of PST has been a brief one, hopefully
it will serve to give the flavour of the effort. Basically, PST
represents an attempt to bring to bear the power of computational
modelling to formulate and test a theory based on the view that
the child acquires a series of grammars for successively more

adult-like languages, and that these grammars arise in a natural manner through the child's attempts to approximate adult sentences. Although it might be said that the child is imitating, his imitation is tied to the grammar that he has previously acquired, and the problem solving modifies this grammar, so that the problem solving method, when coupled with appropriate generalization mechanisms, leads to progressive and productive behaviour. This behaviour is only possible because both surface structure information and semantic information are available to the child, according to PST. In the next section, the extent to which both of these types of information are necessary is examined.

INFORMATION ABOUT SYNTAX AND SEMANTICS

Since the child is learning a grammar that consists of both syntax and semantics, it is appropriate to inquire where he obtains the information about each of them. When we talk about semantics here, however, we shall be referring to the mapping, excluding the basic form of the meaning representation, which is quite possibly innate, or learned by extralinguistic means. A number of hypotheses are presented by the structure of PST.

By 'surface information', we shall mean information regarding word order, and such information as can be obtained by classical immediate constituent analysis (Wells, 1947), without applying any meaning considerations. It has been shown (Gold, 1967) that given appropriate information concerning both which input sentences are grammatically well-formed and which are not, a grammatical inference mechanism can be formulated which will, in time, provide a surface grammar for any algorithmically well-defined set of strings. But this grammar does not include a semantic mapping (of course, since it is not given semantic information).

By 'deep information' is meant information concerning the meaning of sentences, communicated either through language or through context. Neither surface nor deep information is sufficient for a formulation of the sort of grammar that PST seeks to provide. As the 'They ---- fish' example above is designed to show, the maximally simple surface grammar will not provide a reasonable basis for definition of the semantic mapping. (This has to be a judgmental issue, since it can be shown that, in fact, a powerful enough mapping technique could - in principle, though inelegantly - assign proper meaning representations anyway). Similarly, deep information alone, without some analysis of the surface forms, in terms of other surface forms, will suffice only to produce a list of sentences and their meanings, which will not admit of generalization. It is this analysis of the surface material that is provided by the structural induction portion of PST.

PST uses both 'local information', regarding individual utterances,
and 'global information' involving the language of the child at any
particular time. It should be clear that the semantics cannot
reside only in the global information, since this is ultimately
obtained from local information. On the other hand, it is not
mathematically clear that the semantic portion of the global
information is necessary in obtaining the local information
(structure and meaning of individual sentences); but PST
claims that it is, and it appears reasonable (again, a matter of
judgment) that global information is necessary in obtaining local
information. (McCawley (1968) has made some telling remarks
concerning the necessity of global information, though he has
a slightly different model in mind).

A last distinction that might be considered is the distinction
which PST draws between structural information and generalization.
As pointed out above, however, both the surface syntax and the
semantic mapping are necessary in both components. Furthermore,
the structural information is necessary for generalization; and
while generalization is not necessary to the extraction of raw
structural information, it is clearly necessary if any sort of
reasonable grammar is to be available to the structural inference
portion.

It appears that it is not possible to characterize the notions of
'where semantics resides' or 'where syntax resides' by any of the
distinctions above. Furthermore, within each pair of distinctions
- surface, deep; local, global; structure, generalization - one
type of information appears necessary to characterize the other.
It might be added that, given the power of semantic dependency
mappings of the type used in PST (or for that matter, of either
the generative or interpretive versions of transformation theory),
it is not possible to determine a unique surface syntax and a
unique semantic mapping, for a language; or more strongly, that
given a grammar containing both surface syntax and semantic
mapping, it is possible to modify both components, moving
constraints in the semantic mapping into the syntax, and vice-
versa, without changing the language generated. This again
suggests the impossibility of dissevering syntax from semantics.
It also suggests that people's mental grammars quite likely
differ, one from another.

AN EXAMPLE: RELATIVE CLAUSES

To illustrate the application of PST to situations often described
by transformations, and thereby to indicate how the syntax and
semantics interact in the acquisition, we shall consider the

acquisition of relative clauses.   The capability for doing this
type of learning is something new in Instance 3 of PST.

First, a word about transformations: A number of constructions
that are often described in terms of transformations may be
naturally acquired with the mechanisms of PST.   This includes
passive and, of course, adjective preposing. This author's
conviction is that transformations have no psychological reality to
these cases, at least at the early acquisitional level. The
question then arises whether they have any reality in any case.
While it is presently not possible to give a definite answer to
that question, there are economy arguments that are still very
persuasive.   Anyone who has tried to write a phrase structure
grammar for use in a computational situation will appreciate
the economy afforded by transformations.   A transformational
grammar of English would be very large, but a phrase structure
grammar would be orders of magnitude larger.   It seems to this
author that these arguments, which were the original arguments
given for transformations (Chomsky, 1957), have some psychological validity.   But certainly no strongly persuasive evidence
exists.

The transformations that we will consider are perhaps more
analogous to those of early transformational theory (e.g. Lees,
1960) than to later theories.   This is because most 'semantic'
notions are handled within the semantic mapping, rather than
within the syntax.   Nor is there any motivation for finding
the maximally economical formulation, though in the large,
economy is a consideration, as indicated above.

The child must acquire four basic types of relative clauses,
and we will use an example from the interesting paper of
Sheldon (1974).   Our terminology will vary a little from hers,
however.   A relative clause in which the relative pronouns serves
as the object of the clause will be called an o-relative; if the
relative pronoun is the subject, it will be called an s-relative.
A subject o-relative will be an o-relative clause modifying
the subject of the main sentence, and similarly for the other
cases.

Let us consider the case of the object s-relative.   'The pig
bumps into the horse that jumps over the giraffe'.   PST will
immediately reduce this by removing the relative pronoun, which
can be an important marker in complex embeddings, but is relatively
content-free, and therefore acquired fairly late.   The production
portion uses the 'child grammar', which we will assume to
contain such rules as S → NP V NP,NP → the pig, etc. along with
the correlated semantics, and the meaning input (simplified
somewhat, for illustration)

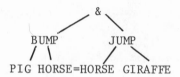

to determine the longest 'child' sentence which matches the input.
This is 'the pig bumps into the horse', which is matched to the
reduced adult sentence, leaving 'jumps over the giraffe'.

A suffix must be added to the child sentence at this point, and
this is 'jumps over the giraffe'. Now the implications for the
child's grammar are seen. Adding the suffix causes the form 'the
pig bumps into the horse', which is an S, to be expanded to S V NP.
The S has the structure NP V NP. As explained elsewhere (Reeker,
1976), the rules in the CC-table refer to the maximal syntactic
classes on the left side, and to the minimal syntactic classes
on the right. The new rule is therefore

$$S \rightarrow NP\ V\ NP\ V\ NP$$

The correlated semantics are obtained by matching the right
hand constituents to the words to which they refer in the meaning
representation of the sentence, i.e.

The pig bumps into the horse jumps over the giraffe

$$NP_{(1)} \quad V_{(1)} \quad NP_{(2)} \quad V_{(2)} \quad NP_{(3)}$$

for the rule

$$\text{Sem}(S \rightarrow NP_{(1)}\ V_{(1)}\ NP_{(2)}\ V_{(2)}\ NP_{(3)}) =$$

The other types of relative clauses are dealt with in an almost
identical manner, using the CC-table in each case to determine
both local and grammatical changes.[1] Notice that the grammar
captures the intuition that the sentence in question has a form

analogous to S → NP V NP, but a meaning that is quite different. Later, if 'that' is added to PST's lexicon, the input sentence will not be reduced, and a grammar will be produced that can account for the whole adult sentence.

CONCLUSION

In this paper, we have sketched the computational model of first language acquisition known as PST, have discussed some issues concerning the blending of syntactic and semantic information, and have outlined the treatment of relative clauses by Instance 3 of PST. It is hoped that the reader will get a feeling for the promise of this model as an explanation of the acquisitional process.

Footnote

1 Several types of transformational phenomena will be dealt with in a forthcoming paper, but the details are too long for this paper.

References

Anderson, J.R. and Bower, G.H. (1973). Human Associative Memory. Winston and Sons, Washington.

Chomsky, N. (1957). Syntactic Structures. Mouton, The Hague.

Fillmore, C. (1968). The Case for Case. In Universals in Linguistic Theory, E. Bach and R. Harms, (Eds.), Holt, New York.

Gold, E.M. (1976). Language identification in the limit. Information and Control, 10, 447-474.

Knuth, D.E. (1968). Semantics of context-free languages. Mathematical Systems Theory, 2, 127-145.

Katz, J.J. (1972). The Philosophy of Language. Harper and Row, New York.

Katz, J.J. and Fodor, J.A. (1963). The structure of a semantic theory. Language, 39, 170-210.

Lees, R. (1960). The Grammar of English Nominalizations. (Now available from Mouton, The Hague).

McCawley, J.D. (1968). Review of Current Trends in Linguistics, vol. 3: Theoretical Foundations. Language, 44, 556-593.

Newell, A. and Ernst, G.W. (1969). GPS: A Case Study in Generality and Problem Solving, Academic, New York.

Reeker, L.H. (1969). The generalization of syntactic specification. Abstract in Meeting Handbook, Annual Meeting, Linguistic Society of America.

Reeker, L.H. (1971). A problem solving theory of syntax acquisition. Journal of Structural Learning, 2, 4, 1-10.

Reeker, L.H. (1974). The problem solving theory of syntactic
    acquisition. Unpublished doctoral dissertation, Carnegie-
    Mellon University.
Reeker, L.H. (1975). Semantic dependency notation. Tech.
    rep. TR - 1 Computer Science Department, University of
    Oregon.
Reeker, L.H. (1976). The computational study of language
    acquisition. Advances in Computers, vol. 15 (M. Yovits,
    Ed.,), Academic, New York.
Schank, R.C. (1969). A conceptual dependency representation
    for a computer-oriented semantics. A.I. Memo No. 172,
    Computer Science Department, Stanford University,
    California.
Schank, R.C. (1973). Identification of conceptualizations
    underlying natural language. In Computer Models of Thought
    and Language, (R.C. Schank and K.M. Colby, Eds.), Freeman,
    San Francisco.
Sheldon, A. (1974). The role of parallel function in the
    acquisition of relative clauses in English. Journal of
    Verbal Learning and Verbal Behavior, 15, 272-281.
Wells, R.S. (1947). Immediate constituents. Language,
    23, 81-117.
Woods, W.A. (1970). Transition network grammars for natural
    language analysis. Communications of the ACM, 13, 591-606.

PREDICTIVE UNDERSTANDING

ROGER C. SCHANK

Yale University

I. THE USE OF KNOWLEDGE

A frequently ignored issue in the disciplines of psychology
and linguistics is the structure of texts. Sentences rarely
come in isolation, but rather in groups, the meaning of whose
combination is more than the meaning of the individual parts.
Such meaning combination can occur within a sentence as well.
Consider sentence (0)

(0)   John burned his hand because he forgot the stove was on.

Clearly an important element in the meaning of (0) is that
"John touched the stove". This element we have come to call
an inference (Schank and Reiger (1974), Wilks (1975)).   In
order for a person to make such an inference, he must know that
something was missing in (0). Without a need to fill in the
empty space in (0), no such inferring would take place.

Work on natural language understanding in Artificial Intelligence
has, by necessity, concentrated on such issues as inference
and prediction from input texts (as for example, Schank et al.
(1975), Wilks (1975), Reiger (1975) and Charniak (1972)).

In order to make a computer understand what it reads, it has to
have some preformed idea of what makes sense, so as to be able
to discover the need to know more, or the need to make sense of
disconnected information.

91

We illustrate with some simple stories:

1a)   John wanted to go to Hawaii. He called his travel agent.

1b)   John wanted to go to Hawaii. He called his mother .

2a)   John was in his history class. He raised his hand.

2b)   John was in his history class. He knocked over the
       garbage can.

3a)   John was walking in the park with his sweetheart, Mary.
       Suddenly, some boys jumped out from behind the bushes
       and held a knife to Mary's throat. They asked John for
       all his money. He gave it to them and they ran away.

3b)   John was walking in the park with his sweetheart, Mary.
       Suddenly some boys jumped out from behind the bushes and
       held a knife to Mary's throat. John laughed. The boys
       killed Mary and ran away. John picked up another girl
       in the park and went to the circus.

In examples 1-3, the "a" stories are more comprehensible than
the "b" stories. That is, they seem to violate processing
expectations that were built up in reading the story less than
do the "b" stories. Any theory of natural language understanding
must be able to explain this fact. In the processing of some
stories, the information that fills in the gaps, or connects
the pieces together is readily available. In other cases, it is
difficult, if not impossible to make the connections. In still
others, while it is possible to make the connection, the
inferred connections violate what one ordinarily would have
expected.   In particular, in 3b, the expectations that one has
are trampled upon to the extent that one is forced to conclude
that John is either crazy, extremely callous, or was in on the
murder in the first place. The fact that one's expectations
boggle, and can be fixed up by making up conclusions such as
those given above, would indicate that there is a capability
that people have in language understanding, that transcends
the language itself. People can make inferences, apply knowledge
about known situations, ascertain actor's goals, predict the
plans that actors' are using to attain these goals, and know when
a prediction has gone awry.

With this view then, let us examine the above stories, with
respect to the kinds of processing necessary to understand
them. In order to understand 1a, it is necessary to know that
"travel agents can make reservations for trips". If one also
has available the facts that "going to Hawaii" is an instance

of a trip and that "if you want to go on a trip, you must make
reservations", then the connections between the two sentences
are easily accomplished. (In fact, we have a computer program,
SAM, that does exactly this kind of connecting up of text). The
connections here come from knowledge about trips which we call
the trip script. A script is a detailed list of the events
that characterize a given situation, complete with options that
are available, what to do when things go wrong and any other
detailed knowledge that is specific to a particular situation.
In order to connect up a sentence with others that might follow
it, we call upon the scripts that it references and make predictions about how and when the pieces of those scripts can be instant-
iated. In the first sentence of 1a, there is a reference to the
trip script (going to Hawaii), and in the second, there is a
reference to one of the predictions made by assessing the trip
script (calling a travel agent). Thus the process of connecting
these sentences is easy. The story is readily understandable
because of the availability of knowledge in a standard form
(a script) to connect it up. The result of understanding 1a is a
representation of the meaning of both sentences connected by the
inferred reason behind John's action - namely, that he intended
to ask the travel agent to help him plan his trip. The story is
entirely sensible because we are able to construct between the
pieces of the text (in this case by reference to a script).

In story 1b however, such knowledge is not as readily present.
We cannot expect to have access to information about people and
their mothers with respect to trips. In any standard model of
the world, "mother" is not a part of the trip script. How do we
connect up the sentences of 1b then?

The connection can be made, but not anywhere near as certainly as
before. To make our connection, the trip script will not help us.
We must consult higher level information about trips and "mothers"
in general. Three possibilities normally occur here. People often
guess that either John's mother will supply the money for the trip,
accompany him on the trip, or give permission for John to go on
the trip.

Some comments are due here. First, it seems clear, given the
above possibilities, that 1b is an ambiguous story (or alternatively that the second sentence of 1b is an ambiguous sentence).
Clearly, this is a non-standard type of ambiguity in linguistics.
We call this ambiguity planning ambiguity which we shall not
discuss here (but see Schank and Abelson (in press)).

Second, in contrasting 1a and 1b, it becomes clear that 1b
requires more work than 1a. The reason for this is what we call
predictive understanding. In 1a the second sentence is predicted

by the script and thus easily handled. In 1b, the predictions of
the script do not help. More processing must be done. This
processing is now bottom-up (as opposed to the previously top-
down predictions). We would claim that in natural language
understanding, a simple rule is followed. Analysis proceeds in
a top-down predictive manner. Understanding is expectation-
based. It is only when the expectations are useless or wrong,
that bottom-up processing begins.

There is no "solution" to handling the second sentence of 1b.
The possible solutions come from matching the prerequisites
and goals for the trip script with the functional meaning of
"mother". One such need is money and the role of the mother is
often a provider of money. In some cases one such need is
permission and a mother is often a provider of permission. One
goal in a vacation-type trip is often companionship. Any person
can fill this role and mother is a person. There are, of course,
other possible interpretations. What is important to recognise
here is that in the understanding process such interpretations
are considered and selected among by the reader (or hearer). In
order to comprehend what you hear then, sufficient knowledge
must be available. The definition of mother available to the
system must contain the above information. Without it,
understanding cannot take place. Similarly, trips must be
defined in detail with respect to what happens in them and why
and how one starts them. Such information is crucial to the
meaning of trip and thus to the understanding of any sequence of
sentences about a trip. Understanding then, is basically an
inferential knowledge-based process.

In sentence 2a, a similar problem is shown. With knowledge
of the "classroom" script, the sequence is readily understandable.
In 2b however, the second sentence (John knocked over the garbage
can) is not normally in the classroom script. Any sentence must
be understood both with respect to how it relates to previous
predictions made by the text, and to how the new sentence itself
sets up predictions. The second sentence of 2b answers no
predictions, it does however, set up predictions.

Since it answers no predictions, there is a sense in which the
second sentence is not understandable in its context. Our
definition of understanding, in this view, is that a sentence is
understandable in context if it can be related to what has
preceded it. If it cannot be so related, although it may be a
perfectly understandable sentence in isolation, it is not
understandable in context.

But ,understandability is also a function of what follows a
sentence. In that sense we may or may not have an

understandable sentence in 2b.   If 2b is a complete paragraph
(that is, with nothing following it) then it is nonsensical because
the second sentence is not understandable.   However, if the
paragraph is still to be expanded, then the second sentence of
2b can be viewed as one that sets up predictions (context) for
future sentences. Predictions that would be set up here, might
relate to the reactions of his teacher, or some other consequence
of the kicking of the garbage can.   Furthermore, sentences that
provide connectivity might be encountered later on in the
paragraph that would concern possible reasons for his actions
(e.g. if it was intentional).

If subsequent sentences were, "John sat down", "later John was
made to stay after school", we would want to use our predictions
to understand that a punishment had occurred, and to tie that
punishment to the garbage can incident rather than to sitting
down.

In order to tie these sentences together correctly, we must
have available to us a theoretical apparatus that includes
within it, normal behaviour (scripts), intentional behaviour
(plans), and prediction about future behaviour (goals and themes).
One way to correctly handle 2b then is:

1)   predict, from the odd event in the classroom script, a
     reaction by the teacher;

2)   infer that a noisy event is a violation of the decorum
     goal of a teacher;

3)   predict a plan that will be created by the teacher to satisfy
     the decorum goal;

4)   understand "staying after school" as a standard scriptal way
     of satisfying such a goal.

What we are saying then is that understanding natural language
sentences requires one to understand why people do what they do
and to be able to place new inputs into a world model that makes
sense of people's actions.   To do this requires knowledge about
what people are liable to do in general, and detailed knowledge
about the standard situations that predicts what they will do in
a given context.

Knowledge about detailed contexts is crucial for understanding
stories dealing with those contexts.   Suppose 1a had been:

4a) John wanted to go to Hawaii.   He called his travel agent.
    He said they took cheques.

What is superficially a simple paragraph to understand is
virtually incomprehensible without a predictive apparatus
that has access to detailed knowledge about contexts (<u>scripts</u>).
The three sentences individually are much less than the sum of
their parts. The first sentence implies that John will need
money for his trip, for example. The second sentence implies
that he is aware of this fact and furthermore knows that the
travel agent can procure tickets and reservations that he will
need. The third sentence, implies that John was worried about
how to pay for his trip (probably he does not have the cash on
hand) and that he asked the travel agent about payment possibilities. The travel agent's response, in isolation, is highly
ambiguous. The sense of "take" used is not one of the first
few that come to mind ordinarily. Yet, in context, the sentence
is readily understandable. Furthermore, it is a reasonable
hypothesis that hearers of that sentence do not have any trouble
at all finding the correct sense of "take". Rather, it seems
likely that they never consider at all the possibilities that
"cheques" had been "cheated", "beaten in a fight", "stolen" or
anything else. Why is it that people just happen to see the
correct sense of "take" first? The answer is prediction. They
are expecting the possibility of "paying" in the first place
because "paying" is a part of the "travel agent script".

Furthermore the ambiguous "he" references are easily dealt with by
people because they know who would be paying and who would not.
People consistently keep the same role in a script. The traveler
in the travel agent script is the actor in the planning scene,
the payer in the paying script and the passenger in the airplane
scene. Similarly, the "they" in this story refers to the travel
agency management, an implicit referent within the story that is
quite crucial for its understanding Scripts are highly important
for resolving such anaphora problems.

II.   PREDICTIONS IN ACTUAL UNDERSTANDING

At Yale we have a computer program, SAM that uses scripts for
top-down predictions to enable understanding (see Schank et al
(1975) and Cullingford (1975)). An example of a typical story
that it has processed follows

John went to Leone's. He ordered a hamburger. The waiter said
they didn't have any. He asked for a hot dog. When the hot dog
came, it was burnt. He left the restaurant.

Our programs are capable of producing long paraphrases, summaries
and answers to questions that indicate an access to information
not explicitly in the story of the type that we have been
claiming is a prerequisite for understanding. The output of

SAM in question-answering mode (see Lehnert (1975) for a discussion of the question answering problem) is reproduced below:

Did John go to a table?
YES

Was John mad?
YES

Why did John get mad?
BECAUSE JOHN REALIZED A HOT DOG WAS OVERDONE

Did John pay the check?
NO BECAUSE HE BECAME UPSET BECAUSE THE HOT DOG WAS OVERDONE
AND SO HE LEFT LEONE'S.

Why did John go to Leone's?
BECAUSE HE WANTED TO EAT A MEAL AT LEONE'S.

As we have said, SAM applies its knowledge of scripts so as to make predictions about what it expects to hear. This predictive information aids in the initial parsing, where the correct senses in context are chosen for ambiguous words. It also aids in the inference process, where information that is left implicit is assumed because it has to be there according to the context. This allows SAM to have no problem with sentences that could refer to many different situations out of context, but are clearly unambiguous in context. Examples (5), (6), and (7) show such ambiguous sentences that in context clearly offer no problem.

5)   John wanted to impress Mary at the banquet. He asked for the check.

6)   John was finished eating. He asked for the check.

7)   John was disgusted at the service he was getting. He asked for the check.

We have no problem understanding the import of "He asked for the check" in (5), (6) and (7). Yet the import is clearly quite different in each instance. Scriptal ambiguity is, in essence, solved beforehand, by predictions based upon knowing what script (or scene of a script) you are in. These predictions cause one to not even notice these ambiguity problems. Our theory here would be that people do not see a number of choices and then select one. Predictions force your hand so that you only see the most obvious choice in each case.

All top-down predictive understanding cannot be explained by

scripts however. We have been developing (Schank and Abelson
(in press)) a theory of higher level knowledge structures that are
used in understanding of which scripts are the simplest kind.
Stories 3a and 3b are best handled in terms of what we term
plans and goals.

Basically, we view an understander to be a mechanism that is
continually constructing a theory about the goals, motivations,
and intentions of the characters he is hearing about. This theory
takes into account default norms of behaviour since we do not
expect to hear complete descriptions of every character as a
prelude to every story. Thus a predictive understander, has
available to it a set of standard goals for any given individual.
We expect then, that any new John will get hungry, lonely and
tired from time to time, and will want to eat, love, talk, walk,
play and sleep with varying frequency. Further we will want
valuable objects and dislike noxious things. All this seems rather
obvious, and it is. Somehow though, it has been ignored by
theorists who worry about understanding. Understanding takes
place in the presence of all this information. What seems obvious
is that any theory of understanding must take account of such
goal information and use it to aid understanding processes.

In story 3, after the first sentence, our default assumptions
about John include that he wants to love, that he will protect
his lover from harm, that he is happy and so on.

When a threat to Mary's health occurs in the second sentence of
3, predictions about John's doing something to prevent it, and
about him being frightened and worried must be brought forward.
These come from a standard default goal which we call π HEALTH
(loved ones). Attached to π HEALTH (loved ones) when it is
activated, are predictions about what one can do in given circumstances to protect someone from harm. Also are attached standard
state changes that occur in the person for whom it is a goal.
Predictions are attached to those state changes with respect to
how they might affect future actions in other areas. All this
information is now brought to bear by an understander so that it
can deal with new inputs accordingly. If the next sentence were
"He called for the police", we would understand this as reasonable
on the level of protecting a loved one in that it might ultimately
help, but foolhardy in that it might change the anger of the
assailant and further threaten Mary's health. New inputs are thus
evaluated in terms of expectations. In 3a when the money request
is made, we consult a rule about the relative ranking of π HEALTH
versus π(Valuable Possessions) and predict that John values Mary
more than money and will act accordingly. 3a is thus entirely
understandable because it meets our expectations fully.

Story 3b is another matter. We are left, after reading it, with a series of unanswered questions about just what was going on. Why did John laugh? How could he be so callous as to go to the circus? Was his only goal to have a date for the evening?

These questions are the residue from the predictive understanding mechanism. We expect a negative reaction by John. "There was moisture on John's face" would be interpreted as satisfying a "sadness-crying" prediction. Without this prediction that sentence would be quite difficult to understand. Without a goal prediction/goal understanding mechanism, we would not worry about why John laughed. A bottom-up program would treat "John laughed" as any other new input for which connecting reasons and results must be found. But, a top-down predictive program would recognise immediately that a prediction had been violated. This recognition comes from the expectation here that John will do something about his π HEALTH (loved ones) goal. This goal has been generated from the stated thematic relations between John and Mary.

The recognition of a prediction that has been violated forces the understander to create solutions to explain the error violation. In other words, it immediately directs processing to the source of the prediction. Here this would be the π HEALTH (loved ones). At this point in the processing we become bottom-up. We ask if there are goals or general states that are naturally relating to the action (laughing). "Laughing" is associated with a happy state. Since we had predicted a sad state the LOVE theme is rewritten and we must come to a new understanding of John's goals. ("Cry" here would have violated an action prediction but satisfied our state prediction. That is, we would wonder why John did nothing to prevent Mary's negative HEALTH change but we would not wonder about John's sadness).

At this point in the processing (after laughing) then, we must find out what goal John actually was operating under that had as its consequent a negative result for Mary. If none can be found directly, we can assume some things about John's mental health or let it remain a mystery. A hierarchy of goal types and information about how such processing proceeds is given in Schank and Abelson (in press).

The point here then is that the language understanding process is heavily reliant on predictions based on understanding the goals and intentions of the actors in a story. Psychological research should be concentrating finding out what assumptions people are making about goals and intentions during the understanding process so as to develop a theory of how predictions that come from those assumptions are used.

From the point of view of building intelligent programs, it
should be clear that story understanding program must be as
top-down predictive as possible.   At Yale we have built a prototype of such understanding system called PAM.   PAM understands
only a few stories right now, because it has only limited
knowledge.

An example of input and output of PAM is:

INPUT

John loved Mary but she didn't want to marry him.   One day a
dragon stole Mary from the castle.   John got on his horse and
killed the dragon.   Mary told John she would marry him.   They
lived happily ever after.

PAM produces the following paraphrase:

JOHN WAS IN LOVE WITH MARY.   SHE DID NOT WANT TO MARRY HIM.
A DRAGON TOOK HER FROM A CASTLE.   HE LEARNED THE DRAGON HAD
TAKEN HER FROM THE CASTLE.   HE MOUNTED A HORSE.   IT TOOK HIM
TO HER.   HE KILLED THE DRAGON.   SHE WAS INDEBTED TO HIM.   SHE
TOLD HIM SHE WAS GOING TO MARRY HIM.   HE MARRIED HER.   HE AND SHE
WERE HAPPY THEREAFTER.

PAM also answers questions such as these:

(1)   Why did John kill the dragon?
      BECAUSE HE WANTED HER NOT TO DIE.

(2)   Why did John get on his horse?
      BECAUSE HE WANTED TO BE NEAR HER.

(3)   Why did Mary tell John she would marry him?
      BECAUSE SHE WAS INDEBTED TO HIM.

PAM works in the way that we have been describing.   It locates
new inputs in terms of its expectations about how characters
will act.   Thus it really knows a lot about what it hears.   While
both PAM and SAM are predictive understanding systems, PAM
works exclusively from general knowledge and SAM works exclusively
from specific knowledge.   They are of course, instruments for
future research.

It seems clear that predictive understanding is an important issue
to which researchers on natural language should be paying
serious attention.   Sentences in isolation are not particularly
good vehicles for researching these predictive problems and
unfortunately such context-free sentences have formed the

back-bone of most linguistic and psycholinguistic research. The
problem of how humans understand text is full of complex issues
that interact with each other in ways that have made many
linguists and psycholinguists back off from them.

Understanding how language functions clearly involves understanding how humans understand. Our assertion here is that they
understand predictively. Figuring out exactly how and when
these predictions work is a serious and important issue for
research.

This work was supported in part by the Advanced Research Projects
Agency of the Department of Defense and monitored under the Office
of Naval Research under contract N00014-75-C-1111.

References

Charniak, E. (1972). Towards a model of children's story
     comprehension. AI TR-266, MIT.
Cullingford, R.E. (1975). An approach to the representation
     of mundane world knowledge: The generation and management
     of situational script. Americal Journal of Computational
     Linguistics. Microfiche 35.
Reiger, C. (1975). Conceptual memory. In R.C. Schank, ed.
     Conceptual Information Processing. Amsterdam: North
     Holland.
Schank, R.C. and Abelson, R.P. (in press). Knowledge Structures.
     Lawrence Erlbaum Press, Hillsdale, N.J.
Schank, R.C. and Reiger, C. (1974). Inference and the computer
     understanding of natural language. Artificial Intelligence,
     5, 373-412.
Schank, R.C. and Yale A.I. (1975). SAM-A story understander.
     Yale University, Department of Computer Science Research
     report 43.
Wilks, Y. (1975). A preferential pattern-seeking, semantics for
     natural language inference. Artificial Intelligence, 6,
     53-74.

SENTENCE CONSTRUCTION BY A PSYCHOLOGICALLY

PLAUSIBLE FORMULATOR

GERARD KEMPEN

University of Nijmegen

Natural language production comprises a variety of processes that may be grouped under two headings. The conceptualization processes select a conceptual theme for expression. They decide which parts of the theme must be actually communicated to the hearer and which can be left unexpressed: the latter are already present in the hearer's memory or can be inferred by him from what the speaker said. And the conceptual content selected for expression must be organized into a linear sequence of messages so that each is expressible as a complete or partial sentence. The psychological mechanism that accomplishes these tasks I will call the conceptualizer. The second main mechanism is the formulator which maps each input conceptual message into a natural language utterance. Formulating consists of two main processes:

(1)   lexical search, for locating and retrieving from memory language elements which express the conceptual information, and

(2)   sentence construction, i.e. assembling a partial or complete sentence from language elements.

It is only the latter aspect of language production this paper is concerned with. I will try to work the various empirical data that are known about human sentence production into a blueprint of a possible sentence construction procedure. The data I have in mind are observations on speech errors (Fromkin, 1973; Garrett, 1975), hesitation and pausing phenomena, (Fodor, Bever and Garrett, 1974), experimental results on semantically constrained sentence production and reproduction (several chapters in Rosenberg, 1976),

103

and some general properties of the human cognitive system. I offer
the blueprint as a point of departure for more detailed and more
formal theorizing on human sentence production and as a source of
meaningful hypotheses for psychological experimentation.

Section I reviews the main empirical facts the sentence construction system tries to encompass. Section II outlines the system
itself. Section III explores some linguistic consequences of the
model. Finally, Section IV contains a few detailed generation
examples.

I.   Empirical observations on sentence construction

I.1 Heavy reliance on multiword units

As building blocks to construct sentences from, the formulator
uses not only single lexemes but also multiword units which span
several lexemes and/or "slots" to be filled by lexemes. Becker
(1975) argues a similar point. Left-to-right order of the
elements of a multiword unit is more or less fixed. I quote
the following observations in support:

(i)   Phraseology linked to stereotype situations.People
         speaking in a stereotype situation often have available
         sentence schemes which they know will enable them to
         express what is on their mind. Especially if rapid speech
         is required, such schemes will actually be put to use
         (example radio reporters doing running commentaries of
         soccer matches or horse races). Although there is little
         empirical data to either underpin or undermine this claim,
         some informal observations by this author support it
         strongly (Kempen, 1976b).

(ii)   Syntactic retrieval plans. In a series of experiments on
         paraphrastic reproduction of sentences,I demonstrated
         the existence of retrieval mechanisms which look up a
         specific pattern of conceptual information in memory and
         directly express it in the form of a specific syntactic
         frame. For details see Kempen (1976a,b) and Levelt &
         Kempen (1975).

(iii)   Speech errors. Garrett (1975), in the most extensive and
         detailed study of speech errors to date, suggests that
         people sometimes use syntactic frames consisting of
         functors(articles, prepositions, inflection morphs, etc.)
         in left-to-right order, with slots to be filled by content
         words. Occasionally, content words are put into a wrong
         slot and speech errors like I'm not in the READ for MOODing
         and she's already TRUNKed two PACKS result (inter-changed

pieces in capital). But Garrett never observed errors like <u>The</u>
<u>boyING shoutS disturbed us</u> with functors interchanged.

I.2 <u>A clause-like unit as the largest unit of planning</u>

The units by which sentence construction proceeds vary from, say,
words to clauses. That is, the segments of speech that get attached
to the output string are neither very small units like single
phonemes nor very large ones like convoluted sentences. (I'm only
concerned with spontaneous speech production, not with writing).

The largest unit, and also a very predominant one, is often thought
to be the surface clause or the phonemic clause. Boundaries between
such clauses frequently attract pauses and hesitations. Boomer
(1965), however found that the highest proportion of pauses occurred <u>after</u> the first word of a phonemic clause. Disregarding this
exception, the other positions showed a gradual decrease towards
the end of a clause.

Many first words of clauses must have been subordinate or coordinate conjunctions. If so, Boomer's data suggest that decisions
regarding conjunctions are rather independent of decisions regarding the other lexical material (verb, noun phrases, etc.) of a
phonemic clause. Two possibilities come to mind. First, the
conjunction may have been selected at a very early stage of the
formulation process. E.g. a speaker who wants to express a
causal relation between two events may very early on decide to
construct an utterance of the form "EVENT2 <u>because</u> EVENT1".
After verbalization of EVENT2 he only needs to put the word
<u>because</u> that is waiting in some buffer store, into the output
stream. After <u>because</u> a pause may develop depending on how much
time it takes to verbalize EVENT1. The second possibility is
perhaps more interesting: the conjunction initializing a certain
clause may be syntactically required by the verb of another clause.
For instance, the <u>if</u> in <u>I don't know if John is in</u> is dependent
on the verb <u>know</u> of the main clause, and can be uttered even
before the formulation process for the subordinate clause has
begun. The line of reasoning of this paragraph.

(i)   I can't help you because *1* I don't know if *2* John is in.

would point at positions *1* and *2* of sentence (i) as likely
places for pausing. The segment between *1* and *2*, that just
misses being a phonemic clause, I will call a <u>verb dependency</u>
<u>construction</u> (VDC). A VDC contains a verb as the "head" or
"governor" and all the phrases that are dependent on it. In
sentence (i), <u>if</u>, <u>don't</u>, and <u>I</u> are dependents in the VDC which
has <u>know</u> as its governor.Two other VDCs in (i) are <u>I can't help</u>
<u>you</u> and <u>John is in</u>.

My motivation for introducing notions of syntactic dependency
here is, first of all, that they nicely represent the relations
between "predicting" and "predicted" sentence elements, even if
the predictions surpass clause boundaries. Other empirical
arguments for dependency grammar are reviewed in sections I.3
and III.2.

Raising a clause-like unit such as the VDC to the status of the
largest unit of planning is also supported by an observation on
the maximum size of idiomatic expressions. Except for proverbs,
there doesn't seem to exist any idiom or phraseology that is
substantially longer than a single clause. All idiomatic expressions which allow for some variation (word order, slots to be
filled) observe this upper boundary. Proverbs sometimes spanning
several clauses, are no counterexamples since they are totally
fossilized and don't need a "formulator" at all. To put it
differently, non-fossilized phraseology never consists of, for
instance, two successive half clauses or two complete clauses.
If an expression spans more than one clause, then only one
clause is variable and all the others are fossilized. A Dutch
example is: "NP LACHEN als een boer die kiespijn heeft" (NP
LAUGH as a farmer who has toothache). This limitation on the
size of syntactic constructions suggests that the formulator
never works on more than one clause at the same time.

To be sure, non-fossilized idioms sometimes do violate clause
boundaries, but apparently only with conjunctions just as VDCs do.
Examples are: My name is not ... or ..., NP TAKE it that ...

I.3   Speech errors which are exchanges within dependency levels

One category of speech errors in the collection studied by
Garrett (1975) are "word exchanges": two complete words which
occupy non-adjacent positions end up at each other's place.
Examples (2) through (13) are all the word exchange errors
(interchanged words in capital) Garrett lists in his paper.

(2)   I have to fill up the GAS with CAR.

(3)   Prior to the operation they had to shave all the HEAD off
      my HAIR.

(4)   She donated a LIBRARY to the BOOK.

(5)   Older men CHOOSE to TEND younger wives.

(6)   ...which was parallel TO a certain sense, IN an experience..

(7) Every time I put one of these buttons OFF, another one comes ON.

(8) She SINGS everything she WRITES.

(9) ... read the newspapers, WATCH the radio, and LISTEN to T.V.

(10) Slips and kids - I've got BOTH of ENOUGH.

(11) I broke a DINGHY in the STAY yesterday.

(12) Although MURDER is a form of SUICIDE, ...

(13) I've got to go home and give my BATH a hot BACK.

In his total corpus there are 97 such cases. Garrett remarks not only that the interchanged words usually belong to the same word class but also that they "come from corresponding positions within their respective structures ... The parallelism of structure is most strikingly evident for the word exchanges that cross clause boundaries, but even the within-clause exchanges show a strong correspondence, usually involving two similarly placed words from distinct phrases. These phrases are quite often, for example, the noun phrases (NPs) of direct and indirect objects, or the NPs from a direct object and an adverbial phrase, or from successive adverbial phrases" (p.155-156).

I would like to add one further constraint that seems operative in such word exchanges. From examples (2) - (13) and some further statistics provided by Garrett one can conclude that the overwhelming majority of the exchanged words belong to the same syntactic dependency level. This is true of 8 out of the 12 examples listed above (5, 8, 10 and 12 seem to be exceptions).

These observations and those of the previous Section provide some empirical basis - although I admit it is very slender - for setting up the sentence construction process as one which roughly proceeds dependency level by dependency level.

I.4   Parts of the syntactic form of an utterance may be given before the formulation process starts.

In certain circumstances the formulator is not completely free in determining the syntactic shape of an utterance. Stylistics includes phenomena such as these: certain themes/contents and certain audiences prefer certain syntactic forms; the syntax of individual sentences is partly controlled by their position in the total text. Apparently, some mechanism prior to the formulator (perhaps the conceptualizer) biases him towards certain syntactic forms.

Another instance of limited "freedom of expression" is provided
by situations of repetitive speech. A radio reporter who has to
read out a series of sports results is tempted to use the same
syntactic scheme for several successive scores. Elsewhere I
have discussed this point in detail (Kempen, 1976b).

I.5   Preferred word orders

Recent experiments by Ertel (1976), Osgood & Bock (1976) and
Jarvella (1976) have uncovered some of the rules underlying
preferred or "neutral" word order in spontaneous speech. For
instance, speakers have a tendency to express ego-related, vivid
and salient concepts early in the sentence. Such tendencies
importantly determine the selection the speaker makes from the
total set of paraphrases he might use to express the content he
has in mind.

In terms of standard transformational grammar, some such
paraphrases have a longer transformational history than others.
For instance, passive sentences are supposed to be more complex
than actives, and subject complement constructions with extra-
positioned (trailing) subject (It amazed me that he went) have a
longer derivation than their counterparts with subjects in
initial position. However, the available evidence disconfirms
the hypothesis that differences in derivational complexity will
show up in actual human sentence production. The experimental
study of James, Thompson & Baldwin (1973) renders very implaus-
ible the hypothesis that passives are more difficult to produce
than actives (except, perhaps, for length). Jarvella (1976)
compares ease of production of subject complement sentences with
the that-clause in leading vs. trailing positions. He concludes
"there was no real indication that subject complements were
effortfully postponed".

I.6   Very limited working memory

Humans have a small working memory and a huge, easily accessible
and very flexible long-term memory (LTM). The opposite is true
of modern computers. Large amounts of data (inputs, intermediate
results of computations, etc.) can be very quickly stored without
"rehearsal" and don't get lost as a function of time or of new
data coming in. On the other hand, LTM lookup of some little
piece of data in a large computerized data base is very cumbersome.
Sentence generators built by both transformational and computational linguists tend to require a large working memory for keeping
intermediate results (typically, tree-like structures). And this
memory can only be cleared at a very late stage, if not after
completion, of the generation process. No part of the content of
the working memory may be released earlier, e.g. put into an
output channel for pronunciation since there is always the chance

for a later "transformation" to be dependent on it or to change
left-to-right word order.

Thus, in order to ease the burden put onto a working memory by
the sentence generation process, it seems wise to first decide
upon the left-to-right order of constituents so that speaking can
start relatively early and need not wait till all details of the
total utterances have been computed.

This line of reasoning, however, also applies to the level of the
conceptualizer. Since the conceptual messages he makes up have
to fit in a small working memory it would be efficient if he could
pass partial results down to the formulator for quick translation
into natural language. Since most people would agree they often
start talking before they have completely worked out what they
want to communicate, I will allow for conceptual messages that
are fed into the formulator in bits and pieces. And the formulator must be enabled to start working on parts available instead
of having to wait until the complete message is in.

Such a system has an interesting consequence as regards naturalness of word orders (cf. previous Section). The speech segments
the various conceptual pieces translate into will show an order
correlating positively with the order in which these pieces were
sent out by the conceptualizer. (Syntactic constraints on word
order will often prevent the correlation from being perfect).
Following a suggestion by Osgood & Bock (1976), we might hypoth-
esize that the order in which concepts are processed by the
conceptualizer is determined by saliency, vividness, ego-
relatedness, etc. Also, the linguistic notions of topic and
comment may be related to order of conceptualization (topic first,
comment later). Consequently, a formulator which is able to
process fragmentary conceptual messages in their order of arrival,
will spontaneously show natural or preferred word order and won't
need any special machinery for computing them. In other words,
the problem of natural word order should be studied at the level
of the conceptualizer, not the formulator.

II. Outline of the sentence construction procedure

II.1 A constructional lexicon

In Section I.1, the importance of syntactic frames, sentence
schemes, standard (canned) phrases and the like was discussed.
Here I will introduce the notion of a syntactic construction - a
notion that I think encompasses most multiword units occurring in
natural language, and one which has proved useful in accounting
for the results of some experiments on sentence (re)-production
(Kempen, 1976a,b). A syntactic construction is a pair consisting

of

(1)   a pattern of conceptual information, and

(2)   a sequence of lexemes and/or syntactic categories.

The latter I will call a syntactic frame, the former a conceptual
pattern.   The syntactic frame expresses the conceptual pattern.

For example, the syntactic frame "NP1 give NP2 to NP3" expresses
a specific form of transfer of possession of NP2 between NP1, the
actor, and NP3, the recipient.

The "passive" syntactic frame "NP1 be given NP2 by NP3" belongs
to a separate syntactic construction whose conceptual pattern is
identical to that of the active give-construction.   The idiomatic
expression "NP shoot Q bucks" is a syntactic frame expressing the
number of dollars NP spends.   So far, the examples all have open
slots (indicated by capital letters) but there are also many
constructions whose syntactic frame is completely closed and allows
no variation at all, not even word order permutations (like as a
matter of fact, other things being equal, proverbs).   Parenthet-
ically, the examples make it clear that I use the term syntactic
frame in a broader sense than Garrett (1975) who only considered
sequences of functors as the body of frames (e.g.The N is V-ing;
cf. Section I.1.iii).

The lexicon contains syntactic constructions as lexical entries.
An individual lexeme (single word) figures as a lexical entry only
if it constitutes a syntactic frame on its own. (14) gives an idea
of what the syntactic frame of a VDC lexical entry looks like, in
LISP notation.

(14)   VDC: ((pp1 {Cat: NDC; Case: Subj})
              (leave {Cat: V})
              (pp2 {Cat: NDC; Case: Obj; Status: Opt}))

It is a list containing three sublists as top-level elements. Each
of the sublists is a pair whose right-hand member is a list of
attribute-value pairs {between square brackets}. The latter provide
syntactic information for the procedures operating on lexical entries
that have been retrieved from the lexicon.   These right-hand members
I will call synspecs (syntactic specifications).   The left-hand
member of the top-level sublists is either a single lexeme or a
"pointer procedure" which computes a pointer to a field of the
conceptual pattern that is being translated.   For instance, pp1
sets up a pointer to the actor field in the conceptual pattern.
It is from this field that the lexical filler for the subject slot
will be derived.   Likewise, the value of pp2 will be a pointer to
the location the actor travels away from. {Cat: NDC; Case: Subj}

means: the lexical realization must be a Noun Dependency Construction (or NP if you wish) in subject case. { Status: Opt} in the
third sublist marks this NDC as optional (leave is a middle verb).

I will now give a more formal definition of a syntactic frame. It
is a list of one or more pairs of the form "(pp synspec)" or
"(1 synspec)", where pp is a pointer procedure (returning a pointer
to a field of a conceptual pattern); synspec is a list of attribute-value pairs (marking syntactic properties that have to show
up in the utterances under construction); and 1 is a lexeme(which
can be put into the output stream after the applicable morphological rules have worked on it). Furthermore, I propose the
following conventions. If the lefthand member of a top-level
sublist is a lexeme and the righthand member is a single attribute-
value pair {Cat: X}, where X is any part of speech, then the
sentence construction procedure will assume this lexeme can be
dumped into the output stream without any modification. E.g.
(because { Cat: Conj}) means that because, a conjunction,
doesn't need any morphological shaping up before it is pronounced.
The part of speech attributes can also be used to decide which
sublist contains the governor of a construction. Each syntactic
frame in the lexicon is explicitly marked as a Dependency
Construction of some sort: VDC, NDC, ConjDC, etc. The governor
of the frame is the sublist which contains the corresponding "Cat:"
mark. (This will work only if the frame contains exactly one such
sublist. Nominal compounds like apartment building or graduation
day which have two nouns in them could not be handled. Since the
first noun cannot be separated from the second one and is not
subject to morphological changes, I propose to treat these compounds as single nouns, as is done in German and Dutch).

II.2.   Sentence Assembly

II.2.1.   General Overview

The formulator starts constructing an utterance with two pieces of
information:

(1)   a conceptual pointer, and

(2)   a synspec which enumerates zero or more syntactic properties
      of the to-be -constructed utterance.

Empirical arguments for (2) were given in Section I.4. I will
first describe the workings of the proposed formulator if it
operates on complete conceptual patterns. In Section II.2.3 the
extra machinery for dealing with fragmentary conceptual patterns
will be outlined.

As for terminology, the two main procedures the formulator uses I
will call LEX (for lexicalization) and FIN (for finalization).
LEX receives as input a <u>formula</u>, which is a pair of the form
"(p synspec)" or "(1 synspec)" where p is a pointer to a conceptual
pattern, 1 a lexeme, and synspec as defined above.

The formulator passes the input, which is a formula, on to LEX which
replaces it by another formula or by a list of formulae. To this
output, LEX is applied again,that is, to each of the formulae,
going from left to right. The result of this "pass" or "sweep" of
the lexicalization procedure is, again, a new list of formulae.
The formulator continues such lexicalization sweeps until all
formulae in the list have the form "(1 synspec)", i.e. until they
all have lexemes in them and no pointers to conceptual patterns
anymore. To this list, the formulator applies FIN which computes
the final form of the lexemes. The left-to-right order of lexemes
in the formula list corresponds to order of words in the final
utterance.

Although this is not clear from the description just given, applying LEX this way enables growing a syntactic dependency tree from
top to bottom, dependency level by dependency level. Consider the
dependency tree in Fig. 1 which depicts dependency relations among
the words of sentence (15).

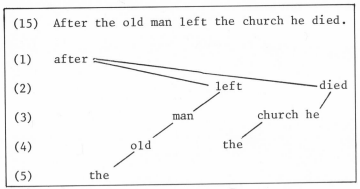

(15)   After the old man left the church he died.

(1)   after

(2)   left   died

(3)   man   church he

(4)   old   the

(5)   the

Fig. 1 Syntactic dependency tree for sentence (15).

Suppose the formulation process starts out with the formula
(p1 {Cat: S}. The first lexical frame LEX finds (N.B. this paper
is not concerned with lexical search itself) is the ConjDC "after
S S" which will replace the earlier formula:

((after {Cat: Conj}) (p2 {Cat: S}) (p3 {Cat: S})).

This list of three formulae, with the conjunction as governor, has
two slots to be filled. Pointers p2 and p3 refer to fields of p1:
the events between the temporal relation <u>after</u> is specified. These

pointers have been set by executing the pointer procedures (pp's)
in the syntactic frame. Now the formulator notices that after,
the leftmost lexeme in the list, has its final shape (cf. the end
of section II.1) and can be pronounced. The remaining two-member
formula list is then lexicalized with the VDC frames leave (see
(14) and die: (((p 4{Cat: NDC; Case: Subj}) (leave {Cat: V}
(p5 {Cat: NDC; Case: Obj})) ((p4 {Cat: NDC;   Case: Subj})
(die {Cat: V}))).

The next lexicalization pass replaces the slots p4 and p5. During
this pass, LEX notices that p4 occurs twice, inhibits lexical lookup for the second token and uses a pronoun instead.

This is a rough description of how lexicalization proceeds if the
complete conceptual pattern is known to the formulator right from
the start. After each lexicalization pass, the formulator checks
if the leftmost top-level element of the formula list has been
completely lexicalized. If so, this element is processed by FIN
which computes the definitive shape of its lexemes, dumped into
the output stream, and finally detached from the formula list.
If the new leading top-level element has been completely lexicalized, too, FIN will work on it; if not, then the whole remaining
formula list is subjected to a new lexicalization pass. FIN will
be mainly a routine for handling VDCs, since top-level elements
of the formula list are either VDCs or unchangeable words like
conjunctions or sentence adverbs. The latter may be uttered as
they are, without any further processing. So FIN's task is to
shape up the constituents of VDCs in accordance with rules of
tense, number, person, case, etc. (To this purpose FIN might use,
among other things, a stock of syntactic frames in Garrett's sense
(see Section I.1.iii); FIN will call itself recursively if a VDC
contains another VDC.

This general setup of the formulator is consistent with the
empirical observations reviewed in Section 1. Since the left-to-
right order of formulae in the formula list is never changed, the
formulator can release leading VDCs, conjunctions and adverbs
very quickly and reclaim the freed working memory space (cf.
Section I.6). The observations on pausing in Section I.2 can be
accommodated too.   For sentence (16), Boomer's rule would
identify the transition between after and he as the place most
likely to attract a pause.

(16) The man died after he left the old church.

This is also true of the proposed model: after FIN has worked on
the first VDC the formulator can just read off the man died and
after. But then he has to complete lexicalization of the second
VDC and to finalize it before he left the old church can be said.
(Lexicalization of the second VDC takes one more lexicalization

sweep than the first one, because of the modifier old.) Since
lexicalization proceeds dependency level by dependency level, the
type of speech errors discussed in Section I.3 become understandable. E.g. in terms of sentence (9), the VDCs watch and listen to
were interchanged during one lexicalization pass, or the radio and
T.V. were during another. (Exactly how such interchanges come
about, I don't know. Here I only appeal to some notion of temporal
contiguity - "within the same pass..." - but other factors may be
involved as well, e.g. similarity between interchanged elements).

II.2.2   Details of the lexicalization procedure

The complicated job LEX has to do for each formula may be divided
into eight tasks. I will discuss them in turn, in the order they
are carried out. Subtasks i through v are mainly concerned with
finding adequate syntactic frames in the lexicon. By subtasks vi
through viii, a selected syntactic frame will be trimmed down to
the format needed for insertion in the formula list.

(i)   Expanding synspec. The two sources of synspec we have
         considered up till now are the conceptualizer and the
         lexicon. Synspecs may be "incomplete", in the sense that
         syntax requires further specifications. For instance, the
         synspec {Cat: VDC} must be expanded so as to contain inform-
         ation about subcategories "main" vs. "subordinate" (at
         least in German and Dutch where they condition certain
         aspects of word order). If the lexicon or conceptualizer
         didn't specify which, the formulator must have a means of
         adding this information; for instance by assuming a
         default value, by looking at neighbouring formulae, or by
         inspecting the current conceptual pattern. If "Subcat:
         Main" is chosen, further information about Mode must be
         added: declarative, interrogative or imperative. The
         value of the Mode attribute can only be determined by
         inspecting the conceptual pattern. Similarly, a synspec
         {Cat: V} requires information about Tense, which can be
         derived from time information in the conceptual pattern.

         LEX must have a set of rules defining completeness of
         synspecs, and mechanisms which execute the tasks illus-
         trated by the examples.

(ii)   Inspecting the conceptual pattern. The pointer in a form-
         ula points to a total conceptual pattern (i.e., the input
         pattern delivered by the conceptualizer) or to a part of
         it (e.g. the actor field of an event describing pattern).
         For simplicity, I call them both "conceptual patterns".
         LEX must know what kind of information to extract from the
         current conceptual pattern. For instance, if it has the

form"EVENT 1 cause EVENT 2" or "EVENT 1 time-relation
EVENT 2", LEX should pick out the connector information
and not, say, the actor of EVENT 2.

(iii)   Lexical search. The information extracted from the
         conceptual pattern guides LEX through the lexicon when
         searching for an adequate syntactic frame. A good example
         of how to set up procedures for both (ii) and (iii) is
         provided by Goldman's (1975) generator.

(iv)   Matching a candidate frame to synspec. Not every syntactic
         frame which expresses the current conceptual pattern can
         be inserted into the formula list. First, a candidate
         frame must be checked for compatibility with synspec. For
         example, if synspec is {Cat: S} , then a NDC frame wouldn't
         match, but a Conjunction-S-S frame would. Also, the syn-
         spec of a formula may impose certain word order restrict-
         ions upon the syntactic frame that will replace it. So
         LEX has to check if these restrictions can be met by a
         given candidate frame. What is needed of course, is a
         system of rules formalizing the notion of "matching" and
         "non-matching" frames.

(v)   Checking for modifiers. Often, lexical search will not be
         able to locate a syntactic frame which expresses all asp-
         ects of the current conceptual pattern. In terms of sent-
         ence (15), suppose a speaker of English doesn't have a
         single syntactic frame expressing the conceptual pattern
         which underlies old man. Lexical search will suggest man
         for part of the conceptual pattern; which leaves the part
         underlying old unexpressed. I assume this second part will
         be lexicalized during LEX's next pass. As a reminder, LEX
         will tag the formula (man {Cat: N}) with a special
         synspec label: (man {Cat: N; NMOD: pl}). The attribute-
         value pair "NMOD: pl" says that the conceptual pattern pl
         will have to be lexicalized in the form of a noun modifier
         (e.g. an adjective or a relative clause). LEX will find
         this tag during the next pass and then come up with old
         as the translation of pl. LEX must be supposed to know
         special rules for modifier placement. (old {Cat: Adj.})
         may simply be placed before the formula containing man,
         but especially in the case of verb modifiers LEX will
         need to consult more complex rules.

(vi)   Executing lexical procedures. A syntactic frame may
         reference procedures that have to run before it is put
         into a formula list. As yet, the only type of lexical
         procedures we have seen are the pointer procedures (cf.
         Section II.1), but other types may very well prove necess-
         ary).

(vii)   <u>Transforming syntactic frames</u>. It is generally recognised
          that humans have a very limited working memory and a large
          and easily accessible long-term memory (cf. Section I.6).
          In line with this, I assume that the lexicon, which is part
          of long-term memory, contains many ready-made syntactic
          constructions which in standard transformational grammar
          would be produced by applying transformations, i.e. by
          real-time computations in working memory. Examples are
          the passive constructions and subject complement construct-
          ions with extrapositioned subject. By assuming that these
          constructions as well as their transformationally less
          complex counterparts are both entries in the lexicon, we
          have an easy way of accounting for the experimental data
          mentioned in Section I.5.

          But I certainly do not hold the view that all transform-
          ations should be dealt with this way.   Consider, for
          instance, interrogative sentences (yes-no questions) in
          Dutch and German. They differ from declarative sentences
          in that the subject follows the tensed verb; e.g. <u>John</u>
          <u>saw Mary</u> ---> <u>Saw Mary John?</u> The problem at issue
          is: does the lexicon contain a separate interrogative
          entry in addition to each declarative VDC entry, or are
          interrogative constructions computed from declarative
          entries? Another feature of German and Dutch is word
          order differences in subordinate and main clauses. E.g.
          "John fainted: for he SAW MARY" turns into "<u>John fainted</u>
          <u>because he MARY SAW</u>". This example raises the same
          question: is the NP1-V-NP2 order in the lexicon, or
          NP1-NP2-V, or both? Whatever is the answer in these two
          concrete cases, I don't think we can do without a limited
          number of transformations which reorder or delete elements
          of syntactic constructions retrieved from the lexicon.

          Pro-forms, too, entail changes of constituent orders. For
          instance, object NPs follow the verb in French, but precede
          it if they are pronouns. In many languages, interrogative
          pronouns occur in initial position, even if the standard
          position of the questioned NP is further down the sentence
          (e.g. <u>John saw ?NP</u> ---> <u>What did John see?</u>

          Such transformations are applied to a syntactic frame
          before it is inserted in a formula list, for at that time
          the relevant synspec information is available. E.g. the
          Question transformation will be applied to a syntactic
          frame if synspec reads {Cat: VDC; Subcat: Main; Mode:
          Y/N-Question} . The transformed syntactic frame is then
          put into the formula list; whereafter synspec is lost.
          (The formulator doesn't keep a generation history of the

formula list, for reasons of efficient management of working memory).

At the present time I don't know which members of the set of linguistically defined transformations should be treated in terms of alterations to syntactic frames (like the Question example) and which members deserve separate lexical entries (like passive constructions). This problem may be experimentally investigated in experiments where subjects spontaneously construct sentences of a specific syntactic format, e.g. while describing a perceived or memorized event or picture.

(viii) Replacing the input formula. Finally, LEX replaces the input formula with the selected and possibly transformed syntactic frame.

II.2.3 Lexicalization of fragmentary conceptual patterns

The conceptualizer often delivers a conceptual pattern in bits and peices, and the formulator must be able to immediately operate on such fragmentary information (cf.Section I.6). There are obviously many ways to divide a conceptual pattern into parts, and many different orders for feeding these parts into the formulator. I will outline here how the present model can handle an interesting subclass of all these cases, namely when a dependent (more precisely: the conceptual pattern underlying a dependent) arrives earlier than its governors, so that the natural top-down order of lexicalization cannot be followed.

By way of example, the conceptualizer delivers the nominal concept "Mary" to the formulator before embedding it in a conceptual pattern as the recipient of a transfer-of-possession action. The formulator prepares and utters "Mary..." without knowing what kind of VDC it will have to fit in. Then, after receiving the conceptual action, it is forced to look up a VDC which expresses the recipient in leading position. The passive give-frame "NP 1 be given NP 2 by NP 3" will do, as well as the active get-frame "NP 1 get NP 2 from NP 3", but neither the active give-frame nor "NP 1 is given NP 2 by NP 3".

To permit the formulator to handle such cases it has to be extended along the following lines. The conceptualizer provides his output messages with a delimiter symbol, informing the formulator when messages start and finish. The input formula - which is either a complete or a fragmentary conceptual pattern - is passed along, not to LEX directly but to a monitor function MON. On the one hand, MON watches the lexicalization sweeps and prepares a "syntactic summary" for the utterance part LEX is constructing

currently. On the other hand, MON registers any new parts the
conceptualizer adds to the current message. Suppose, at a given
moment, LEX has finished his last pass for a fragmentary conceptual pattern and, in the meantime, the conceptualizer has added a
new part to it. The latter implies there is a new input formula
with a new pointer and a new synspec, MON will now

(1)   append the syntactic summary for the last-produced
        utterance part to the synspec of the new input formula,

(2)   add a tag "DONE" to the part of the conceptual pattern which
        has just been lexicalized, and

(3)   register the role played by the DONE part in the new
        conceptual pattern.

As a result of this, LEX will receive a new input formula whose
synspec tells him what kind of partial utterance the formulator
has committed himself to already, and which part of the conceptual
pattern need not be expressed anymore. This is enough information
for LEX to construct a good continuation, if any (!), of the
utterance.

In terms of the above example, if the synspec of the new input
formula would simply have said "produce an S for this conceptual
pattern", then MON would change it to "produce an S which expresses
the recipient as an NDC in leading position" (assuming here that
"NDC" is the syntactic summary for "Mary ...").

Finally, MON will hand over the modified input formula to LEX,
monitor and summarize the lexicalization process and, if it sees
no delimiter symbol, repeat its operation for still other fragments coming in. This facility for handling fragmentary conceptual
patterns requires only one modification to LEX: rules for treating
DONE parts of conceptual patterns.

III   The formulator viewed from a linguistic point of view.

In this Section attention is shifted from psychological and
computational to lingusitic aspects of the proposed formulator.

(i) Except for cases discussed in Section II.2.3, lexicalization
proceeds top-down (dependency level by dependency level, and from
left to right within dependency levels). Is this regime compatible with the bottom-up principle of the transformational cycle,
as discovered by transformational grammar?

A definite answer to this question cannot be given as long as many details of the model remain unspecified. But the examples I have worked out show that the proposed formulator is indeed able to handle some sentence types which need cyclically applied transformations in a standard transformational grammar. Consider sentence (17) (cf. Fodor et al., 1974, p. 121-131), whose deep structure contains a sentoid "doctor-examine-John" as part of the verb phrase of the matrix sentoid "Bill-persuade-John".

(17)   Bill persuaded John to be examined by the doctor.

Two cyclical transformations are applied to the subordinate sentoid:

(1)   passivization, resulting in "John-be-examined-by-the-doctor", and

(2)   equi-NP-deletion, deleting the first NP of the subordinate sentoid ("John") which is referentially identical with the object of the matrix sentoid.

The proposed formulator can use synspecs to make such transformations superfluous. The active persuade-frame in the constructional lexicon looks (informally) like (18):

(18) VDC: ((ppl {Cat: NDC; Case: Subj})
                (persuade {Cat: V})
                (pp2 {Cat: NDC; Case: Obj})
                (to {Cat: Prep})
                (pp3   {Cat: VDC; Subcat: Infinitive-construction;
                        Detail: ppl of this VDC must deliver the
                                same value as pp2})   )

This frame is selected during one lexicalization sweep; during the next, candidate frames for the subordinate VDC are matched against the synspec following pp3. The active examine-frame wouldn't do because the value of its pointer procedure ppl (the concept "Doctor-such-and- such" in the field referenced by pp3) is not the same as the value computed by pp2 of the persuade-frame (the concept "John"). But the passive examine-frame would match. The attribute-value pair "Subcat: infinitive-construction" has two consequences as regards the final shape of the subordinate VDC:

(1)   it influences later lexicalization sweeps in such a way that no verb tensing will occur, and

(2)   it will delete the subject NDC.

The general idea seems to be:

(1)   to make the synspecs for subordinate VDCs maximally specific
       so that they are only matched by frames which approximate
       the required syntactic shape as closely as possible:

(2)   if any transformations to a selected frame are still needed,
       to execute them before inserting the frame into the formula
       list. The effect of doing this will be similar to the effect
       of cyclically applied transformations, but the computational
       processes are very different.

(ii) My chief motivation for using syntactic dependency as
generative mechanism is computational efficiency. The only
psychological evidence consists of the observations discussed
in sections I.3 and I.2. Linguistic evidence, be it of an
indirect nature, is provided by the study of linguistic intuitions. Levelt (1974) summarizes several studies of so-called
cohesion judjments. They strongly favour dependency grammar over
other grammar types (e.g. constituent structure grammars). Schils
(1975) has confirmed this finding.

Levelt remarks that dependency trees do not represent the difference between endocentric and exocentric constructions. Interestingly, the difference is brought out by the lexicalization
procedure. Exocentric constructions (like verb with subject,
object, etc.) are the ones that are retrieved as a whole from the
constructional lexicon. Endocentric constructions result from
modifiers (see Section II.2.2 (v)): LEX notices it can only
find a syntactic frame which expresses part of the current conceptual pattern, marks this frame with MOD, and lexicalizes the
remainder of the conceptual pattern at a later stage.

IV   Two generation examples

(i) In Section II.2.1, the first lexicalization steps for sentence
(15) were discussed. Here I will follow the remaining passes.
The formula list after LEX's second pass looks like (19).

(19)   (((p4 {Cat: NDC; Case: Subj})
          (leave {Cat: V; Tense: Past})
          (p5 {Cat: NDC; Case: Obj}))
          ((p4 {Cat: NDC; Case: Subj})
          (die {Cat: V;   Tense: Past})))

It differs from the formula list given in Section II. 2.1 in that
it contains tense properties for the verbs. These were added by
LEX because, in English, verbs dominated by S (finite verbs) have
obligatory tense markers. This is done by the procedure which

expands synspecs (Section II.2.2.(i)).

The third lexicalization pass works on the three noun dependency
constructions. The synspecs of (20) lists four properties.

(20) (man {Cat: N; Case: Subj; Number: Sing; Mod: p6})

Of these, the first two were simply copied from (19), the other
two were added by the synspec expanding routine (Section II.2.2(i))
and the modifier checking routine (Section II.2.2(v)) respectively.
The last NDC becomes (21), with the appropriate Personal Pronoun
instead of the noun.

(21) (he {Cat: PP; Case: Subj; Number: Sing})

The fourth lexicalization pass adds one level of modifiers to the
lexeme string man-leave-church-he-die. (20) is changed to (22)
by looking up a frame for (part of) p6 and consulting rules for
placement of noun modifiers.

(22) ((old {Cat: Adj})
      (man {Cat: N; Case: Subj; Number: Sing; Mod:p8}))

The fifth pass only leaves p8 to operate on. The article is
inserted before old. Others might prefer to determine the
article during the same pass as the governor noun. Speech errors
like (2) and (4) which have the articles at the correct place even
though the nouns have been interchanged, might be taken as evidence
for that alternative. Here I have strictly followed the dependency hierarchy. The end result is formula list (23) to which FIN
is applied.

(23) ((((the {Cat: Art})
        (old {Cat: Adj})
        (man {Cat: N; Case: Subj; Number: Sing})))
      (leave{Cat: V; Tense: Past})
      ((the {Cat: Art})
        (church {Cat: N; Case: Obj; Number: Sing})))
      ((he {Cat: PP; Case: Subj; Number: Sing})
        (die {Cat: V; Tense: Past}))   )

The first top-level element is processed, resulting in the
utterance the old man left the church. Finally, FIN treats
the remaining formula list (((he {...}) (die{...}))).

(ii) the second example has to do with fragmentary conceptual
patterns (Section II.2.3). I will demonstrate how sentence
(24) is produced

(24)   The old man left the church and then he died.

if the conceptual pattern underlying (15) comes in in two
fragments: first EVENT1, then" ... time-relation EVENT2".

The input formula for the first fragment is, I assume, (p1{Cat:S}).
The translation into English proceeds in exactly the same way as
the first VDC of (15). The syntactic summary prepared by MON
reads simply "Cat:S". Assuming that the input formula for the
complete conceptual pattern is (p2 {Cat: S}), MON changes it to
(25).

(25)   (p2 {Cat: S; Order: {EVENT1 {Cat: S} rest{ } })

MON has also figured out that p1 plays the conceptual role of
EVENT1 in the event sequence delivered by the conceptualizer.
The notation between curly brackets specifies order and form of
expression of the various conceptual parts: first EVENT1 is an S
(which has been DONE already), then the rest in any form LEX
wishes (this synspec is empty). This order prescription excludes
"After S S" as a matching frame, but frame (26) is alright.

(26)   ConjDC: ((pp1 {Cat: S})
                  (and {Cat: Conj})
                  (then {Cat: Adv})
                  (pp2 {Cat: S}))

Procedures pp1 and pp2 set pointers to EVENT1 and EVENT 2
respectively. During his first pass, LEX will simply detach
the first top-level element, that has been expressed already,
and select the die-frame for EVENT2. Subsequently, FIN will put
and and then into the output stream. Then the one remaining
VDC is lexicalized and finalized.

Notes

This paper was written when the author was a Postdoctoral
Researcher at the Department of Computer Science of Yale
University. His stay there was made possible by a grant from the
Netherlands Organization for the Advancement of Pure Research.
(ZWO).

I'm indebted to Dick Proudfoot, Department of Computer Science,
Yale University for setting up a tentative computer implementation
of the model outlined in this paper and for commenting on earlier
drafts of the paper.

References

Becker, J. (1975). The phrasal lexicon. In R.C. Schank & B.
    Nash-Webber (eds.) Theoretical issues in natural language
    processing. Cambridge, Mass.:MIT.

Boomer, D.S. (1965). Hesitation and grammatical encoding.
    Language and Speech, 8, 148-158.

Ertel, S. (1976). Where do the subjects of sentences come from?
    In Rosenberg (1976).

Fodor, J., Bever, T.G. & Garrett, M. (1974). The Psychology
    of Language. New York: McGraw Hill.

Fromkin, V.A. (1973) (ed.). Speech Errors as Linguistic Evidence.
    The Hague: Mouton.

Garrett, M. (1975). The analysis of sentence production.In G.
    Bower, (ed.). The Psychology of Learning and Motivation,
    Vol. 9. New York: Academic Press.

Goldman, N. (1976) Conceptual generation. In R.C. Schank
    Conceptual information processing. Amsterdam: North
    Holland.

James, C.T., Thompson, J.G., & Baldwin, J.M. (1973). The
    reconstructive process in sentence memory. Journal of
    Verbal Learning and Verbal Behaviour, 12, 51-63.

Jarvella, R. (1976). From verbs to sentences: some
    experimental studies of predication. In Rosenberg (1976).

Kempen, G. (1976a). Syntactic constructions as retrieval plans.
    British Journal of Psychology, 67, 149-160.

Kempen, G. (1976b). On conceptualizing and formulating in
    sentence production. In Rosenberg (1976).

Levelt, W.J.M. (1974). Formal Grammars in Linguistics and
    Psycholinguistics. The Hague: Mouton.

Levelt, W.J.M. & Kempen, G. (1975). Semantic and syntactic aspects
    of remembering sentences. In R.A. Kennedy & A.L. Wilkes
    (eds.) Studies in long-term memory. New York: Wiley.

Osgood, C.E. & Bock, J.K. (1976). Salience and sentencing: some
    production principles. In Rosenberg (1976)

Rosenberg, S. (1976).(ed.) Sentence production: developments in
    research and theory. Hillsdale, N.J.: Erlbaum.

Schils, E. (1975). Internal Report, Department of Psychology,
    University of Nijmegen, The Netherlands.

A COMPUTATIONAL MODEL OF DISCOURSE PRODUCTION

A.C. DAVEY   and   H.C.   LONGUET-HIGGINS

University of Sussex

I.   Introduction

A central problem in the psychology of language is to explain as
precisely as possible how the speaker of a natural language
expresses himself in words. The problem is really twofold. In
the first place, how does the speaker decide what to say? Secondly, how does he find the words in which to say it? Existing
grammatical theories (Chomsky 1965; Lakoff 1971) address the
latter problem by positing a semantic representation for each
sentence, and attempting to prescribe, for a given language,
a set of rules for mapping such semantic representations on to
syntactic structures, or vice versa. This approach, however,
does not cast any light upon the first problem, namely how the
semantic representations are created in the first place; the main
difficulty here is in specifying the message which is to be
conveyed independently of the linguistic structures which must
be created in order to express it.

This paper describes an attempt to deal with both aspects of
the problem, by specifying effective procedures for the production of whole paragraphs of English in response to a well-
defined communicative need. These procedures have been cast in
the form of a computer program (Davey 1974) which generates a
post mortem, in English, on any given game of noughts and crosses
(tic-tac-toe). The choice of this particular universe of discourse
was motivated by the desire to keep the subject matter of the
message as simple as possible without reducing it to complete
triviality; as will be seen later, even for such a simple

125

universe quite subtle linguistic problems arise about the choice
and ordering of the words in which the messages are couched. The
present enterprise complements, in a sense, the work of T.
Winograd, whose book "Understanding Natural Language" (Edinburgh
University Press 1972) describes a computer program capable of
understanding and responding to quite a wide range of English
sentences typed in by a human operator. In Winograd's work the
main emphasis was on linguistic comprehension, the sentences
produced by the program itself being of relatively limited syntactic complexity, and being closely constrained in form by the
operator's own questions. It therefore seemed worthwhile to try
to develop a program which would produce coherent discourse
spontaneously, without prompting by a human operator, and in the
following sections we outline the way in which the program works,
and the general implications of the ideas on which it is based.

II.   An overview of the program

In order to give the program something to talk about, the
operator begins by playing a game of noughts and crosses with
the program, whose grasp of the tactics of the game was made
relatively unsophisticated, so that occasionally it loses. Once
the game is over a suitable instruction will then cause the
program to type out a paragraph of English describing the progress
of the game and its outcome. An example of such a game (in which
the program's moves are symbolised by noughts, and its opponent's
moves by crosses) is shown in Figure 1, and Figure 2 gives the
commentary which was subsequently typed out.

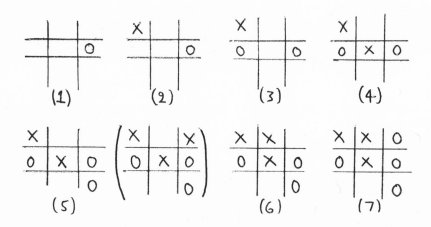

Fig. 1

"I started the game by taking the middle of an edge,
and you took an end of the opposite one. I
threatened you by taking the square opposite
the one I had just taken, but you blocked my
line and threatened me. However, I blocked
your diagonal and threatened you. If you had
blocked my edge, you would have forked me,
but you took the middle of the one opposite
the corner I had just taken and adjacent to
mine and so I won by completing my edge."

Figure 2

It is important to realise that none of the sentences or
phrases in the above commentary had any previous existence
in the program's memory; they were all constructed de novo
by the systematic application of pragmatic and syntactic
procedures to the situation in hand. It will be convenient
to comment first on the grammatical component of the program.

The grammatical theory embodied in the program is basically
the same as that used by Winograd, namely a systemic grammar
in the tradition of Halliday (1961, 1967, 1968), but owing
a particular debt to work by Hudson (1972a & b). Essentially,
the grammar specifies the program's linguistic resources, by
enumerating systematically the various constructions which are
available in the relevant subset of English. The grammar
is generative in that it has complete and explicit rules of
formation, which are here used to govern the production
of grammatical items. One part of the grammar specifies the
options which may or must be selected for a given grammatical item; a major clause, for example, must be past or
present tense, but cannot have gender or number. These options
are set out in the network of "systems" in which each system
is a set of simultaneous exclusive alternatives, and the
network structure exhibits the logical relation of each system
with the rest. The systems network thus sets out exactly what
grammatical decisions must be taken in order adequately to
characterise any item under construction. The other part
of the grammar comprises sets of rules which state how
the available options may be constrained by the role of
the item in the grammatical environment. For example,
the grammar contains a rule which constrains a pronoun such as
"I" to have the accusative form "me" in an environment in
which the pronoun is dominated by a preposition.

By itself however, the grammar is powerless to determine
what sentences should actually be produced, as it
merely lays out the options which are available to the
speaker and controls the production of sentences in
accordance with those options which he selects. The
selection of the options is made by the pragmatic component of the program, which has the task of specifying
how to decide what has to be put into words, how to
divide this information into sentences, how to arrange
each sentence so that its parts fit their context and
are easy to understand, and then how to pick words and
combine them into phrases to mean the right things. It
also specifies - and this is perhaps the most interesting pragmatic problem - what can be left unsaid: it
attempts always to avoid telling the hearer anything
he knows already, anything more than he needs to know,
or anything he might reasonably be expected to work
out for himself.

The general principle underlying the design of the pragmatic component was that a speaker must accommodate his
remarks to the hearer's presumed state of knowledge, and
that any speaker who ignores this maxim runs the risk
of baffling or boring his addressee. One reason for
selecting a simple game as subject matter was that a
game offers a suitable context in which the speaker can
keep up to date with the hearer's presumed state of
knowledge - presumed because in the absence of responses
from the hearer the program must assume that he both
understands the rules and point of the game, and follows
the developing description of the particular game in
hand. The program's discourse is thus accommodated to
what the hearer's state of knowledge ought to be, rather
than to what it actually is.

It is impossible in such a brief space to explain the
workings of the program in detail, so we shall content
ourselves with analysing the commentary quoted in Figure
2, in the hope that this example will illustrate the
problems which the program has to face, and the manner
in which it handles them. It should however be pointed
out that there are about 20,000 tactically distinct
games of noughts and crosses upon which the program
might be asked to report, and that for any one of these
it should produce an equally coherent description, so
that our observations on this particular commentary are

of general application.

III An example

The first obvious point to note about the game commentary
quoted in Figure 2 is that it consists of a number of
sentences each of which reports upon one or more actual or
possible moves. In order to construct such a paragraph
the program must decide how to allocate the semantic material
to individual sentences in a sensible fashion. The general
principle adopted was to group together the descriptions
of moves which are tactically related - with certain quali-
fications to be mentioned in a moment. The first two moves
are both tactically neutral, and can be conveniently described
in a single sentence of coordinate construction.

> (i)   "I started the game by taking the middle
>   of an edge, and you took an end of the
>   opposite one".

Moves 3 and 4 belong together tactically in that 4 is, among
other things, a reply to 3.   For this reason the description
of 3 is associated, in a new sentence,with that of 4 rather
than being appended to the sentence describing moves 1 and 2.
A coordinate conjunction is again appropriate, but since the
expectations raised by describing 3 as a threat are deflated
by learning that 4 was a defence, the required conjunction
is contrastive:

> (ii)   "I threatened you by taking the square
>   opposite the one I had just taken, but you
>   blocked my line and threatened me".

Because move 4 has an offensive as well as a defensive aspect,
and the same applies to move 5, one might be tempted to add an
additional "but" clause to sentence (ii). The result would,
however, be somewhat confusing -

> "Move 3, but move 4, but move 5".

- because the hearer would not receive a clear signal from
the second "but" as to the element with which move 5 is to
be contrasted. The program therefore places move 5 in a
fresh sentence, prefaced by "however" to make it plain that
expectations aroused by the description to date are not to
be fulfilled:

> (iii) "However, I blocked your diagonal and
>   threatened you".

Moves 6 and 7 call for special treatment, because the threat
posed by move 5 was not countered by move 6. The program can
recognise this fact by referring to its own game-playing
routines, which assign a tactical evaluation to any move made
in any board position. This way of integrating the move-
making and move-describing functions is designed to suggest a
way of thinking about semantic representations. The effect
here is to treat move 6 as a mistake, in that the program
would not have made move 6 in the situation created by move 5.
In such circumstances the program draws attention to the mistake
by first describing, not what actually happened, but what
would have happened if the operator had adopted the program's
own tactics. For this purpose a counter-factual hypothetical
construction is appropriate:

> (iv)   "If you had blocked my edge you would
>   have forked me ..."

The actual course of events, however, contrasts with the
previously favourable position of the operator, and so the
appropriate conjunction is "but":

> (iv b) "...but you took the middle of the one
>   opposite the corner I had just taken and
>   so I won by completing my edge".

As the winning move was the immediate consequence of the
operator's mistake, the causal connective "so" prefaces
the description of the final move.

The hypothetical construction just illustrated is used by the
program when the hypothetical move has both an offensive and
a defensive aspect, requiring two separate clauses. If the
best available alternative were merely defensive, the program
would produce

> (v a)   "You could have blocked my edge, but you..."

or

> (v b)   "Although you could have blocked my edge, you.."

In both cases, as in (iv b), there is a contrast between the
favourable hypothetical move and the less favourable actual one.
In (v a) the contrast is signalled by "but"; in (v b) the
hypothetical is expressed in a subordinate clause and the
expectations thereby aroused are disappointed in advance by
the introductory "although".

In setting about the construction of each sentence, the

program makes a tactical evaluation of the next few remaining
moves in deciding how many of them to describe in the forthcoming
sentence. Its next task is to distribute the move-descriptions
between major and minor clauses. Normally each move is described
in one or more major clauses, linked by conjunctions in the ways
just explained. This is appropriate where the moves are of comparable interest, but when a move is futile or vacuous, its description may be relegated to a subordinate clause. For example, if the
program had posed a double threat - a "fork" - the operator cannot
defend himself, and his final move is of no avail. The program
may then produce:

>   (vi)   "...I forked you. Although you blocked my
>   edge, I won by completing my diagonal".

The subordination of the operator's final move is appropriate
only if the program did in fact go on to win; otherwise the
operator's move is tactically important, and is therefore described
in a main clause. This resource of the program is no more than a
hint at one of the many ways in which clause subordination may be
motivated in richer universes of discourse.

Within each major clause describing a move, there may be major
and minor clauses conveying the various tactical aspects, if any,
of the move. If there is both an offensive and a defensive
aspect, both aspects are likely to be significant, and so a coordinate structure is appropriate:

>   (vii a) "I blocked your line and forked you",

while a subordinate structure is distinctly odd:

>   (vii b) "I forked you by blocking your line",

or even odder:

>   (vii c) "I blocked your line by forking you".

If, furthermore, a coordinate structure is employed the program
must conform to the likely train of thought of the hearer and must
mention the defensive aspect of the move before indicating its
offensive aspect. It would not do to say

>   (vii d) "I forked you and blocked your line",

both for this reason and because the second clause is necessary
for the identification of the move referred to in the first
clause.

Sometimes two clauses are required to describe a move, but one is
plainly not as significant as the other. For example "taking a
square" is tactically non-committal while "threatening" is not.
In such a case it seems equally acceptable either to subordinate
the less significant clause:

(viii a) "You threatened me by taking a corner",

or to present the less significant clause first and then enlarge
upon it:

(viii b) "You took a corner and threatened me".

The alternatives

*(viii c) "You took ... by threatening"

and

*(viii d) "You threatened ... and took ..."

are plainly unacceptable as confounding the end in view and the
means adopted for achieving it.

IV Referring expressions

A noteworthy feature of the program is the way in which it
identifies the moves of the game. As the procedures for doing
this illustrate several important aspects of the use of English,
it seems worthwhile drawing attention to some of the underlying
principles.

First, it may not even be necessary to identify explicitly the
square which was taken in a particular move; clauses such as
"I blocked your line" or "by completing my edge" serve to
identify unambiguously the square which was taken provided
the hearer follows the commentary.

Secondly - a point which is implicit in what has just been said -
the construction of the commentary takes for granted that the
description of each event is to be interpreted in the light of
what has been said so far. In order to achieve this effect
of "a train of thought" the program must tacitly replay the
game which it is describing, so that the identity of the square
can be established with reference to the board position which
has been defined by all the statements so far enunciated. In
order to achieve this the program must arrange the sentences,
and the main clauses within them, in an order corresponding
to that of the events which they describe. This necessity
is not, we suggest, a peculiar feature of the chosen universe
of discourse, but a direct reflection of the way in which in
any ordinary situation the hearer's state of mind is progress-
ively modified by each sentence or clause which the speaker
utters.

Thirdly, the reader will notice the subtle - and correct

- distinctions which the program makes in its use of
indefinite as opposed to definite noun groups. The use
of indefinite referring expressions is appropriate for
identifying certain moves because of the symmetry of
certain board positions, and the corresponding tactical
equivalence of some alternative moves. Thus in sentence
(i) it is of no consequence which of the four edges
had its middle square taken in the first move, or which
end of the opposite edge was taken in the second move.
By contrast, a definite noun-phrase such as "the one
I had just taken" or "my line" is used when the entity
referred to is in a class by itself, and is uniquely
indentifiable by such a description.

Fourthly - and this is a pervasive feature of natural
language use - the program identifies squares, edges
and so forth by the relations in which they stand to
other identifiable entities. The program has a repertoire of such relationships, defined by such words as
"opposite", "middle", "end" and so on, and works
through these systematically in deciding which one to
use for specifying any particular move. According to
which relation is chosen, it will construct an appropriate qualifier which may take the form of a relative
clause, a prepositional phrase or a relational adjective; the relevant grammatical decisions are again
governed by the principle of economy of expression,
through the details are too complex to describe in this
account.

In employing the device of relational reference the
program sometimes relies on its assumption that the
hearer is keeping abreast of events. For example, if
a particular square has just been mentioned and the
program now needs to refer to the only one of its
neighbours which is free, it produces

     "... took the adjacent square".

The referent is unambiguous despite the fact that more
than one square is adjacent.

Fifthly, the program leans heavily on anaphoric reference, especially on the use of pronouns. The program's
use of the pronoun "one" is of special interest. In
contrast to "it", which is co-referential with a definite noun phrase, "one" refers to a type of entity
previously specified by a noun of appropriate significance. In the given universe of discourse "types"
include such entities as edges, corners and so forth;
the use of the word "one" is deemed appropriate if

reference has recently been made to such a type and
there is need to refer to it again.   There is a
"pronoun specialist" which has to decide, in the light
of what has been referred to so far, what type of entity
will be uppermost in the hearer's mind on the basis of
both its recency of mention and its depth in the existing constituent structure.

The pronoun specialist must not be thought of as operating solely within given constraints arising from the
existing context.   Just as the part of the program
which allocates conjunctions to major clauses does so
by manipulating sentence content in order to make the
fullest use of the conjunctions available, so too the
pronoun specialist may manipulate the planned output
to make full use of its anaphoric resources.   Consider
in particular the construction "one of...the other"
occurring in the sentence

> "although you blocked one of my edges, I
> won by completing the other".

By the time the hearer reaches "...completing the other"
one of the two equivalent edges referred to in the
"although" clause has been eliminated from consideration
by the operator's blocking move (see Figure 3).   None-
theless, "other" implicitly refers to a pair of items
so that the program must suspend its usual practice
of replaying each move as soon as it is described, and
treat the two moves in question as a composite event.
The point of general interest to emerge from this is
that the pragmatic procedures involved in the production of discourse must be sensitive to the grammatical
resources of the language.

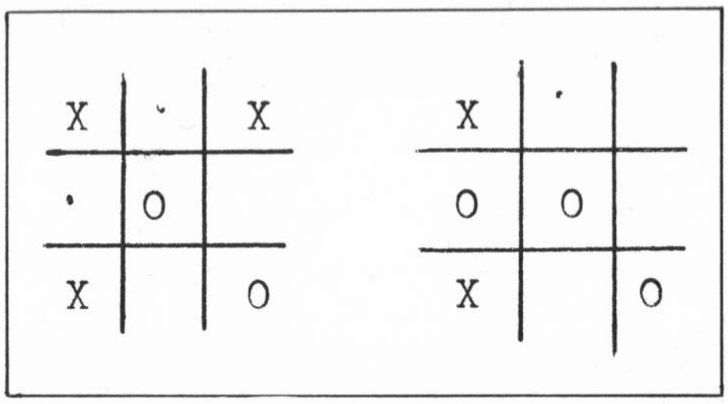

Figure 3

V Discussion

Our main purpose in developing this particular model
of discourse production was not to demonstrate, which

would be impossible, that human beings do produce language in the way described, but to present a fully
explicit theory of how this task might conceivably be
accomplished in a limited universe of discourse . The
suggestion which the program explores is that the
generation of discouse is an ongoing process in which,
far from merely translating a pre-existing structure
into words, the speaker must proceed in a stepwise
fashion, first deciding what requires to be said and
then creating a succession of linguistic structures
whose interpretation by the hearer may be presumed to
depend upon exactly what has been said so far. It is
of course essential for the speaker to have a working
knowledge of the language in order to make himself
clear to the hearer, and this knowledge must be expressed, at least in part, in the form of a generative
grammar of some kind. A crucial question, which
remains open, is how, given a particular language and
a particular universe of discourse, the speaker's pragmatic knowledge is to be related to the one and to the
other. Our program was designed in such a way that
its pragmatic component would be fully adequate for
describing games of noughts and crosses in English,
but we hope that our suggested solutions to some of
the problems we encountered may have application to
richer universes of discourse. But whatever the merits
or demerits of our detailed suggestions, we feel that
a full understanding of discourse production is likely
to require the formulation of psychological hypotheses
at least as detailed as those which our program offers
and, furthermore, hypotheses which specify, in full
procedural detail, the logical processes which may
occur in a speaker's mind when he is engaged in the
production of an intelligible and informative utterance
or sequence of utterances.

Acknowledgements

We are specially indebted to Stephen Isard, who jointly
supervised this project, for invaluable suggestions
and criticisms, and would also like to thank the Royal
Society and the Science Research Council for financial
support and computing facilities.

References

Chomsky, N. (1965). Aspects of the Theory of Syntax.
   Cambridge, Mass.:MIT Press

Davey, A.C. (1972). A Computational Model of Discourse Production. Unpublished Ph.D. thesis, Edinburgh University Library.

Halliday, M.A.K. (1961). Categories of the theory of grammar. Word, 17, 241-292.

Halliday, M.A.K. (1967). Notes on transitivity and theme in English. Parts 1 and 2. Journal of Linguistics, 3, 37-81 and 199-244.

Halliday, M.A.K. (1968). Notes on transitivity and theme in English. Part 3. Journal of Linguistics, 4, 179-215.

Hudson, R.A. (1972a). Systematic Generative Grammar. (Privately communicated.)

Hudson, R.A. (1972b). English Complex Sentences. Amsterdam: North Holland.

Lakoff, G. (1971). On generative semantics. In Steinberg, D.D. and Jakobovits, L.A. (eds.) Semantics. Cambridge: Cambridge University Press.

Winograd, T. (1972). Understanding Natural Language. Edinburgh: Edinburgh University Press.

STRUCTURE AND USE OF VERBS OF MOTION

W. J. M. LEVELT, R. SCHREUDER, & E. HOENKAMP

NIJMEGEN   UNIVERSITY

I   Introduction

This article outlines in a summary fashion aims, present
status, and further plans of a research programme[1] on
the structure and use of motion verbs.

One of the ultimate aims of a semantic theory is the
specification and explanation of the relations between
semantic representations and cognitive structures. In
linguistic theory semantic representations are formal
characterizations of the information conveyed by
sentences. But linguistic theory is not self-contained:
a theory of <u>what</u> we understand should be part and
parcel of a theory of <u>how</u> we understand.

Verbs of motion form an attractive domain for the study
of such relations between structure and use. Situations in which verbs of motion are used have been
widely studied in the psychology of perception (Michotte,
1946, Heider, 1944, Johansson, 1973). Especially
Michotte's work is highly relevant for our semantic
purpose. Coming from a neo-Kantian tradition Michotte
proposed that our innate notions of space and time, such
as substance, permanence, causality have their genetic
origin in the innate structure of perception. Study of
the perception of motion and locomotion could therefore
lead to the roots of these concepts. Michotte's
experimental method consisted of systematically varying
the visual motion patterns, and analysing the subjects'
description thereof. These analyses centered around the

use of certain verbs or classes of verbs, a major
instance being the class of causal verbs. In spite of
the fact that Michotte was fully conscious of his experimental dependence upon the verbal reactions of his
subjects (Michotte, 1962), he never undertook a truly
linguistic analysis of his subjects' verbs of motion.
That part of his work remained intuitive.

Such analyses, however, are available in the linguistic
literature. Though there are several older sources
(e.g. Collitz, 1931), it seems to have been Gruber's
(1965) work which has reopened the interest in the
structure of verbs of motion. Like Michotte, Gruber
was not interested in motion per se, but in more general
notions which resemble, and are probably derived from,
concrete concepts of physical motion. A recent extension of Gruber's work is to be found in Jackendoff
(1976). Other important linguistic analyses are
Miller's (1972), and Schank's (1972).

It is not surprising to find that Michotte's perceptual
categories, such as causality, direction, velocity,
return as semantic components in linguistic analyses,
in spite of the fact that these latter are not based on
perceptual arguments. Our research programme is an
effort to bridge the gap: it is on the one hand concerned
with a more systematic analysis of linguistic intuitions
about verbs of motion, whereas, on the other hand, it
tries to link these intuitions to the actual use of
such verbs in perceptual situations, as well as situations in which inference is required. Again, it is
hoped that some of the main results of this study apply
more widely than to the field of verbs of motion alone.

II Linguistic intuitions

II.1 A coincidental classification of verbs of motion

If one is interested in the use of motion verbs, one
would like to know the conditions under which such
verbs can be used. The semantic representation of a
particular verb should in some way or another express
the information which, if present in the interpretation
domain of the use, makes the verb, or better the
sentence containing it, a true statement. This interpretation domain can be a perceptual situation, but
also a conceptual structure which is less directly
related to the real world.

Gruber (1965) has proposed that some of the essential
information expressed by verbs of motion is about the
moving theme, the source it comes from and the goal it
goes to. For a subset of these verbs there is further
information about the agent which causes or permits the
theme to move. Verkuyl (1976) elaborates these notions
in much more detail. Here we will limit ourselves to
making a gross classification of motion verbs, which
will be mainly based on relations of co-reference
between the just-mentioned entities figuring in Gruber's
system of thematic relations. But before we go into
this, it should be noticed that not all verbs of
motion involve the change of location which is expressed
in Gruber's schema.

Verbs like tremble, shrink, mix do not have Miller's
(1972) travel-component, or Schank's (1973) PTRANS, but
it would be counter-intuitive to exclude them as verbs
of motion. Therefore the following preliminary distinction is made:
(i) Transposition vs. non-transposition verbs
    Non-transposition verbs express that the theme is in
motion at a certain fixed location. This is, of course,
not a mathematical point, but a region which is conceptually not further partitioned. This latter criterion
is sufficiently vague to allow for some doubtful cases.
An instance is the airplane circles over the town, where
one might consider such a region as unpartitioned,
making circle a non-transposition verb, or a differen-
tiated area marked by towers, high buildings, or clouds.
If under this latter conceptualization the verb could
still be used, circle would (also) be a transposition
verb.

We decided to devise a linguistic test to determine
whether a verb is a transposition verb. Since transposition verbs involve the change of one location to
another, it should be possible to conceive of a third
location where the theme can be in the mean time. The
test can, therefore, be the following simple completion
task: "They verbed (X) via ...", where the subject has
to invent a location at the place of the dots. There
is an optional X for transitive verbs.

This so-called via-test was applied in an experiment,
where twenty subjects were asked to find completions
for 157 Dutch verbs of motion. They were told that
this would not always be possible, but they were invited
to try.

Figure 1 summarizes the results. For most verbs (nearly)
all, or (nearly) none of the subjects were able to find
an appropriate completion. That is, the test rather
neatly dichotomizes the set of verbs into a transposition
and a non-transposition class. Some typical examples of
both classes are given in the figure. It should be
noted that these are translations from Dutch, and it is
more the rule than the exception that no straight one-to-
one translation is possible.

TEST   OF   TRANSPOSITION   (PTRANS) :

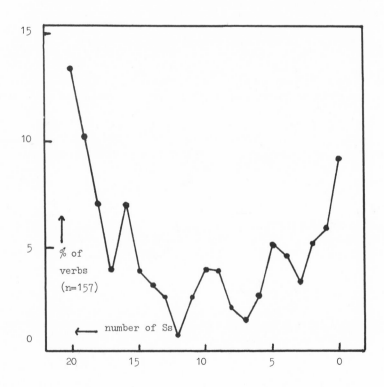

+PTRANS: -PTRANS:
approach, bring, climb,   brake, kneel, separate,
creep, escape, fly, go,   shiver, shudder, shock,
lead, proceed, push, reach, shrink, split, stoop,
roll, row, run, rush, sail, stretch, swell, tremble,
shuffle, travel, trudge,   wrap, yawn.
walk.

Figure 1 : Percentage of verbs judged as transposition verbs as a function of number of subjects.

(ii) Agentive vs. non-agentive verbs

Transposition verbs can be agentive or not. In
John threw the ball, John is the agent causing the
theme's (ball)motion. In the ball fell in the water
there is no such agent. In most linguistic studies
two forms of agentive action are distinguished. One is
always called causation, the other is denoted by
permission, allowance, or the like. In John threw the
ball, John is taken to be a causative agent: John by
some action generates the motion of the ball. In John
released the bird, however, John stops preventing the
bird's own motion. Release is a permissive verb, like
drop. Jackendoff (1976) introduces a function LET in
the semantic representation of permissive verbs. (In
Dutch drop has to be translated by laten vallen, i.e.
let fall).

At this point it should be noticed that Michotte (1947),
on convincing experimental grounds, makes a distinction
between moving objects which are perceived as "being
displaced", and objects which seem to have "proper
motion". Only the first perceptual structure allows for
a causative agent, i.e. for perceived causality.
Michotte's classical case is entraining (i.e. pushing),
where object A moves towards a stationary object B; at
reaching it A continues moving, "pushing" B along. Here
B is not seen to have proper motion: it is simply displaced, participating in A's pre-existing motion. Here,
according to Michotte, there is genuine perception of
causality. But if an object has "proper motion" all the
time, i.e. does not participate in another object's
motion at any time, causality is not perceived. A case
is the perception of braking (Levelt, 1962) where a
speedy moving object gets "stuck" in a certain differently
coloured region of the field. Here subjects do report
braking of the object by the coloured area, but Michotte
(1963) gives arguments to suppose that in these and
similar cases like releasing, and triggering, there is
only immediate perception of dependence, not of causality. If subjects use causal verbs in these cases it
can be the language itself which is to blame: "We should
bear in mind that ordinary language totally lacks
precision in this point, and that in ordinary discourse
we continually confuse cause and condition." (p. 367).
Whatever the truth in this statement, Michotte's distinction between perceptual causation and mere perceptual dependence seems to parallel the linguistic
distinction between causative and permissive agents.

Only agentive verbs allow for instruments. It is the

agent which uses an instrument in bringing about the
theme's motion. Non-agentive verbs like fall will never
carry an instrument in their semantic representations.
It is less clear whether both causative and permissive
verbs allow for instruments. Jackendoff (1976) gives
an apparently positive example for the permissive
verb release: David released the bird from the cage with
a coat hanger. However, Jackendoff gives arguments for
the supposition that the coat hanger is not an instrument for releasing, but for an unexpressed causative
action, namely opening the cage. This causative action
is the means (not the instrument) by which the bird is
released.

We now turn to a further classification of agentive
verbs of motion. If source and goal are different
locations, and if the theme is going from the one to
the other, it is excluded that any two of theme, source
and goal can coincide. However, there is no a priori
reason why agent could not coincide with theme, with
source, or with goal. Of course, there is the final
possibility that none of these coincidences hold.
Together, these four possibilities make a fourfold
subclassification of agentive verbs:

(iii) Agent ≡ theme verbs.[3]

A major subclass here is formed by the intransitive verbs
of locomotion such as run, walk, skate, swim, etc.

There are also transitive verbs in this class, but the
agent/theme will never be in the direct object position.
Examples are leave, enter, and pass. For these examples
at least, it should be obvious that the agent is
optional. In the ball passed my head, there is no agent.
This cannot happen with verbs of auto-locomotion. In
order to prevent confusion in a later section (III.1) we
must make a further remark on pass. If pass is used in
the agent ≡ theme sense, it is certainly causative.
Michotte, however, would never call pass a causative
verb. The good reason is that there is no causal
relation between the activity of the agent and the motion
of the object being passed. It is in this sense that
we will use pass as a non-causative verb in III.1.
Finally, it is our impression that all verbs in the
present class are causative, and allow for genuine instruments. The subclass apparently excludes permissive
verbs.

(iv) Agent ≡ source verbs
      Here the moving theme displaces away from the agent.

Examples are throw, fling, kick, drop. This subclass
contains both causative and permissive verbs.

(v) Agent ≡ target verbs
    For these verbs the theme should move in the direc-
tion of the agent. It is hard to find examples in this
subclass; Attract is one, fetch may be another. Causing
motions "from a distance" so to say seems to be conceptually hard, as will also appear in the next subclass.
Agent ≡ target verbs are more natural in the related
semantic field of ingest-verbs, analyzed by Schank (1973);
examples are swallow, drink, etc.

(vi) Non-coincidental verbs (Agent - external verbs)
    Agent is different from theme, source and goal in
this subclass. Push, transport, drive (transitive),
carry are examples. It seems to be the case that for
most or even all of these verbs the agent moves with
the theme. Again, it seems hard to imagine causation
of motion from a distance. Maybe this will change in
our era of space travel.

As a summary of this section, Figure 2 depicts classification of verbs of motion in a schematic fashion.

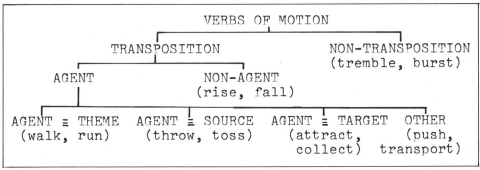

Figure 2 : Classification of verbs of motion in terms
           of thematic relations.

The classificatory properties of motion verbs will
appear to be important for the way in which subjects use
these verbs in inference tasks, as will be discussed
in Section III.3.

II.2 Specificity of verbs and the principle of minimal
      negation
      The schema in Figure 2 is an admittedly very rough
classification. Much more subtle distinctions can be

made among verbs of motion. Take, for example, the subclass of agent ≡ theme verbs, and compare run, hop, and
skate. It seems that each of these verbs over and above
its general meaning of autolocomotion, has a further
much more specific meaning component: run is used for
indicating speed, hop for a particular use of the legs,
and skate for a particular instrument. Moreover, this
specific, or salient component is probably the one
thing which is in the language user's centre of attention
when he or she is using the verb. One does not say the
children are skating if the information to be conveyed
is only that the children are in locomotion. The sentence is used where locomotion is background information
and where the new information is the particular instrument of motion.

Linguists use different means for representing such
components. Jackendoff (1976) uses "restrictive modification", i.e. the simple addition of a marker to the
semantic representation. Others would prefer to add
such components in the form of higher order predicates.
We are not in a position to judge the merits of these
different formalisms. Here we only want to argue that
these representations should allow for the expression
of what we will call "hierarchies of saliency", a
notion which will be worked out in some detail here.

Verbs, also verbs of motion, vary in complexity. Compare
move and rise. Rise has all components of move, but in
addition marks upward directionality. Rise, therefore
is more complex than move. This additional component,
moreover, seems to be the salient one, which is in the
language user's centre of attention when he or she uses
or understands the verb. If one moves from simple very
generic verbs such as move and travel (in one of its
readings) to very complex verbs like bounce or deceler-
ate, more and more components are involved, and the
obvious question then is: which of these components is
the specific one for this verb, or better: is there a
hierarchy of saliency among the meaning components, one
being more "typical" for the verb than another? A
partial answer to this question seems to be the
following: In some cases one meaning component entails
another: braking has a component of deceleration, which
implies a component of velocity, which in its turn
implies some form of motion. Such a chain of reduncancy
can probably be interpreted as a hierarchy of saliency:
the more specific components are probably more available
to the language user than the implied less specific
components (see for a similar argument especially Miller,

1969). However, this is very partial answer indeed. In
many cases components do not have implication relations
at all, Compare swim: on the one hand the verb expresses
that locomotion takes place in the water, on the other
hand it conveys that the locomotion takes place by
means of body parts as instrument. There is no reduncancy between these two pieces of information, but one
could still ask which one is the more salient meaning
component for the language user. There is an empirical
way to go about this question. It is based on what
Noordman and Levelt have called the "principle of
minimal negation" (see Noordman's paper at this conference) which has been independently called the "principle
of minimal change" by Seuren (1976). The basic idea,
however, is certainly older, and can for instance be
found in Miller (1969).

Seuren shows that, normally, when a listener is given a
negative statement he will only make minimal changes in
the knowledge structure which is relevant for understanding the sentence (its "interpretation domain").
The listener who is in an appropriate context confronted
with John didn't give the book to his brother, will
probably infer that actually no transfer took place, or,
alternatively - and very much dependent on prosodic
features of the utterance - that it was not his brother
to whom it was transferred. Whatever the negated element, the important observation is that it is a single
one: The listener will not at the same time infer that
it was in fact neither John nor the book that were
involved, or similarly for other combinations of elements.

Noordman and Levelt's experiments show that also for
lexical negation there is a principle of minimal change.
The experiments, involving the negation of kinship terms,
clearly show that subjects change one meaning component
at a time.

In inference tasks where they may correctly conclude
from not father to either uncle, mother or aunt, they
never give aunt as a response. Aunt differs in two
components from father: sex and parency, whereas uncle
and mother involve only a one-component change. Similarly, uncle is evaded as a response in inferences from
not mother; here father and aunt are the preferred
responses.

How can this principle of minimal negation be used for
determining relative saliency of meaning components?

The answer is based on the assumption that it is the
most salient component which is the probable candidate
for change under negation. We have described a salient
component as a component which is typical, specific,
highly available; it easily gets into the foreground
of attention. The assumption adds that it is this
foreground component which is most likely the one to
be affected by negation.

Applying this to the kinship terms, Noordman and Levelt
have found that for father parency is far more salient
than sex (most subjects gave uncle rather than mother as
a response in the above mentioned inference task). For
mother, however, sex and parency are about equally
salient (father and aunt are about equally frequent
responses).

The empirical procedure, then, which we propose for
determining the most salient meaning component of a
motion verb is to negate the verb and to register the
subject's interpretation. More specifically, we
presented subjects with the incomplete sentence "They
do not verb (X), but they ..." and they were requested
to find an appropriate completion. (X stands for an
optional direct object.)

Take again ski as an example. Presented with They do
not ski, but they ..., most subjects reacted with
skate, indicating that the most salient component is
the instrument 'ski', which is changed into instrument
'skate(s)'.

We have applied this procedure to our 157 verbs of
motion. Twenty subjects did the completion test for each
of these verbs. For each of the verbs the twenty
completions were categorized in (near-)synonymous
groups. The complete results will be reported elsewhere, here we will limit to a few observations:
Figure 3 gives the frequency distribution of the
largest group size. For instance, for travel 9 subjects
complete with stay home, all other reactions, like
wander are produced by smaller numbers of subjects.
The largest group for travel, therefore, has size 9.
This same size 9 is reached by 17 verbs in the sample,
which can be read from the figure. It appears from the
figure that the median major group size is 7. For the
median verb one third of the subjects give the same
reactions under negation. The completions, therefore,
are far from random, and in fact fairly systematic. A
typical median verb is wenden (to turn, especially a

car, a ship, etc.). The 7 completions are to go (or)
drive on straight. It is the directionality component
which is affected by negation here: change of direction
seems to be the salient component of turn. There are
verbs at both sides of the median. At the right side
are the verbs which we will call specific: their most
salient meaning component is much more salient than
any of the others.

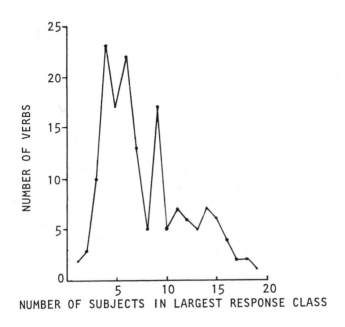

Figure 3 : Frequency distribution of largest response
           class in negation test.

Examples are ascend (19 completions descend), come (18
go), open (18 close), arrive (16 depart), seesaw (16
swing), as well as all the inverses of these. In many
of these cases the salient component is directional in
some way or another, but also instrument is often
involved (like in seesaw), or size (swell - shrink).
At the other side of the median there are the less
specific verbs: two or more of their components are
close in saliency. Examples are swim (6 dive, 4 row),
throw (5 catch, 3 fling), follow (6 lead, 5 stay).

It should be added, however, that there may also be
specific verbs at this side of the median. If the

salient feature is an instrument or means, like in <u>pour</u>,
the subject may replace it by a variety of other instru-
ment or means. For <u>pour</u> we find <u>sprinkle</u> (6), <u>spirt</u> (3),
<u>spray</u> (3). <u>Pour</u>, therefore is a rather specific verb.
However, in order to make this inference, one must have
explicit ideas about the component involved. In other
words, one cannot at the same time use the negation
test as a discovery procedure, and as a means for
determining the saliency of components, except where
subjects give equal or about equal reactions. A
similar problem arises when we ask the question whether
the principle of minimal negation works in this test.
The decision whether one or more components are changed
under negation depends on the definition of components.
However, at scanning the most frequent type of completion for the different verbs, we have not found a
single case where, on intuitive grounds, that completion differed in more than one component from the
original verb meaning. Less frequent reaction types
do show multiple feature changes in certain cases. An
example is <u>drive</u>, where we find <u>walk</u> (6) as first
reaction type. This is a change of instrument. The
second reaction type is <u>sail</u> (5), also involving change
of instrument, but moreover a change of medium.

Turning back to semantic representations which allow for
treatment of hierarchies of saliency, it should be
noticed that none of the existing linguistic systems
are very natural in this respect. In Jackendoff's way
of representation one could, of course, give special
marking to the modifier(s) or function(s) which is
(are) salient, but that is a trivial solution. Also a
solution in terms of predicate hierarchies seems to be
somewhat forced since, mostly, components do not have
implication relations which are strong enough to
determine the predicate hierarchy. We would welcome
representations where the thematic structure of the
verbs integrates naturally with its more specific
semantic aspects.

III   Use of motion verbs

In this section two uses of motion verbs will be
discussed. The first is the use in perceptual verification tasks, i.e. in tasks where the subject is
presented with visual motions, and where the reaction is
either the choice of a verb out of a set, or a yes/no
reaction with respect to a single verb. Here some
experiments and further plans will be discussed. The
second is the use of motion verbs in inference tasks.

Here only plans are available, no experimental results.

III.1   Perceptual verification tasks

In the introduction it was observed that there
exist noteworthy correspondences between Michotte's
perceptual categories and semantic components in
linguistic analyses of verbs of motion. The experimental programme started out with a set of verification
experiments with Michotte-type motions such as push,
launch, pull, pass, brake, bounce, etc. These experiments are reported in van Jaarsveld's (1973) thesis.
Here one of his experiments will be reported in some
detail since it is one of the typical tests of the
theory we started out with:

Theory 1   Sequential elimination

The theory describes the decision process for the situation where the subject is presented with a visual
motion and has to make a choice among n alternative
verbs, one of which is a correct description of the
motion picture. The theory can be extended to the one-
verb yes/no decision task.

The theory states that the underlying decision process
has a tree-like structure: the subject is supposed to
apply a series of tests to the perceptual trace. Each
test involves one semantic component. If the component
is m-valued, there are m possible outcomes of the test.
For each of the outcomes a further test may be
applied, etc. Testing proceeds until a single verb-
alternative is left, i.e. the process is selftermin-
ating.

Take as example the set of verbs which was actually used
in the experiment to be described: Meenemen (a rather
generic term for Michotte's entrain, we will translate
it by take along), passeren (pass), wegstoten (launch),
ophalen (pick up, collect). Pretests had shown that
subjects prefer to use these verbs for describing two-
object motions of the four types shown in Figure 4.
Table 1 gives a componential analysis for the four verbs
in terms of two semantic components: causality, and
(change of) direction. Causality is two-valued, direction three-valued. If it is supposed that the subject
successively applies perceptual tests related to these
components, and if the procedure is self-terminating,
there are two possible decision trees. They are
presented in Figure 5. They only differ with respect
to the order of testing, but they predict rather

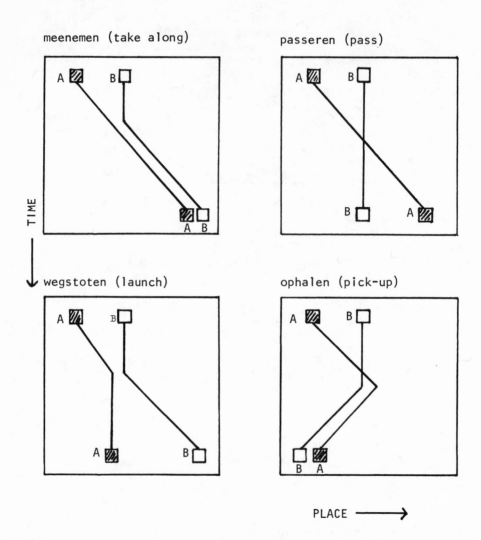

Figure 4 : Motion patterns with their preferred
    verbal description as used in van Jaarsveld's (1973)
    experiment.

Table 1 Componential analysis of four verbs

VERB	CAUSAL	DIRECTION OF A AFTER IMPACT
1 take along	+	+
2 pass	-	+
3 launch	+	O
4 pick up	+	-

different patterns of reaction times: the first tree
predicts that pass will be decided quicker than the
other three verbs. The second tree predicts that
launch and pick up are the fastest to decide on. This
should suffice to exemplify the theory.

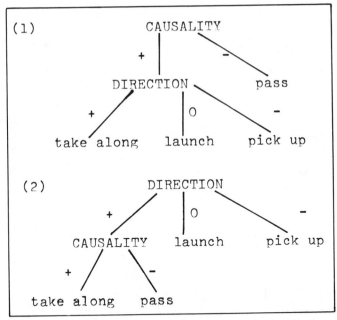

Figure 5 : Decision trees for the verbs: "take along",
"launch", "pass" and "pick up".

Apart from the two decision trees derived from Table 1,
van Jaarsveld derived four more by adding a second
directional feature involving object B. All six
theories were put to test in an experiment where subjects

had to make a forced choice between the four verbs.

Each verb corresponded to a prototypical motion (see
Figure 4); all motions were identical up to the moment
of impact, and all differed from then on. Of each
motion 8 variants were created: four variants were
slight spatio-temporal variations, four other variants
were the mirror images of these (i.e. starting from
right, going to left). The 4x8 different motion patterns
were presented in random order on the screen of a
Vector-General, connnected to a PDP 11/45. The objects
A and B were differently striped rectangular forms. The
subject was seated before the screen with his index
finger on a rest-button. After each presentation he or
she left the button for one of the four reaction keys.

In order to prevent lifting the finger before a
decision was made, catch trials were introduced. They
consisted of motion patterns also starting out with A
approaching B, but then A simply stopped. Subjects were
instructed not to lift their finger in such cases.
Reaction times were measured from the moment of impact
to the lifting of the finger. Responses were recorded;
error rate was smaller than 5%. The experiment was
repeated 10 times in succession for each subject.
Twenty subjects served in the experiment.

Figure 6 presents the average reaction times for the
four verbs (correct reactions) over the ten repetitions
of the experiment. The pattern is fairly consistent:
the mean RT's of the four verbs are all highly significantly different. Ordered from long to short they are
take along, pick up, pass, launch. The result could
not have been worse for the theory: none of the six
patterns of reaction times are in agreement with this
rank order. This result, combined with several other
experimental failures, forced us to reconsider the
theoretical starting point of these experiments.

Theory 2   Testing the salient component

There were several findings in van Jaarsveld's experiments which indicated a different direction of theorizing. Take along (meenemen) is a rather generic verb
in Dutch which can be true for a large variety of
motion patterns. The interesting finding was that in
the above experiment take along produced the longest
reaction time, but not only there, also in free naming
experiments. Moreover, it appeared that also other
generic verbs like meet and disappear gave rather long

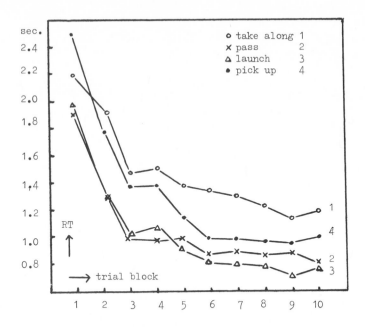

Figure 6 : Average reaction times for the four verbs
  over ten repetitions of the van Jaarsvedd (1973)
  experiment.

reaction times in free naming situations. This does
not seem to be a frequency affect as these more generic
verbs have a higher frequency count than more specific
verbs as pick up, pass and launch. (Uit den Boogaart,
1975) Could it be possible that subjects find it
easier to produce and verify specific verbs than non-
specific verbs?

At this point in the project, international cooperation
came to our help. Dr. Johnson-Laird asked us to help
him run an experiment on motion verification which
could easily be done on our system. His interest was
in possible differences in verification reaction times
for simple and complex verbs. The background of that
interest is not at issue here; it will be discussed,
together with the details of the experiment, in an
independent publication. Here only one result of the
experiment will be mentioned which is essential for the
just mentioned theoretical question. In the experiment
19 pairs of simple and complex verbs were used. The
meaning of the complex verb always implied the meaning

of the simple verb, so for instance rise and move, fall
and travel. That is, the complex verb always had an
additional feature. If both verbs applied to a
certain perceptual event, the complex verb could therefore be a more specific description of that event. The
way in which it would then be possible to understand
that specific verbs could be easier to verify than
generic verbs is depicted in the flow diagram of
Figure 7. It describes the process involved in verify-
ing a verb which is presented to the subject right after
presentation of a motion pattern. The idea is that the
perceiver stores the perceptual event, but has his
attention directed to the most pregnant feature of that
event: here any striking perceptual Gestalt quality may
figure: symmetry, force of impact, speed, etc. Also
the perceiver stores the verb and directs his attention
towards the salient meaning component (in sense described
above).

The subject then makes a comparison between perceptual
feature and salient component. It is first decided
whether the perceptual feature is in the domain of the
meaning component, and if yes, whether their values
match. This latter is not the case, for instance, if
there is high speed perceptually whereas the salient
speed component of the verb has the value 'low'. What
happens if the perceptual feature is not in the domain
of the salient meaning component? Then the subject has
to check back and find another feature of the perceptual
event to which he has not yet attended, but which corresponds to the salient component of the verb.

Let us now take move and rise as examples of simple and
complex verbs. The subject is presented with an object
moving up from the bottom of the screen, and then with
the verb rise. Pregnant perceptual feature is upward
direction, salient component of rise is directionality,
with value 'upward'. The component corresponds to the
feature, the values match, and the response is "true".
Now consider move. The same motion pattern is presented;
the attention getting feature is again upward direction.
The salient component of the verb, however, does not
correspond to this, it is something like '+ displacement',
and the perceiver has to redirect his attention to this
displacement aspect of the motion. Only then it is
possible to judge whether '+ displacement' is true for
the perceptual event. Prediction, therefore, is that
the complex verb will take less reaction time to verify
than the simple verb. This prediction is somewhat
counter-intuitive in the light of what is known about

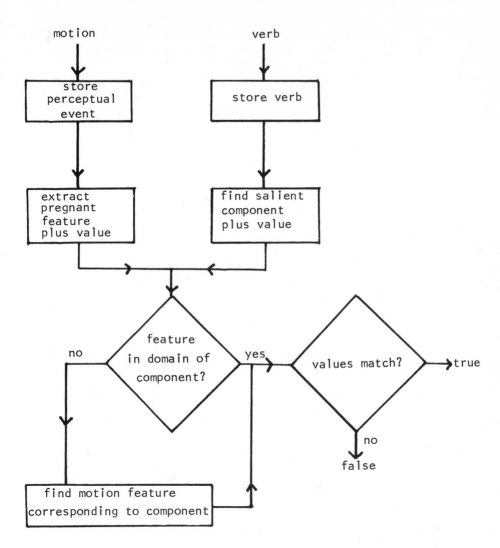

Figure 7 : Flow diagram of verification process of
          motion verbs.

verification times for simple and complex <u>sentences</u>.
So far verification times for "true" responses.

What about "false"? Imagine the subject to be presented
with a stationary object (e.g. an O). If the verb is
<u>rise</u>, there will necessarily be a mismatch between
salient meaning component (directionality) and attention-
getting feature of the percept (certainly not directionality): redirection of attention is needed, and the
reaction time will be relatively long. But now for <u>move</u>.
The critical question is whether the displacement component of <u>move</u> corresponds to the dominant perceptual
feature of the stationary object. This would be the
case if the stationary object would have as its most
pregnant perceptual feature 'no displacement'. This,
however, is very unlikely. Clark & Chase (1972) argue
on convincing grounds that perceptual coding will
normally be positive. The pregnant Gestalt property of
our object will be something like 'round' or 'circular',
and this produces a mismatch for the displacement component of <u>move</u>. Here also, then, redirection of
attention to another aspect of the percept (its
stationarity) is necessary in order to verify the verb.
For "false" responses, therefore, the prediction is
that reaction times will not differ for simple and complex verbs.

The experiment was run in Nijmegen with 20 native
speakers of English as subjects. For each of Johnson-
Laird's 19 pairs of verbs one perceptual event was made
where they were both true, and one perceptual event
where they were both false. So, in fact, each subject
made 76 true or false judgements.

Of the results we only mention the one which is relevant
for the present considerations: in agreement with the
predictions there was a significantly ($p < 0.01$) longer
reaction time for simple verbs than for complex verbs
in case of "true". For "false" there was no significant
difference between simple and complex verbs. But it
should be remarked that these results have to be taken
with caution: different pairs of verbs behaved rather
differently, and there is a general, though non-significant tendency for complex verb reaction times to be
shorter also in case of "false". Further experiments
are necessary (and are in the planning stage) to sub-
stantiate these results.

III.2 Human locomotion

At the beginning of the project, when a sampling
of Dutch motion verbs was made (see Schreuder, 1976),
it was at once clear that most of these verbs could
indicate human patterns of motion, and that a substantial part of the verbs specifically referred to human
locomotion. Distinctions here are highly subtle:
shuffle and shamble, limp and hobble, jog and trudge.
The extent of this vocabulary suggests that variations
in gait are extremely important in daily life. They
may signify moods, intentions, individual styles, etc.

These considerations led to the decision to give
special attention to human locomotions in the project.
For this it would be necessary to generate and manipulate
human locomotions as stimuli for verficiation and naming
tasks. Perceptual studies of human motion hardly
exist, and for a simple reason: locomotion patterns are
highly complex natural stimuli which are very hard to
control experimentally.

A marked exception is Johansson's (1973) beautiful study
on the visual perception of biological motion. Johansson
developed various techniques to register the motion
patterns of humans. One of them consisted in attaching
light spots to the joints, and then making film registrations in the dark. If shown to subjects such patterns
were immediately recognized as human gait, even if the
presentation time was no more than 100 ms. We have
adapted this method in order to be able to manipulate
the registered motions by computer. (Johansson himself
developed a rather different method for the same purpose.) It consisted of using infrared light sources
(LED's actually) to attach to the joints (shoulder,
elbow, wrist, hip, knee, and ankle of the right side,
and elbow, wrist, knee and ankle of the left side).

These LED's switched on and off in turn in a very
rapid cycle (1000 Hz.). A so-called Selspot camera
registered the coordinate of each of the LED's at each
successive cycle, and x and y signals were digitalized
and fed in the PDP 11/45 for further processing.
Further technical details can be found in Hoenkamp
(1976). A list of 46 verbs was made (like stroll, trip,
limp, hop, race, etc.), and an actress was provided [5]
with this list some two weeks before the registration.
Because of the infrared system, the registrations could
be made in daylight, which made it possible to simultaneously make video recordings of them. In this way
we could get an indication of the quality of performance.

Five subjects watched the video pattern on monitors
during the performance. Each of them was provided with
a list of ten verbs, one of them being the correct one.
The 45 incorrect ones were distributed over the five
subjects. If three or more subjects were able to detect
which verb was being played, the Selspot registration
was kept in the "Motion Library" (i.e. on disc). If
not, the verb was played again later on. Only five
verbs could not be recognized even at repeated
performance.

Various transformations were made of these patterns.
The most important being the so-called "conveyor-belt"
transformation. This consists of taking one full phase
out of a periodic motion pattern, choose a tracking
frame of reference, and connect end and beginning of the
movement. Looking at such a pattern gives the impression of the camera following the actress, so that the
(invisible) background moves while the actress keeps in
the middle of the screen. In this way the motion
pattern can be presented to the subject for any length
of time. Figure 8 shows the original registration and
the conveyor belt transformation of hopping. Other

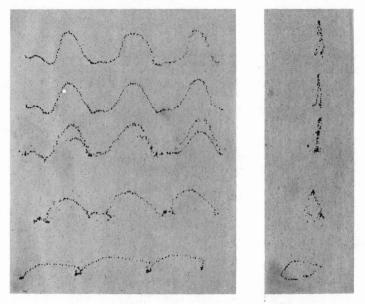

Figure 8 :   The verb "hop"
left : the original pattern
right : the same pattern transformed to a conveyor-
        belt image (no noise reduction).

transformations are easily made. Joints can be connected by limbs, joints and limbs can be removed, or
changed in their motion patterns. One can have walking
feet and running arms, etc. The technique being
available now, we will soon start verification experiments with human locomotions as stimuli.

III.3 Inference tasks
       There are many tasks involving verbs of motion
where the salient component has to be ignored by the
user. His or her attention has to be directed towards
other aspects of the verb's meaning in order to cope
with the task requirements. Good examples are various
sorts of inference tasks. Take for instance the situation where the subject is presented with sentence (1):
(1) John just kicked the ball in the canal.
and the subject is then asked where the ball is just
after this event. This can be done by means of sentence (2):
(2) Where is the ball?
The answer should be "in the canal", but the answer is
independent of the salient component of kick ("by force-
ful foot movements" or the like). It would have been
the same for the sentence (3):
(3) John threw the ball in the canal.
Actually, the answer would be the same for all agent ≡
source verbs and it seems to be only dependent on
relations in the thematic schema of section II.1. This
is especially apparent if one questions the position
of the ball just before the event, e.g. by sentence (4):
(4) Where was the ball?
Since the source is identical to the agent, the answer
should be "with John" or some synonymous expression.
Again, this does not depend on the verb, but on the
thematic class.

Locative inferences of this sort show an interesting
interaction between thematic class and tense of the
locative question.

There are two basic semantic questions at stake with
inferences of this sort. The first is to show how the
inference follows from the semantic representations of
the sentences involved. Rules specifying an inference
in terms of semantic representations are called meaning
rules. Jackendoff presents two sets of meaning rules.
The first set he calls rules of (logical) inference.
The two rules required to answer questions (2) and (4)
are of this sort. They basically say that if a theme
goes from source to goal, it must have been at the

source at some time t_1 and at the goal at some time t_2, and t_1 is before t_2. [1]Jackendoff does not build tense conditions into his rules, which makes it rather compli- cated to handle the just-mentioned tense-dependency of the inference. Tense-dependency is nicely treated by Stillings (1975) in his account of the meaning rules involved in inferences about borrow and loan. The second set of meaning rules are rules of invited inference, or "implicatures". An invited inference would for instance go from John didn't kick the ball to the ball didn't move. Here the first sentence could very well be true while the second is false. But the implicature should hold if, on the principle of minimal negation only the highest predicate is affected by negation. Both sets of inference rules may have prag- matic conditions to them. An example is the inference of the authors went from Holland to Scotland from the authors went from Nijmegen to Stirling on the pragmatic condition that Nijmegen is part of Holland and Stirling part of Scotland.

The second basic semantic question is how to relate the actual inference behaviour of the language user to these meaning rules. Stilling's (1975) paper is an exemplary study of this question. He uses the subject's reaction times for solving inferences on borrow and loan to test alternative sets of meaning rules in combination with so-called control programmes which control the inference procedure by choosing appropriate meaning rules at each step in the inference.

It is our intention to use Stilling's paradigm to ana- lyse meaning rules for the different classes of motion verbs discussed in section II.1.

Footnotes

1 Sponsored by the Netherlands Organization for the
   Advancement of Pure Research (ZWO).

2 The selection of these verbs is discussed in
   Schreuder (1976).

3 ≡ stands for "coincides with".

4 This cooperation was possible under the twinning
   arrangement between the Sussex Laboratory of
   Experimental Psychology and the Nijmegen Unit of
   Experimental Psychology sponsored by the ETP.

5 We are very grateful to Evelyn Schippers who studied
   these motion patterns, and patiently performed them
   during a whole afternoon of registration.

References

Clark, H. H. & Chase, W. G. On the process of comparing
     sentences against pictures. Cognitive Psychology,
     1972, 3, 472-517.
Collitz, Klara H. Verbs of motion in their semantic
     divergence. Language Monographs, 8, March 1931.
Gruber, J. S. Studies in lexical relations, 1965,
     unpublished doctoral dissertation, M.I.T.
Heider, F. An experimental study of apparent behaviour.
     American Journal of Psychology, 1944, 57, 242-253.
Hoenkamp, E. C. M. A computer registration of human
     locomotion. Report 76FU02, Dept. of Psychology,
     Univ. of Nijmegen, 1976.
Jaarsveld, H. van. Het gebruiken van bewegingswerk-
     woorden, Unpublished doctoral thesis, Nijmegen,1973.
Jackendoff, R. Toward an explanatory semantic repres-
     entation. Linguistic Inquiry, 1976, Vol. 7 nr. 1,
     89-150.
Johansson, G. Visual perception of biological motion
     and its analysis. Perception and Psychophysics,
     1973, Vol. 14, no. 2, 201-211.
Levelt, W. J. M. Motion braking and the perception of
     causality. In: Michotte, A., Causalité, permanence
     et réalité phénoménales, Louvain, 1962.
Michotte, A. La perception de la causalité, Louvain,
     1946.
Michotte, A. Causalité, permanence et réalité phénomén-
     ales. Louvain, 1962.

Michotte, A. The perception of causality, London, Methuen, 1963.

Miller, G. A. A psychological method to investigate verbal concepts. Journal of Mathematical Psychology, 1969, 6, 169-191.

Miller, G. A. English verbs of motion: a case study in semantics and lexical memory. In: Melton, A. W. & Martin, E. (Eds.), Coding processes in human memory. Washington, Winston, 1972.

Schank, R. C. Conceptual dependency: a theory of natural language understanding. Cognitive Psychology, 1972, 3, 552-631.

Schank, R. C. Identification of conceptualizations underlying natural language. In: Schank, R. C. & Colby, M. C., Computer models of thought and language, San Francisco, Freeman, 1973.

Schreuder, R. Een lijst van Nederlandse bewigingswerkwoorden, Rapport 76FU01 Dept. of Psychology, Univ. of Nijmegen, 1976.

Seuren, P. A. M. Echo, een studie in negatie, to be published, 1976.

Stillings, N. A. Meaning rules and systems of inference for verbs of transfer and possession. Journal of Verbal Learning and Verbal Behaviour, 1975, 14, 453-470.

Uit den Boogaart, P. C. Woordfrequenties, Utrecht, Oosthoek, Scheltema & Holkema, 1975.

Verkuyl, H. J. Thematic Relations and the Semantic Representation of Verbs Expressing Change, Report 76FU07, Dept. of Psychology, Univ. of Nijmegen, 1976.

VERBS OF JUDGING AND ACTS OF JUDGING

IVANA   MARKOVA

University of Stirling

In her recently published book Presupposition and Delimitation of
Semantics Kempson (1975) argues that in the work of linguists
such as the Lakoffs, the Kiparskys and Fillmore, problems concerning the semantic and the pragmatic nature of presuppositions
have been conflated, and, moreover, that the notion of presupposition
has no place in semantics. The phenomena labelled by linguists
and logicians as presuppositions can, according to her, be explained
within the framework of Gricean pragmatics. In Kempson's own
treatment of a pragmatic theory, there is, however, no suggestion
as to what to do with lexical presuppositions.

I shall be concerned with, in this paper, lexical presuppositions
of the group of verbs which Fillmore (1971) and McCawley (1975)
have called verbs of judging and verbs of bitching respectively
and which include such verbs as accuse, criticise, blame, apologise,
forgive, condemn, impeach and so on.   The purpose of this paper is
two-fold: firstly, to consider the inadequacies of Fillmore's and
McCawley's semantic descriptions of verbs of judging; secondly, to
argue that the lexical presuppositions of these verbs can only be
discovered by a psychological analysis of acts of judging.

Fillmore's exercise in "semantic description" of verbs of judging
is an attempt to offer an alternative to the feature analysis of
the meaning of words. Having distinguished between meaning and
presupposition Fillmore focussed on such questions as what do we
know about various aspects of the situation and how do we evaluate
it in terms of the Affected, Defendant, Judge, Judgement, Statement
or Addressee. In more detail: a situation has either a favourable or
an unfavourable effect on an individual; someone is responsible

for the act: someone may make a moral judgement about the state
of affairs;   or the situation may be factual rather than assumed,
and so on.   For example, the semantic descriptions of 'accuse'
and 'criticise' are as follows:

ACCUSE (Judge, Defendant, Situation (of)) (Performative)

 Meaning: SAY (Judge, 'X', Addressee)

X = RESPONSIBLE (Situation, Defendant)

Presupposition: BAD (Situation)

 CRITICISE (Judge, Defendant, Situation (for))

 Meaning: SAY (Judge, 'X', Addressee)

X = BAD (Situation)

Presupposition: RESPONSIBLE (Defendant, Situation)

Presupposition: ACTUAL (Situation)

What Fillmore says is that when we use the verb 'accuse', we
presuppose that the situation was bad and we claim that X is
responsible for such a bad situation;   on the other hand, when
we use 'criticise', we presuppose that X is responsible for the
situation and we claim that the situation is bad.

Kempson criticises this analysis on the grounds that it is, in
fact, no more, than a decomposition of lexical items into semantic
features in the style of Katz and Fodor (1963) and Bierwisch (1970).
McCawley, on the other hand, is not critical of the very basis of
such an analysis but points out certain problems: for example
that the differences between verbs cannot be attributed to
differences in presuppositions in the way suggested by Fillmore;
that the presuppositions of, for example, 'accuse' are more
complex than claimed by Fillmore;   and that 'situation' in Fillmore's
analysis refers to several different things.   McCawley's criticism
thus focusses on some important difficulties with Fillmore's
analysis, though neither he nor Kempson tackle what seems to me to
be the crucial problem.

Let me say firstly that in my opinion Fillmore's semantic
description is an important step away from feature analysis.
However, although he correctly asks "what do I need to know in
order to use this form appropriately?" (p.94) his appeal to
reality, and in this particular case to the social reality of
acts of judging, is only a common sense appeal to FACTUALITY,

ACTOR, RESPONSIBILITY, rather than an attempt to go beyond common
sense and analyse problems such as "what do I need to know about
the actor's role in order to use this form appropriately?" or
"what do I need to know about an actor's responsibility in order
to label his involvement appropriately?" and so on. Let us, for
example, consider the notion of RESPONSIBILITY in more detail.

RESPONSIBILITY in Fillmore's analysis is a unitary notion, i.e.
the person either is or is not responsible. In the analysis of
action, however, we distinguish between different forms of
responsibility, each of which differs as to the relative contribution of environmental and personal factors. At the lowest level
RESPONSIBILITY is a global notion and refers to any sort of
connection between the person and his environment. For example,
a person may be congratulated for the victory of his school team.
On the next level the person is judged according to the result of
what he does rather than according to his intentions. This type
of responsibility in fact corresponds to the Piagetian subjective
responsibility stage of a child's development of morality. This
is followed by a level at which a person is felt to be responsible
for an after-effect that he might have foreseen even though it
was not a part of his own goal. The next level refers to intentional responsibility: a person is responsible for what he intends
to do. Finally, a person may in certain circumstances be seen as
not responsible for his action and even his motives may be ascribed
to the environment in which he lives.

Although in English there are not different words for different
types of responsibility, it seems to me that different verbs of
judging do take into account at least some of these different
types. The RESPONSIBILITIES connected with 'accuse' and
'criticise' are different types of responsibility and this may be
why 'accuse' is more often used with a human subject while, on
the other hand, we criticise states of affairs as well as people
or their neglect, i.e. failure to act when action was required
by a situation. To take McCawley's examples

> The principal criticised Rocky's hair for being untidy.

> *The principal accused Rocky's hair of being untidy.

Although it does not seem to me that sentences can be labelled
as semantically anomalous on their own (as Rommetveit (1974) has
shown, contexts can be found in which the anomaly disappears), we
do not often accuse the river of something, accuse stones of something and so on. Such distinctions between different types of
inbuilt 'responsibility' may also account for the following
difference: 'accuse' usually refers to an action:-

I accuse the chemical factory of polluting the river and
not to an outcome of an action:-

*I accuse the river of being polluted while 'criticise' may
refer to both an action and to an outcome of an action:-

I criticise the chemical factory for polluting the river

I criticise the pollution of the river

Responsibility in 'accuse' is the moral type of responsibility -
and we attribute morality to people and not to inanimate objects
- while 'responsibility' in 'criticise' encompasses different
types of responsibility. While the accused,if he proves to be
guilty, deserves punishment, the one who is criticised does not
necessarily deserve punishment, and criticism is often understood
as a sort of help.

I hope that this example demonstrates the necessity to go beyond
Fillmore's common-sense notions. This, however, cannot be achieved
by detaching verbs from the actions to which they refer or by
establishing whether a sentence is well- or ill-formed, but only
through a systematic exploration of actions in terms of the
differences between the actor's roles, the social evaluation of
actions, the moral involvement of the participants and so on, and
only then can we assess what is presupposed by the use of
different verbs. Murdering Mary or splashing Mary with red paint
or kissing Mary not only all have different consequences for the
victim but also imply different reactions from the observers. If
it is Peter who was the actor and someone comments:

"The father condemned Peter for murdering Mary"

we might enquire as to the WHY of the murder but would probably
be quite satisfied as to the WHY of the father's reaction because
it is socially expected that a murderer should be condemned. If,
however, the father condemns Peter for splashing Mary with red
paint - our WHY would refer not only to Peter's silly act but also
to the inappropriateness of the father's reaction. Finally, if
the father condemns Peter for kissing Mary we would have even more
WHYs, as kissing is not normally associated with badness, unless
of course, it carries some additional meaning, such as the biblical
example 'Jesus was kissed by Judas', or kissing in order to pass
on a fatal disease and so on. The point I am trying to make is
that in order to account for the appropriateness of the use of
verbs, rather than making verbs the independent variables and
situations the dependent variables, we must treat the situations
as independent variables and the verbs as the dependent variables.

Any psychological process of attribution of guilt or virtue can be

considered to be a problem-solving situation consisting of a
series of sub-problems of a hierarchical order such that a solution
to a sub-problem which is lower in the hierarchy will provide
presuppositions for the next stage. Such a problem is defined as
the gap between an action (= the initial data) and the final
attitude of the judge towards the actor. The process of the
attribution of guilt might proceed as follows:

Sub-problem 1:

ATTITUDE TOWARDS THE OUTCOME OF THE ACTION

Sub-problem 2:

WHO IS RESPONSIBLE?

degree of certainty that X did it

role in which X did it

social setting

power relationships

Sub-problem 3:

X DID IT

social evaluation of the action

type of responsibility:did he intend to do it?

did he intend something else?

Sub-problem 4:

X's REACTION

degree of acceptance of guilt

degree of acceptance of social evaluation

degree of acceptance of responsibility

degree of regret

Sub-problem 5:

VICTIM'S REACTION

attitude to X's reaction

level of forgiveness

It is not necessary, of course, to proceed through all the stages
in all acts of judging. In the case of an event, rather than an
action we do not go beyond stage 1. If the culprit is a child
we may stop at a lower level, depending on what we think the
youngster intended and is responsible for. Or we may know from
the very beginning who was the actor, why he performed the act,
and so on.

This type of procedure can hopefully divert our attention from
the pre-determined semantic structures of words and their fixed
components to what Rommetveit calls operative semantic competence.
Mastery of verbs of judging presupposes the mastery of basic
categorizations and social attributions, such as GOOD/BAD, ACTUAL/
ASSUMED, RESPONSIBLE/NOT RESPONSIBLE and so on. Mastery of the
meanings of these verbs, however, cannot be reduced to the mastery
of these categories. Our operative, that is, our dynamic
semantic competence can be assessed only in terms of our ability
to elaborate these general categorizations and attributions in
the particular contexts in which verbs of judging are used. Our
problem-solving analysis of acts of judging grasps the basic categorizations, which are necessary, though not sufficient aspects
of their meaning potential. Thus we cannot understand the difference between 'criticise' and 'accuse' unless we master categories
such as GOOD/BAD, FACTUAL/ASSUMED across different contexts.

We all differ with respect to the neatness of our categorizations.
A child may, from the point of view of an adult, have a very
simple system of categories. For example, for the child the
difference between 'hope' and 'fear' may be reduced to differences
between 'good' and 'bad'. A poet, on the other hand, will have a
very complex and subtle system of categorizations, although he
will share with the child the distinction between 'good' and 'bad'.
For example, for the philosopher Marcel, 'hope' is contrasted
with 'desire' and incorporates categorizations of active and
passive waiting, faith and impatience.

Lexical presuppositions are, therefore, in our account, basic
categorizations which can be discovered from a psychological
analysis of action. In communication we become aware of them
only when we realize that we do not share them with the other
participants in a social encounter. For example:

A: "This is not appraisal this is criticism".

B: "What do you mean?"

A: "What I mean is that what John did was good and you are
     saying it was bad".

Or on a more subtle level Marcel might say:

M: "It is not hope, it is desire".

X: "What do you mean?"

M: "What I mean is that "it is precisely here, in the relation
     to Time ... that the essential difference lies between
     Desire and Hope. The latter involves, indeed, a waiting,
     and we will have to recognize the existence of a range
     which goes from inert waiting to active wating"" (p.280).

Mastery of the meanings of words generally and of verbs of
judging in particular is revealed through creative optional
elaborations of basic categories in particular contexts and are
the means by which we achieve inter-subjectivity. In one of our
interviews with the parents of handicapped children the mother said:

    "Our neighbours are shocked by our high moral values".

A linguist would probably say that the sentence is semantically
anomalous but it might also be considered as a sincere attempt
to elaborate the basic acquired categories and express something
not adequately communicable by words such as 'surprise', 'admire'
or 'amaze'.

To sum up, lexical presuppositions of verbs cannot be revealed
by breaking down lexical items into components. Neither can they
be accounted for by referring to common-sense notions in Fillmore's,
the Lakoffs', Kartunnen's and others' fashion. They can be
discovered only by psychological analysis, in our case of acts of
judging and more generally, by the psychological analysis of the
appropriate types of action.

References

Bierwisch, M. (1970). Semantics, In Lyons, J. (Ed.)
     New Horizons in Linguistics. Harmondsworth, Middlesex:Penguin.

Fillmore, C.J. (1971). Verbs of judging: an exercise in semantic
     description. In Fillmore and Langendoen (Eds.) Studies in
     Linguistic Semantics. New York: Holt, Rinehart and Winston.

Fodor, J. and Katz, J.J. (Eds.) (1963). The Structure of Language.
     Englewood Cliffs, N.J.: Prentice-Hall.

Kempson, R.M. (1975). Presupposition and the Delimitation of
    Semantics. London: Cambridge University Press.

McCawley, J.D. (1975). Verbs of Bitching, In Hockney, Harper and
    Freed (Eds.) Contemporary Research in Philosophical Logic and
    Linguistic Semantics. Dordrecht: D. Reidel.

Marcel, G. (1967). Desire and Hope. In Lawrence, N. and
    O'Connor (Eds.) Readings in Existential Phenomenology.
    Englewood Cliffs, N.J.: Prentice-Hall.

Rommetveit, R. (1974). On Message Structure. New York: Wiley.

A PROGRAMMATIC THEORY OF LINGUISTIC PERFORMANCE

M.J. STEEDMAN and P. N. JOHNSON-LAIRD

University of Sussex

Introduction: Towards a Programmatic Psychology

The aim of this paper is to present a unitary model
of some of the mental processes involved in speaking
and listening. The model is cast in the form of a
computer program, but the program is not intended as a
contribution to either Artificial Intelligence or
Computer Simulation. We regard it as an exercise in
programmatic psychology, and perhaps a few words are
necessary in order to explain the nature of this claim.
On the one hand, proponents of Artificial Intelligence
(AI) aim to develop machines capable of intelligent
behaviour and, in particular, to devise computer programs capable of such tasks as interpreting visual
scenes, providing theorems, and understanding natural
language. Although the methods implemented in these
programs are likely to interest a psychologist, any
resemblance to human performance may be entirely coin-
cidental. AI is concerned with intelligence in general,
not merely its embodiment in living organisms. On the
other hand, proponents of Computer Simulation aim to
understand human behaviour by simulating it. Turing's
(1950) test has proved to be most influential: a program should perform with a sufficient degree of verisimilitude so as to be indistinguishable from a human
being performing the same task. Of course, the point
of the enterprise is not merely to fool a naive obser-
ver - an exercise that turns out to be relatively easy
(Weizenbaum, 1966), but to use such a test as a method

171

of assessing the adequacy of the psychological principles embodied in the program.

AI programs are very large, and growing larger, and
there is a danger that their size and complexity will
outrun their deviser's ability to comprehend them.
Simulation programs are similarly growing in complexity
and in consequence encounter two serious problems.
First, it has become difficult to compare their performance with human protocols in an informative way:
where there is a divergence it can be hard to assess
its significance and still harder to modify the program in order to reduce it. Second, such programs
inevitably embody a large number of ad hoc decisions
for which there is no psychological motivation and no
possible empirical test.

Over the last three years we have struggled with the
problem of finding a rapprochement between computer
programs and experimental results, and it is from our
common concern to understand linguistic performance
that the idea of a programmatic psychology has emerged.
Its goals, as the term itself suggests, are more
modest than those of AI or simulation. Its philosophy
is, in epitome: small is beautiful. Its precepts are
simple to summarise. The program should be the theory:
ideally, it should contain no ad hoc patches, and its
principles and structure should be easy to grasp. The
purposes of developing a theory in a programmatic form
are to ensure its consistency, to develop new ideas,
to formulate experimental tests of its critical assumptions, to discover both the scope and limitations of
the theory, and to demarcate the set of phenomena that
one does not understand. In our experience these advantages accrue as an almost inevitable consequence of
this mode of theorising (see Steedman, 1976; Johnson-
Laird, Steedman, and Huttenlocher, in preparation).
In short, the aim is to encapsulate fundamental theoretical principles in a program, keeping to the smallest
scale possible, and not to succumb to the temptation
to insert unmotivated patches into the program purely
to engender a spurious appearance of verisimilitude.
A poor simulation is more informative than a dissimu-
lation.

Semantics and Augmented Transition Networks

The overriding impression that has emerged from recent
empirical studies of linguistic performance is that
there is evidence for the psychological reality of

entities analogous to the surface and deep structures
of sentences, but there is no reliable evidence for
the psychological reality of grammatical transformations
(Fodor, Bever, and Garrett, 1974). One device that
resolves this apparent paradox is the Augmented Transition Network (ATN). ATNs were originally invented
to retrieve the surface and deep structures of sentences and they did so without recourse to operations
correpsonding directly to transformations (Thorne,
Bratley, and Dewar, 1968). They have subsequently
been used to parse sentences according to a variety of
different conceptions of grammar (Woods, 1970; Winograd
1972), to model various perceptual heuristics for parsing (Kaplan, 1972), and to codify alternative processing strategies (Wanner and Maratsos, 1975).

The essentials of an ATN can be described intuitively
by way of an example. Consider how an active sentence
such as "The boy hit the ball" might be parsed. The
first task, granted a system restricted to simple
declarative sentences, is to find a noun phrase. The
system relies on a component specifically assigned to
analysing noun phrases. Whenever this component is
called upon it checks to see whether the first word it
encounters is an article, next it tests for the optional
presence of an adjective (or a string of them), and
finally it tests for a noun. Alternatively, it simply
tests for the presence of a noun alone. Once it has
established that the required elements are present, it
assembles them in an appropriate syntactic analysis of
the noun phrase. In so doing it relies upon the fact
that information is stored about the outcome of each
test. Thus, for example, the label Article is assigned
to The and the label Noun to boy, and in turn the
analysis of the noun phrase can be assembled:

(Nounphrase: Article = the Noun = boy). This information is returned to the main system, which specifies
that the noun phrase is the subject of the sentence
and then tests whether the next word is a verb. If it
were to find an intransitive verb, it would proceed
immediately to assemble an analysis of the sentence as
a whole, provided there were no other words to be
parsed. In our example, however, the verb hit is transitive and so, having stored this fact, the system
once again calls upon the services of the noun phrase
analyzer. This component successfully parses the ball
and returns its analysis to the main system for labelling as the object of the sentence. It is now possible to assemble a complete parsing of the sentence.
Typically, this analysis is represented in the following

sort of syntactic notation:

(Sentence:Subject = (Nounphrase:Article = <u>the</u> Noun =
    <u>boy</u>)
                    Verb = <u>hit</u>
              Object = (Noun phrase: Article = <u>the</u> Noun =
                        <u>ball</u>)).

Each component in the above process can be represented
pictorially as a network, and the ensemble of networks
constitutes an augmented transition network (ATN) as
depicted in Figure 1.

The upper network is the main component for the analysis of declarative sentences; the lower network is the
component that analyses noun phrases. The numbers in
the nodes are simple mnenonics denoting different
states of the device. A transition is made from one
state to another along an arc, provided that the conditions specified above the arc are satisfied; when a
transition is made the action specified in parentheses
beneath the arc is executed. It is obviously necessary for the processor operating such a system both
to have access to a dictionary in order to look up the
syntactic categories of words, and to keep track of
where it is working, so that when it jumps from one
network to the other it can ultimately return to the
appropriate place. If no transition can be made but
there are still words in the sentence to be parsed, the
system "blocks" and the sentence is rejected. The system in Figure 1 is, for example, unable to accept
sentences with auxiliary verbs, though it is a simple
matter to extend the network in order to accept them.
What is remarkable is that an ATN can analyse the deep
structure of a sentence without recourse to transformational rules. The point can be established by considering how the simple device in Figure 1 can be enhan-
ced in order to accept passive sentences.

A sentence network that accepts both active and passive
sentences obviously requires an arc that tests for the
auxiliary verb <u>be</u> and an arc that includes a test for
a verb with the past participle affix.   On discovering
both these elements, it will generally be necessary to
take the following steps: the noun phrase currently
labelled as the subject of the sentence must be relabel-
led as its object. With a sentence such as "The ball
was hit by the boy", an arc tests for <u>by</u> and then
assigns the label, subject, to the subsequent noun
phrase. With a sentence such as "The ball was hit",

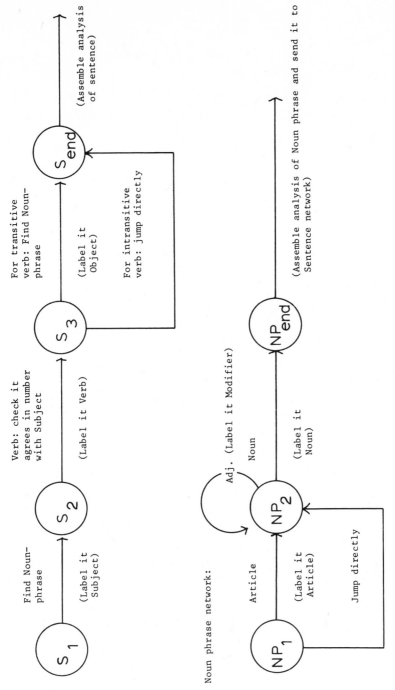

Figure 1   A simple augmented transition network (ATN).

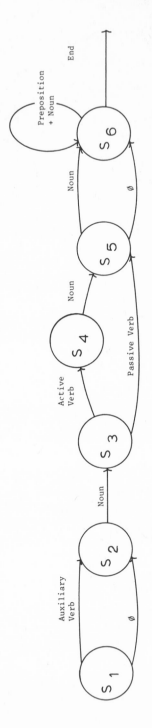

Figure 2   The transition network used as the basis for
the STN.

a dummy element, <u>something</u>, is assigned to the role of
subject. Hence, the fundamental grammatical relations
of a sentence - its deep structure - can, in principle,
be specified by the operation of an ATN.

Although an ATN can be devised to deliver the deep
structures of sentences (see Levelt, 1974), it is
natural to wonder whether listeners actually set up an
explicit representation of deep structure as part of
the ordinary process of understanding a sentence. Most
people have the intuitive impression that they understand a sentence very nearly as they hear it. This
intuition has been bolstered by Marslen-Wilson's
(1976) finding that subjects shadowing speech with a
delay of only about a quarter of a second spontaneously
correct errors which the experimenter has deliberately
inserted into it. These corrections reflect the subjects' sensitivity to grammar and meaning. Yet, an
explicit representation of deep structure cannot be
completed until the whole of a sentence has been perceived; hence, if such a representation is a prerequisite for comprehension, the performance of the
rapid shadowers should not be affected by meaning.
Furthermore, it has proved impossible to find any evidence for the retention of deep structure where it is
uncorrelated with meaning (Fillenbaum, 1973).   Even
Fodor, Bever and Garrett (1974, p.270) remark that
"...so long as the deep structure is assumed to be
that syntactic representation of a sentence which is
semantically interpreted, hypotheses about the psychological effects of deep structure features tend to be
confounded with hypotheses about the psychological
effects of semantic relations." Indeed, it can be
argued that there is no need to derive an explicit
representation of deep structure, and that an ATN can
be devised to return semantic representation directly,
constituent by constituent. Such a <u>semantically augmented transition network</u> (STN) utilises the same information that would be represented in deep structure, but
it does so as a direct signpost to the semantic representation rather than as an explicit map that would
then, in turn, require an interpretation. The STN can
also be used as the basis for a model of speaking by
the simple expedient of reversing the tests and actions
on each transition arc. Thus, for example, a listener
understanding a noun phrase tests for a noun and then
attempts to retrieve its semantic representation from
the mental lexicon, whereas a speaker producing a noun
phrase tests for a certain sort of semantic representation and then attempts to retrieve a corresponding

noun from the mental lexicon. The principles of the
STN were sketched informally in Johnson-Laird (1976);
we shall try to develop these principles in a programmatic theory.

The Given-New Contrast

Although corresponding active and passive sentences
are normally considered to be equivalent in terms of
truth or falsity, there is a subtle distinction between
them. Different suppositions are associated with them
when they are uttered with the "unmarked" intonation
which places the tonic stress at the end. In particular, the question:

      Did Y hit Z?

normally suggests that <u>Y hit something</u>, whereas the
question:

      Was Z hit by Y?

normally suggests that <u>Z was hit by something</u>, which
is a very different matter.

Traditional linguists, such as Jesperson (1924), members of the Prague school (e.g. Firbas, 1964) and more
recently, Halliday (1967), and psychologists such as
Clark and Haviland (1974), distinguish between what is
assumed, the "given" information in a sentence, and
what is asserted or questioned, the "new" information
in a sentence. It seems to be a general principle of
conversation that the given information in a sentence
should be mentioned before the new information. Thus
the exchange:

      Did Y hit X at 2?
      No, at 3 Y hit X

seems awkward in comparison with:

      Did Y hit X at 2?
      No, Y hit X at 3

However, the general rule that given information precedes new information applies only when there is no
explicit indication to the contrary. In the previous
examples, the tonic stress in uttering the sentence
was taken to be in the unmarked final position. The
stress falls in the constituent conveying the new information and so for these examples the given-new order
is directly reflected in the surface form of the sentence. However, when the tonic stress occurs elsewhere,

it locates the new information in something other than
the last constituent of the sentence. Similarly, the
cleft construction marks the fact that the given-new
order is different from the surface structure order.
Hence, the sentences:
         X hit Y
         It was X that hit Y
         Y was hit by X

are similar in that they all take for granted that
something hit Y and assert that it was X.

The given-new contrast is important, as Kay (1975) has
emphasised, in constructing a reversible system that
can be used for both generating and interpreting sentences. Kay suggested that the ordering of constituents from given to new should be represented as a
simple list. However, we shall interpret this ordering
as one that concerns processing rather than simply as
a list of elements within a structure. The order of
processing arises as an implicit consequence of the
system's operation (Johnson-Laird, 1976). Put simply,
the idea is that in answering a question such as,

         Did Y hit X at 3?

the hearer first finds events in memory which involve
Y in hitting something, and that involve X as the
object, and then tests whether one such event occurred
at time 3. The given information does not correspond
to a single proposition, but to several, and accordingly
the computation outlined above may break down either
because Y didn't hit anything, or because Y did not hit
X. It is a virtue of this kind of process model that
the given information is evaluated in successive steps,
and that in the case of a breakdown due to part of it
being in error, the false assumption is immediately
apparent. The order of steps in computation is different for a passive sentence. In the case of

         Was X hit by Y at 3?

the first step is to find events which involve X as the
object of a hitting. As long as the answer to the
question is "Yes", the difference in processing will
not be immediately apparent. But if the question is
based on a false assumption, then its answer will reflect the different division and processing of given
and new information in a passive sentence.

Similar ideas underlie the generation of sentences, but
here the order of the steps in the computation determines the form of the utterance. The extension of the

theory to sentences involving stress or cleft-type
constructions has not yet been attempted, but there
seems to be no reason why it could not handle utterances in which the surface order of constituents is
not the same as the given-new order.

A Programmatic Theory of Linguistic Performance

In keeping with our methodological aims,the program
was kept as simple as possible. It exhibits a minimal
range of linguistic behaviour and has an extremely
restricted universe of discourse. Its basic capability
is to understand questions about the movements and
interactions of a small number of definite entities.
A typical history that it deals with might consist of
three objects (referred to by name as "X", "Y","Z"),
four locations(referred to as "A", "B","C", "D"),and
five successive moments of time (referred to by the
integers 0 to 4.)The system understands only simple
active and passive questions with or without prepositional phrases such as:

        Did Y hit Z at 2?

Since the STN is designed only for single clause sentences it is not necessary to allow for recursion
betweeen a set of networks. The program is based on
the transition network shown in Figure 2.

The program operates in the following way. When a
question is typed in, it is parsed by the STN which
builds up a list of goals as it makes transitions from
node to node. For example, the question

        Did Y hit X at 2?

gives rise to the following ordered series of goals:

        EVENT HIT SUBJ Y   Find an event that has Y as
                            the subject of the hitting.

        EVENT HIT OBJ X   Check that X is the object
                            of the hitting.

        EVENT HIT AT 2   Check that the time of the
                            event is 2.

The goals represent successive stages in the computation
of the answer to the question. This computation is
carried out on the "history of the world", which is
represented in a data-base as a set of assertions
about the positions and elementary movements of the
objects in the world. Higher-level descriptions of
events, such as <u>hitting</u> are not directly represented

in the data-base, and the truth of propositions involving them must be inferred from the simpler propositions
in the data-base. A typical history in the data-base
has essentially the following form (where the square
brackets are used to indicate assertions):

[Y AT A O] :   Object Y is at place A at time O

[X AT B O] :   Object X is at place B at time O

[Z AT C O] :   Object Z is at place C at time O
[E1 Y MOVE FROM A TO B START 1 FINISH 2] :
               Event E1 consists in Y moving from
               place A to place B from
               time 1 to time 2

[E2 X MOVE FROM B TO C START 2 FINISH 3]

[E3 Z MOVE FROM C TO D START 3 FINISH 4]

etc.

This represents a history in which Y moves to where X
is and stops, X then moves to where Z is and stops, and
Z finally moves to place D. It might describe part of
a game of shove-ha'penny, or the events in one of
Michotte's experiments.

We have chosen to use a version of the PLANNER programming language (Hewitt, 1969; Winograd, 1972) to embody
the data-base and to set up and achieve goals. The
version of PLANNER used is PICO-PLANNER (Anderson,1972),
an extension to POP-2 (Burstall, Collins & Popplestone,
1971).

PLANNER is a 'goal-directed" language that permits
goals to be specified, and whose interpreter controls
a search of a data-base to achieve them. Since it is
usaully impossible, or at least expensive, to represent all of the true statements about a world directly
in the data-base, PLANNER also allows the specification
of procedures that achieve goals by setting up sub-
goals, which may, in turn, be achieved by a direct
search of the data-base, or by a further invocation of
procedures and sub-goals. The present program uses
these PLANNER procedures as a representation of the
meaning of verbs such as "hit", which describe events
that are not directly represented in the data-base but
must be inferred from those which are. The meaning of
a verb is distributed over several procedures, since
the goal to check, say, the subject of an event is
evaluated separately from the goal to check its object,
and these two goals require two different procedures.

Such a representation of the meaning of words is
rather unlike the normal conception of a lexicon, but
it has several interesting properties. By not having
a single monolithic definition of a word the system
avoids the problem of what to do if some "argument" of
a verb is unknown or unstated. Moreover, a word may
occur directly in the data-base and as preocedure(s)
to search the data-base. For example, the goal to
check that something is at a particular place may be
achieved either by finding an assertion to that effect
in the data-base, or by using a procedure to infer
that it must be at the place because it moved there.

In order to understand a sentence, the STN turns it
into a list of goals which the PLANNER mechanisms
evaluate in serial order, using the procedures which
define the meanings of words. The system keeps a
record of the series of assertions that satisfy the
goals. Thus, for example, in answering the question:
       Did Y hit X at 2?

the system produces the following goals:

       EVENT HIT SUBJ Y
       EVENT HIT OBJ X
       EVENT HIT AT 2

If the system succeeds in achieving each of these
goals then it produces a series of assertions satisfying them:

$$[E2\ HIT\ SUBJ\ Y]$$: Event E2 has Y as the subject
                        of a hitting,

$$[E2\ HIT\ OBJ\ X\ ]$$: X as its object,

$$[E2\ HIT\ AT\ 2\ ]$$: and occurred at time 2.

In such a case, the system simply replies, "Yes".

What happens if one of the goals cannot be satisfied?
A simple-minded solution would be merely to reply,
"No". However, such an answer is not very informative.
In a world where it is the case that Y hit X at 3, a
speaker is more likely to respond with the helpful
answer:

       No, Y hit X at 3.

A still more helpful answer is called for when the
assumption behind the question is itself false. The
question, "Did Y hit X at 2?" suggests that Y hit X at
sometime. If this assumption is false, then in order
to be helpful an answer must deny it:

Y didn't hit X.

A way to produce helpful answers emerged in the process
of developing the programmatic theory. The system
keeps a record of the series of assertions that satisfy
the list of goals. It is possible that an event which
satisfies the initial goals on the list fails with
a later goal. Y may have hit X several times, and so
in answering the question, "Did Y hit X at 2?", the
system may investigate events which turn out to involve
Y hitting something other than X, or hitting X at some
other time. Such events cause PLANNER to fail at some
stage, and to start again in order to find another
event. All of these false starts are kept on record
as "partial answers" in order to generate helpful
replies just in case there is nothing in the data-base
that satisfies all the goals on the list. Consider,
for example, the case where the question was:

        Did Y hit X at 3?

the answer might contain the following partial answers:

            (1)   (2)
    [E4 HIT SUBJ Y]   [E1 HIT SUBJ Y]
          (FAIL)   [E1 HIT OBJ X]
                                       (FAIL)

which reflect the fact that Y was the subject of two
events, one of which did not involve hitting X, and
the other of which involved hitting X, but not at time
2. Clearly only the latter is of any interest in constructing a helpful reply, and in general the program
need only consider the most complete (i.e. the longest)
of the partial answers.

When the longest of the partial answers above is compared with the original list of goals, it is clear that
it failed only on the final goal, the one representing
the new information on the sentence: all of the goals
representing the given information in the sentence were
satisfied. In this case, the reply is "No", but in
order to be helpful, the program adds the statement,
"Y hit X at 2". It does so by first inspecting the
goal that failed, and then generating a new goal to
find the value that succeeds.

If the given information in a question is false, then
in a helpful reply it must be denied. The program
detects the failure by noting that even the longest

of the partial answers has failed at some earlier
goal than the one corresponding to new information.
In this case, the reply is generated by taking, in
order, all the goals that succeeded, together with
the one that failed, and generating the corresponding
negative sentence. For example, the question,

> Did Y hit Z at 2 at B?

gives rise to the goals

> EVENT HIT SUBJ Y
> EVENT HIT OBJ Z
> EVENT HIT AT 2
> EVENT HIT AT B

The most complete answer that this list evokes for the
scenario described earlier is:
$[E^1$ HIT SUBJ Y$]$

> (FAIL)

By comparing this answer with the original list of
goals, the program can work out what is wrong with the
question. It is the assumption involving the first
two goals:

> EVENT HIT SUBJ Y
> EVENT HIT OBJ Z

They correspond to the erroneous assumption that Y hit
Z, and to deny it the program must generate the sentence:

> Y did not hit Z

To be even more helpful, it finds out what Y did hit,
and adds this information to its reply:

> Y did not hit Z, Y hit X.

The program generates its responses, both to questions
to which the answer is "No", and to those which involve
a false assumption, by using an STN as a generator.
This technique is feasible because the semantic representation of the sentence as a list of goals represents
all the information in the original sentence including
the information conveyed by active and passive voice.
In understanding a passive sentence, the first goal
put on the list involves the first noun phrase of the
sentence as the underlying object of an event. In
generation, if the list of goals begins with one that
involves the object of an event, then the sentence is
couched in the passive voice. Certain aspects of the
process are not reversible in this way. The decision

to generate a declarative sentence rather than an interrogative one is determined by factors outside the STN,
as is the decision about whether it is positive or
negative. Similarly, there may be an ambiguity in
parsing a sentence, whereas in generating a sentence
there can be no ambiguity. The current program accordingly has two similar STNs, one for production and
the other for comprehension, rather than two interpretations of the same network.

The Shortcomings of the Programmatic Theory

Any comprehensive list of the linguistic phenomena
that the programmatic theory cannot accommodate would
be too long to recount without tedium. Its syntactic
component is the simplest possible that could be
devised without abandoning grammar entirely. It is so
restricted that many questions simply cannot be raised
about its performance, e.g. the parsing of complements,
relative clauses, conjunctions, and subordinate clauses.
What is more worrying, however, is that the restrictions on syntax have prevented us from analysing illo-
cutionary force and the ways in which it is signalled
and recognised. We have likewise been unable to investigate the construction and interpretation of referential phrases. However, it is part of the strategy
of a programmatic psychology to begin with a very
simple theory of performance.

The shortcoming in the present theory that is most
amenable to treatment concerns the general problem of
"selectional restrictions". Relational terms such as
verbs and prepositions often impose semantic constraints
on the values of the variables they interrelate. The
preposition "at", for example, is used to introduce
either a temporal or a spatial location. When asked
the question, "Did Y hit X at 2", where 2 refers to a
time, the system should not respond, "No, Y hit X at B"
where "B" refers to a place. Indeed, it does not make
such responses, since it utilises selectional restrictions of the sort advocated by Katz and Fodor (1963);
it ensures that a helpful reply meets the restrictions
imposed by the sense of the preposition in the original
question. However, there are reasons to doubt the
psychological plausibility of such a system since it
is unclear how it could be acquired by children, and it
is constantly violated by metaphorical usage in every-
day conversation (see McCawley, 1968, and Savin, 1973).
Inference based on general knowledge seems to be required
as a basis for conventional selectional restrictions

(Johnson-Laird & Miller, 1976).

A problem which we are still investigating in the light
of the model is that of assigning the various noun
phrases in a sentence to their appropriate semantic
roles in the procedures corresponding to the meaning
of the verb. This assignment depends on the particular
verb occurring in the sentence - a phenomenon that was
responsible in part for the rise of case grammar
(Fillmore, 1968). We have preferred to follow a more
orthodox transformational approach and to adopt a
general principle advocated by Miller and Johnson-Laird
(1976). The basic idea can be illustrated in a simple
example. Suppose that the verbs give and receive had
identical semantic representations, then the difference
between them would be in the way in which the variables
in the semantic representation mapped onto deep structure functions. The critical information would be
captured in lexical entries relating the common meaning
to the different syntactic frames:

Syntax

$$((X)_{NP}(\underline{give}\ (Y)_{NP}(\underline{to}\ Z)_{PP})_{VP})_S$$

$$(((Z)_{NP}(\underline{receive}(Y)_{NP}(\underline{from}\ X)_{PP})_{VP})_S$$

Semantics

DO(X,E) & CAUSE
(E, POSSESS (Z,Y)):
X does something that
causes Z to possess Y.

Our programmatic theory represents this sort of information about the meanings of verbs in their associated
procedures. However, we have yet to solve the problem
of using such definitions in the opposite direction in
order to select a verb to describe an event. The
program's replies are all generated using the same
verb that occurred in the original question.

Evaluation and Conclusions

In psychology, computer programs and experimental data
all too often pass one another by. Hence, perhaps the
strongest point in favour of our programmatic theory
is that there is a considerable amount of empirical

evidence against it. A number of psychologists have
begun in recent years to argue for the autonomy of
syntactic and semantic processing, and to suggest that
the semantic interpretation of one constituent in a
sentence plays no role in the syntactic interpretation
of the next, or any other constituent in the sentence.
This hypothesis is hardly consistent with the programmatic theory. Since each goal represents an independent discrete step in the computation, there is no
reason why it should not try to achieve the goals as
soon as they are set up, i.e. carry out semantic interpretation immediately. It was this intuition, of
course, that led to the development of the programmatic
theory.

The bulk of evidence in favour of autonomous processing
comes from studies involving phoneme monitoring, the
rapid serial visual presentation (RSVP) of sentences,
and the classification of sentences as meaningful and
well-formed (Forster & Olbrei, 1973; Forster, 1976;
Garrett, 1976). Forster and Olbrei point out that
early studies, by Slobin (1966) and others, showing a
facilitation of syntactic processes by semantic cues,
failed to exclude the possibility that the semantic
cues were facilitating an autonomous semantic stage of
processing. To what extent such tasks reflect the
normal processes of comprehension is not yet clear.
In devising the programmatic theory, however, ways in
which it might be put to experimental test did emerge.
There is evidence (Dewart, 1975) that constructions
involving an indirect object are easier to understand
in the form:

The woman gave the baby to the man

than in the form in which the indirect object is preposed:

The woman gave the man the baby.

Presumable the difference is due in part to the lack
(in the second case) of the explicit prepositional cue
to the underlying roles played by the two noun phrases.
However, in a system that takes semantics into account
the problem should be considerably eased where the
meanings of the noun phrases themselves provide a
strong clue to the appropriate interpretation. Thus,
we predicted that the difference in difficulty between
the two sorts of sentence should be markedly reduced
with such sentences as:

The woman gave the book to the man.

and:

        The woman gave the man the book.

In a preliminary experiment carried out by our colleague
J.G. Quinn, such an interaction emerged. The subjects
were presented tachistoscopically with a series of
sentences. They perused each sentence until they
understood it, and then pressed a button. This response led to the presentation of a question such as,
"Who gave the book?", which they then attempted to
answer. On the assumption that the time spent examining
the sentence is directly related to the difficulty of
understanding it, the latencies of the button-pressing
response provide an index of comprehension difficulty.
In a preliminary study, each subject received four
trials on each of four sorts of sentence. The mean
latencies of comprehension are shown in Table 1.

Table 1.   The mean comprehension latencies for four
           sorts of indirect object construction (a
           preliminary study involving 12 subjects).

	Type of sentence	Mean Latency (Secs)
No Semantic Cue	The man took the boy to the girl	3.13
	The man took the girl the boy	3.40
Semantic Cue	The man took the coat to the girl	2.87
	The man took the girl the coat	2.96

Note: The examples are used for illustrative purpose.
        No subject encountered any given lexical content
        more than once during the experiment.

The interaction is significant: the difficulty of the
preposed construction disappears when a semantic cue
to underlying roles is present in the sentence.

A defender of the autonomy thesis may wish to argue
that these results show no more than that semantic
cues facilitate semantic processing. However, one
would then be entitled to ask: what is the nature of a
purely semantic process that resolves the issue of
whether a noun phrase is a direct or indirect object,

and that is affected by surface order? If the autonomy
thesis means no more than that the route through the
STN is governed by syntactic tests on the incoming
material, then it is a thesis to which we subscribe.
However, the thesis seems to imply that a listener can
not use semantic information to eliminate unnecessary
syntactic processing (personal communications from M.
Garrett, K. Forster, and W. Murray), and is accordingly
incompatible with these data. Obviously, the results
of a single experiment are hardly conclusive, and a
sceptic is entitled to wonder whether a task requiring
some three seconds to perform is an accurate reflection
of ordinary comprehension. However, the crucial point
is that the programmatic theory has an easily derivable
empirical content.

The use of semantic cues in syntactic analysis allows
the syntactic component of the STN to be simplified.
It need not distinguish grammatically between such
pairs of sentences as:

> The car was stared by hand.

> The car was started by Hans.

The device is thus constructed in the spirit of Bresnan's
(1976) proposal that all the cyclical transformations
(e.g. passive, dative movement) can be replaced by
lexical redundancy rules. Such transformations are
structure-preserving, that is, they do not move constituents to positions that cannot be generated using
the phrase-structure rules for deep structure (Emonds,
1976). The passive does, indeed, locate the underlying
object of the sentence in a position corresponding to
a prepositional phrase; and Bresnan's proposal, as she
herself points out, is particularly amenable to parsing
by an ATN. The introduction of an STN to do the job
obviates the need for an explicit representation of
deep structure, captures the given-new contrast in
terms of an implicit order of processing, and provides
a uniform treatment of the production and comprehension
of sentences.

Acknowledgements

This research was supported by a grant from the Social
Science Research Council of Great Britain (HR2351/1).
The computing facilities were provided by the Science
Research Council grant B/RG/6975/0. We are greatly
indebted to Geprge Miller. Arnold Zwicky and our col-
leagues, Stephen Isard, Christopher Longuet-Higgins,

and Stuart Sutherland have contributed many hours of provocative and instructive conversation. They also criticized an earlier draft of the paper.

References

Anderson, D.B. (1972). Documentation for LIB PICO-PLANNER. School of Artificial Intelligence, Edinburgh University.

Bresnan, J. (1976). Towards a realistic model of transformational grammar. Paper presented at the Convocation on Communications, M.I.T., April, 1975.

Burstall, R.M., Collins, J.S., and Popplestone, R.J. (1971). Programming in POP-2. Edinburgh: Edinburgh University Press.

Clark, H.H. and Haviland, S.E. (1974). What's new? Acquiring new information as a process in comprehension. Journal of Verbal Learning and Verbal Behavior, 13, 512-521.

Dewart, M.H. (1975). A psychological investigation of sentence comprehension by children. Unpublished Ph.D. Thesis, University College London.

Emonds, J.E. (1976). A transformational approach to English syntax: root, structure-preserving, and local transformations. New York: Academic Press.

Fillenbaum, S. (1973). Syntactic factors in memory? The Hague: Mouton.

Fillmore, C.J. (1968). The case for case. In E. Bach and R.T. Harms (Eds.) Universals in linguistic theory. New York: Holt, Rinehart and Winston.

Firbas, J. (1964). On defining the theme in functional sentence analysis. Travaux Linguistiques de Prague,1, 267-280.

Fodor, J.A., Bever, T.G., and Garrett, M.F. (1974). The psychology of language. New York:McGraw Hill.

Forster, K.I. (1976). The autonomy of syntactic processing. Paper presented at the Convocation on Communications, M.I.T., April, 1975.

Forster, K.I., and Olbrei, I. (1973). Semantic heuristics and syntactic analysis. Cognition, 2, 319-347.

Garrett, M.F. (1976). Word perception in sentences. Paper presented at the Convocation on Communication, M.I.T., April, 1975.

Halliday, M.A.K. (1967). Notes on transitivity and theme in English, II. Journal of Linguistics, 3, 177-244.

Hewitt, C. (1969). PLANNER: a language for proving theorems in robots. Proceedings of the International Joint Conference on Artificial Intelligence. Bedford, Mass.: Mitre Corporation. Pp. 295-301.

Jesperson, O. (1924). The philosophy of grammar. London: Allen & Unwin.

Johnson-Laird, P.N. (1976). Psycholinguistics without linguistics. In N.S. Sutherland (ed.) Tutorial Essays in Psychology. Hillsdale, New Jersey: Erlbaum. In press.

Johnson-Laird, P.N., and Miller, G.A. (1976). Procedural semantics. Paper presented at the Conference on Philosophy and Psychology, Cornell University, April, 1976.

Johnson-Laird, P.N., Steedman, M.J., and Huttenlocher, J. The psychology of syllogisms. (In preparation).

Kaplan, R.M. (1972). Augmented transition networks as psychological models of sentence comprehension. Artificial Intelligence, 3, 77-100.

Katz, J.J., and Fodor, J.A. (1963). The structure of a semantic theory. Language, 39, 170-210. Reprinted in J.A. Fodor and J.J. Katz (Eds.) (1964). The structure of language. Englewood Cliffs, N.J.: Prentice-Hall.

Kay, M. (1975). Syntactic processing and functional sentence perspective. In R.C. Schank and B.L. Nash-Webber (Eds.) Theoretical Issues in natural language processing. M.I.T., June, 1975. Supplement, pp.12-15.

Levelt, W.J.M. (1974). Formal grammars in linguistics and psycholinguistics, Vol. III. The Hague: Mouton.

McCawley, J.D. (1968). The role of semantics in a grammar. In E. Bach and R.T. Harms (Eds.) Universals in linguistic theory. New York: Holt, Rinehart and Winston.

Marslen-Wilson, W. (1976). Linguistic descriptions and psychological assumptions in the study of sentence perception. In E.C.T. Walker and R. Wales (Eds.) New Approaches to language mechanisms. Amsterdam: North Holland. In press.

Miller, G.A., and Johnson-Laird, P.N. (1976). Language and Perception. Cambridge, Mass.: Harvard University

Press. Cambridge: Cambridge University Press.

Savin, H.B. (1973). Meaning and concepts: a review of Jerrold J. Katz's Semantic theory. Cognition,2, 213-238.

Slobin, D.I. (1966). Grammatical transformations and sentence comprehension in childhood and adulthood. Journal of Verbal Learning and Verbal Behavior,5, 219-227.

Steedman, M.J. (1976). Verbs, time and modality. Cognitive Science. In press.

Thorne, J., Bratley, P., and Dewar, H. (1968). The syntactic analysis of English by machine. In D. Michie (Ed.) Machine Intelligence, 3, Edinburgh: Edinburgh University Press.

Turing, A.M. (1950). Computing machinery and intelligence. Mind, 59, 433-460.

Wanner, E., and Maratsos, M. (1975). An augmented transition network model of relative clause comprehension. Mimeo, Harvard University.

Weizenbaum, J. (1966). ELIZA - A computer program for the study of natural language. Communications of the Association for Computing Machinery, 9, 36-45.

Winograd, T. (1972). Understanding natural language. New York: Academic Press.

Woods, W. (1970). Transition network grammars for natural language analysis. Communications of the Association for Computing Machinery, 13, 591-606.

ON THE RE-INTEGRATION OF LINGUISTICS AND

PSYCHOLOGY

BRUCE L. DERWING and WILLIAM J. BAKER

University of Alberta

I. Introduction

The title of this paper is intended to suggest four claims related
to the history of linguistics, psychology and their inter-relations:

(1)   that there was once a time in the past when language study
and psychology were closely integrated;

(2)   that at some time in the more recent past a split occurred and
the two disciplines went more or less their own way;

(3)   that, despite proclaimed attempts to "psychologize" linguistics
in recent years, the rift between the two disciplines has not yet
been genuinely healed – and that, as a consequence, linguists and
psychologists continue to view language phenomena in fundamentally
different ways;   and, finally,

(4)   that we believe that it is important, even essential, for the
future development of language study that a full-scale reconciliat-
ion take place – and we shall eventually have a few suggestions to
offer as to how such a re-integration might be brought about.

II. The Linguistic Perspective of the Past: An "Autonomous" View of Language "Structure"

In order to show that language study was once considered a proper,
even central, concern of psychological enquiry, we need only refer
to Blumenthal's readable (1970) account of the work of Wilhelm
Wundt, the generally acknowledged "father" of modern experimental
psychology, and his contemporaries.   These psychologists were concerned to understand language performance, that is, "to trace the
mental processes that precede, accompany and follow utterances"
(p.16).

It is also not difficult to show, however, that this tradition did
not last in psychology and that a small group of outsiders (now-
adays called "linguists") were eventually allowed to take full
possession of the "language" toy and run off with it (cf. Derwing,
Prideaux & Baker, to appear).   Thus a monolithic linguistic
tradition developed which was less influenced by legitimate psychological concerns than by the concepts and methods of nineteenth-
century historical philology, which had viewed language "as one
of the natural organisms of the world,...that independently of its
speakers' will or consciousness has its periods of growth, maturity and decline" (Robins, 1969, p.181).   In Europe, of course,
it is Saussure who is generally regarded as the legitimate "founder
of modern linguistics" (Lyons, 1968, p.38) - and who was also the
one most influential in propagating the view of language (i.e.,
la langue) as a kind of supra-individualistic "sociological reality"
which maintained a separate "existence" all of its own, independent
of any real speakers of that language (cf. Saussure, 1959, p.18).
But the competing psychological orientation did not die overnight,
either, and at least one powerful voice was raised up against the
separation movement.   That voice belonged to Otto Jespersen, who
felt strongly enough about the matter to place a noteworthy warning in the opening paragraph of his 1924 book, The Philosophy of
Grammar.   His warning was that "the speaker and the hearer, and
their relations to one another, should never be lost sight of if
we want to understand the nature of language and of that part of
language which is dealt with in grammar", and that the grammarian
should thus not be tempted to treat language forms "as if they were
things or natural objects with an existence of their own" (p.17).

Speaking to you as outsiders, it is not completely clear to us
the extent to which the Saussurian tradition won out in Europe,
though we gather that its influence has been at least a very
strong one.   But in North America, to be sure, its victory was
total and absolute.   The Thomas Jefferson of this particular independence movement was, of course, Leonard Bloomfield - the
same Leonard Bloomfield, in fact, who had based his original,
unabashedly psychological orientation "both general and linguistic
entirely on Wundt" (1914, p.vi) and whose first book, An Introduction to the Study of Language, was concerned explicitly "with
those mental processes which most immediately underlie the use of
language" (1914, p.56).   By 1933, however, as is well known,
Bloomfield had changed his tune: "In the division of scientific
labour," he wrote, "the linguist deals only with the speech signal...: He is not competent to deal with problems of physiology
or psychology" (p.32).

Oddly enough, at about the same time - and largely as the result
of the same radical behaviourist notions which had changed Bloomfield's thinking - the psychologists lost interest in the kinds of
questions which the early pioneers like Wundt had raised about lan-

guage, and so were perfectly willing to allow the linguists to
monopolise the field.   Adequate proof of this can be found by inspection of Kling & Riggs (1971), which is intended to represent a
summary of all major research in experimental psychology through
the 1960's.   The subject index of this work contains only a
single reference to "language" - and even that one actually refers
to a section on "speech" (pp.250-9).

Now we come to the question of what actually happened in linguistics
in the late 1950's. Numerous accounts have presented the view that
a veritable "scientific revolution" of Kuhnian proportions occurred
in that discipline, initiated by the publication of Chomsky's
Syntactic Structures in 1957 (cf. Lyons, 1970, p.9).   For reasons
which are developed more fully in Derwing (1973), we regard this
as something of an overstatement.   Yet most certainly there was
at least a terminological revolution. Language "descriptions"
which were "convenient fictions" in the parlance of a Bloomfield
or a Hockett were redubbed "theories" of languages by Chomsky
(1957) - and (in time) came even to be called models of "the competence of the ideal speaker-hearer" (Chomsky, 1965).   And linguistics too, under Bloomfield an admittedly "independent" discipline, is sometimes referred to these days as "a branch of cognitive psychology" (Chomsky, 1968).[1]   Yet while much exaggerated
emphasis has been placed on all that supposedly changed in the
"new" linguistics (see Derwing 1973, pp.25-43 for a full discussion), it is more interesting and cogent for our present purposes
to take a look at some of the important things which did not
change.   In his 1926 "Postulates", for example, Bloomfield defined the term "language" as follows: "The totality of utterances
that can be made in a speech-community is the language of that
speech-community" (p.155).   In 1957 (and on many occasions since)
Chomsky has defined a language as a "set of sentences" (1957, p.13).
Once we understand that "utterances" in the structuralist tradition
were always taken to be "acts of speech" (Bloomfield's Definition 1)
as represented in some standardised "phonetic" transcription - and
the notion "can be made" taken to exclude various non-communicative
vocal noises (coughs, sneezes, etc.) as well as obviously "defect-
ive" utterances (such as false starts, lapses, utterances which
were "non-native" or which "didn't sound right" to the native
speaker, etc.) - it is hard to see that Chomsky's "new" orientation
amounts to anything more than a (laudable) tightening-up of an old
set of definitions.   The fundamental AUTONOMY PRINCIPLE implicit
in the original Saussurian concept of la langue is still maintained,
according to which a "language" is viewed as a set of formal objects,
namely, the set of "good" or "grammatical" utterances, represented
in some "generally accepted" or "universal" mode of transcription.
By the same token, I see no fundamental difference in kind between
the notion of a "simplicity metric" in generative grammar and any
of the various principles of "economy" proposed by the earlier
structuralists: by one device or another, the implicit goal of

either a structuralist "linguistic description" or a transformationalist "generative grammar" seems still to be to attempt to describe
a "language", as characterised below, in some optimal, parsimonious
form.

One obviously new twist of the Chomskyan era, of course, was the
explicit adoption of a supplementary GENERATIVE PRINCIPLE as part
of the attempt to achieve this goal of an economical characterisation of sentences.[2]   But all this amounted to was to take those
"things" called "natural languages" to be just special cases of
the kind of formal languages which are the proper concern of that
branch of mathematics called automata theory (cf. Rosenbloom, 1950;
Davis, 1958; Minsky, 1967; etc.).   The form of a linguistic description thus shifted from one in which the analyst simply listed
various recurring syntactic or morphological patterns to one in
which these patterns (or others such like them) were instead
"recursively defined" from an algorithmic or generative source.
This procedure had a number of descriptive advantages, as its ad-
vocates have not failed to point out: it forced linguists to be
a lot more explicit about their statements (i.e., the grammar had
to "work"), it allowed for some new kinds of economies of description not previously conceived (such as deriving a number of
"surface" syntactic patterns from the same "deep" or "base" structure - a natural development of Harris' idea to "equate" various
surface syntactic patterns (1957)), and, above all, perhaps, it
provided one means for accounting for the open-ended character of
languages (i.e., it provided one possible descriptive analogue for
the notion of linguistic "creativity").   But we construe all of
these changes as attempts to incorporate various improvements in
the formal apparatus of grammatical description, without changing
the fundamental notions either of a language as a "thing" (a
"subset of sentences") or grammar as a concise and precise "specification" of that thing(cf. Wall, 1972, p.166).   Apart from the
later terminological "psychologising" already alluded to, therefore,
the only basic element which we have been able to find in this revised orientation which is "fundamentally new" (cf. Jenkins, 1968,
p.540) has been the degree to which it has allowed its practitioners
to take full advantage of the freedom from psychological constraints
which the Bloomfieldian declaration already made possible in principle and to allow their analytic imaginations to run completely
wild.[3]

As for the psychologising itself, the literature reveals only a
single, original warrant in support of the claim that the modern
grammarian has somehow managed to negotiate the great ontological
leap from describing language forms to describing some kind of
"psychological reality".   The classical statement of this warrant
can be summarised as follows: since TGG's go beyond previous
structuralist accounts in that psychological data related to certain
kinds of "linguistic intuitions" are taken explicitly into account

in their construction and evaluation, such grammars become _ipso_
facto psychological theories (cf. Bever 1968, p.483).[4]   But Derwing
(1973), Spencer (1973), Linell (1974), Botha (1975) and others have
identified a host of difficulties which serve to reduce this warrant
to tattered shreds, bringing into question not only the restrictive
nature and limited significance of the data base involved, but also
questioning the reliability and very validity of much of the "intro-
spective" evidence which has actually been employed by linguists.
It is clear, of course, that the availability of reliable data
concerning the linguistic intuitions of ordinary language users im-
poses a potential challenge which any fully satisfactory psychological theory must eventually meet(see Section III below).   It is
not so clear, however, that the mere imposition of formal constraints
on grammars, only a few of which, after all, have any claim to be
based on established psychological principles (cf. Prideaux, 1971;
Derwing, 1973), must necessarily change the essentially "descriptive", "taxonomic" or "self-confirming" character of these grammars
(cf. Schank & Wilks, 1974, p. 315; Hymes, 1974, p. 450 etc.).

In summary, therefore, "autonomous grammars", as we see them, are
not valid psychological constructs at all, but rather mere descriptive artifacts which have arisen from the linguists' myopic
view of the language product as an isolated phenomenon and hence
a "natural domain" for a scientific theory (cf. Sanders, 1970).
In truth, such "natural languages" do not occur spontaneously in
nature, and the burden of proof must therefore rest on those theorists who believe that anything even remotely resembling a "formal
grammar of a language" plays any role whatever in the mental life
of the source of the language product, who is the human language
user himself.

III. A Psychological Perspective for the Future: An "Information
     Structure" View of Language Use [5]

Having restated some of the main reasons why we are not doing what
we are not doing at Alberta, it is time now to give you a better
idea of what it is that we are trying to do. As already indicated,
our ultimate goal is to understand language behaviour or linguistic
performance; that is, we would like to be able to build and to
test a psychological model of what the speaker does when he speaks
and what the hearer does when he comprehends. We are a long way
from achieving that goal, as you might well expect. We are also
being very cautious, for if the transformationalist era, in particular, has taught us anything, it is that a powerful lot of
heavy intellectual effort can be wasted in the construction of
hypothetical models whose empirical foundations rest on shifting
sand.   Before we progress too far in our development of an explicit, formalized model, therefore, we want to be reasonably
certain that our feet are firmly planted on one ledge before we
risk trying to transport the whole expedition up to the next one.
As a result, our entire research effort is built around an ambitious

program of controlled experimentation all down the line - experimentation concerned primarily with the linguistic abilities of
ordinary language learners and users(see Prideaux, Derwing and
Baker, to appear). Some of our early experiments, to be sure,
had an unavoidable "cage-rattling" character, since we sometimes
have had only the vaguest idea of what it was that we were looking
for at the outset.   By this time, however, our thinking has
sufficiently crystallised that I can present to you a general list-
ing, at least, of all the major factors which we conceive as necessary for the construction of an explicit model of the sort desired.
The result is not yet, of course, a "model" in any sense worthy
of the term, but is rather merely a "way of viewing the problem"
which attempts to lay out in black and white what the critical
experimental variables are, as we conceive them[6].

The first thing we have attempted to do is to "put semantics back
into the driver's seat" (to borrow Osgood's phrase, 1968, p.509),
recognising fully the correctness of Cofer's observation that "our
interest in communication arises in the first place from our desire
to say something, rather than from the pleasure that exercising
syntactic structures may provide us" (1968, p.535). Every act of
communication begins with the intention of the speaker: the speaker
wants to communicate some intended message to one or more hearers
and he ordinarily knows what that message is. It is also obvious
that every act of communication takes place in some particular
communicative situation, a factor which has been largely ignored
in the narrow North American linguistic tradition (but cf. Hymes,
1972), though much less so in Europe (cf. Firth, 1957; Uhlenbeck,
1963; Slama-Cazacu, 1973; etc.); this factor is represented in a
diagram as the large, outer box.

The mind of the speaker is represented as the upper large rectangle
within this box, and the mind of the hearer as the lower large rectangle.   The smaller rectangle in the centre represents the utterance itself (i.e., an acoustic wave form), together with any physical factors which may tend to interfere with its transmission from
speaker to hearer.[7]   Within our conception of the problem, one
critically important aspect of the speaking or "encoding" phase of
linguistic performance is the current state of the speaker's mind.
As indicated in the diagram, this involves such things as the
speaker's general knowledge of the world and his awareness of
certain relevant aspects of the specific communicative situation
(e.g., the number and character of any other people present, the
physical setting and social milieu, etc.), as well as his knowledge
of any previous communication which might already have taken place.
His general motivation (including his knowledge that he "doesn't
know something that he would like to know" - the motivation for
asking a question) and his attitudes (which help determine whether
casual conversation, persuasion, cajolery, insult, etc. are intended) also play a role. The speaker is also, of course, restricted

to those specific communication devices which he has available to
him (cf. the problem of the small child or the non-native speaker)
and by his own particular skill at deploying them effectively
(cf. the skilled orator vs. the incoherent rambler).

In our conception of the problem, a basic task which the speaker
must next achieve is to break down his intended message (m) into
its various information components for purposes of eventual linguistic encoding and acoustic transference.[8]   One important consideration which sharply distinguishes this conception with previous "purely linguistic" views is that only a relatively small
portion of the speaker's intended message will ordinarily have to
be broken down in this way.   In particular, the speaker may well
construe that a substantial part of his "meaning" will be clear
on the basis of information provided by the communicative situation
itself, including the speaker's assessment of the hearer's current
state of mind. (We might have chosen to label this component Ie
(for "environmental information"), had we not decided to restrict
the notion of "information structure" to those components of an
intended message which are to be linguistically transmitted.) The
significance of this situational class of information is that it
relates to factors over which the speaker has only very limited control (such as the choice of his audience, their background knowledge,
and the physical and social environment in which communication must
take place). It is significant that some linguists (notably the
so-called "generative semanticists") have attempted to incorporate
even this kind of information directly into the "underlying structure" of sentences themselves (cf. Langendoen, 1969; Lakoff, 1971
etc.).   This is only a natural consequence of a recognition that
such factors play a very important role in communication, combined
with a conceptual orientation which fails to recognise that sentences themselves do not "have" inherent "meaning" or "structure",
but are rather invested with information by speakers, on the basis
of which hearers can interpret them as meaningful in the context
of everything they know both about a particular communicative
situation and about the world in general (as illustrated by several
of the papers presented at this conference).   In our view, these
latter aspects of "meaning" not only fail to constitute "an inherent part of the structure of sentences", they are not even communicated by means of linguistic devices!

We see the need of at least four types of information, however,
which can be linguistically encoded, and hence for which specific
linguistic devices must be provided in a language.   These are as
follows:

(1) Information which relates an intended message to any of a
variety of aspects of the preceding linguistic context.   This is
the component labeled Ic (for "contextual information") in Figure 1.
As "discourse analysts" have long recognised, even so simple a

Current State of Speaker's Mind

Knowledge, general and of specific situation
Motivation
Available linguistic skills and devices
Then:
  Intended Message (m)
  Information Structure $\left[I_c+(I_s+(I_r+(I_d)))\right]_{Spkr}$

  Linguistic Structure of Utterance (x)
  Motor Plans and Production

Utterance + Its Physical Environment   (y)

Basic Sensation and Perception
Perceived Linguistic Structure   (x')
Inferred Information Structure
  $\left[I_c + (I_s +(I_r +(I_d)))\right]_{Hr}$
Construed Message (m')
Evaluation of m'   (m")

Current State of Hearer's Mind

Knowledge, general and of specific situation
Motivation and attention
Available linguistic skills and devices

COMMUNICATIVE SITUATION

Figure 1

matter as the choice of a word like he presumes a recognition by
the speaker that the intended referent or antecedent must already
have been established (or, in some cases, can reasonably be anticipated), either from something previously said (e.g., "My brother
is really a strange one") or from some other feature in the communicative situation (e.g., only one other male present, speaker
points at some male nearby, etc.), or both.

(2) Information which relates to the speaker's motivation for
wanting to initiate a speech act. This is the component labeled
Is (for "sentential information", so named because the kind of
information involved is often encoded linguistically in a way
which serves to distinguish among a set of basic formal "sentence
types" of a language). Does the speaker want to give information
(declarative) or request it (interrogative) - or does he want to
issue a command (imperative)?   And if it's a question which is
to be asked, is it content-seeking (Wh-question) or merely con-
firmation-seeking (Yes/No-question)? etc.

(3) Information which has to do with the relationships among the
participants in the intended message. This component is labeled
Ir in the diagram (for "relational information").   Which participant is intended to be the "subject" of the action, and which
the "object"?   (See below for further discussion.)

(4) Finally, and most basically, there is also information related to the participants and actions themselves. This is the
component labeled Id (for "denotational information").   This information might vary from as little as that provided by the word
he in English, in an appropriate linguistic context (or even by
nothing at all, as when answering "Yep" to the query, "You mean
that John really beats his wife?"), to something at least as complex as "that skinny gray-haired old man who used to drop by and
visit our next-door neighbours once in a while when we lived in
upstate New York before the war."   Thus Id includes all of the
information which is to be linguistically encoded not merely by
lexical classes such as nouns and verbs in English, but also by
their modifiers - and so includes perhaps the bulk of the recursive linguistic devices required in a language.

Notice also that the diagram does not merely list these four major
information types, but also orders them hierarchically; that is,
Is is embedded within Ic, Ir is embedded within both Ic and Is etc.
This has been done for a number of reasons, all ultimately empirical
in nature. First of all, in constructing a particular utterance so
as to convey a particular intended message, the speaker needs to
find a suitable linguistic device only for those aspects of information which he feels have not already been provided for at some
higher level. Thus, in the example just cited, once the speaker
determines that the hearer has ascertained what or whom the conversation is about (e.g. the same old man who used to visit the neigh-

bours, etc.), he can take advantage of this Ic and make use of the
convenient lexical device he (or one of its morphological variants)
thereafter, so long as the topic of conversation is not changed or
complicated by reference to other males. Similarly, in answer to
the question, "Who did you say beats his wife?" the single lexical
item John may often serve to encode all of the information about
which there is any doubt.   Under ordinary circumstances, in fact,
the "full answer" (say, "I said that John Smith beats his wife")
would be bizarre and quite inappropriate - unless, of course, the
speaker wanted to signify his impatience with his interlocutor's
stupidity, for example (in which case he would likely also employ
an appropriate set of suprasegmental devices, as well).   This is
all the explanation we require for such so-called "elliptical"
sentences as "Yep", "He did", or "John" (cf. Rommetveit, 1974,
pp. 29ff.).   Certainly we need not say that such responses are in
any sense "derived" from other, more "well-formed" syntactic strings
by various rules of "transformation" or "deletion".   There is
simply no place for this kind of process in our kind of model - and
there is no need for it, since we are no longer viewing the language
product as something which must be exhaustively described in isolation from its original source, function and environmental setting.

There are other reasons for bracketing the information components
the way we do, as well. One is to allow us to characterise certain
of the various levels of "paraphrase" or "sentence relatedness"
upon which the transformationalists have laid so much heavy methodological emphasis. How is it that various "sentence families" are
readily construed as "paraphrase sets" by native speakers?   The
general answer, of course, relates to the interpreted meaning of
these sentences (cf. Derwing, 1974, p.30), although most such relationships can be characterised at some particular level of information structure, as well. Sentence pairs such as "The boy hit the
girl" and "My sister hates your brother", though they differ in
their Id (and hence in their lexical content), still do share the
same Ir (and hence have the same surface syntactic structure).
Other sentence pairs, such as declaratives and their corresponding
Yes/No interrogatives (e.g., "The boy hit the girl" and "Did the
boy hit the girl?", etc.), share common information at both the
denotational and relational levels and differ only at the sentential
level.[9]   Since relationships like these are already expressed at
one or another level of the information structure of the sentences
involved, it would be completely redundant to re-express them syntactically: hence no syntactic rules of "transformation" or the
like are required.[10]

It is also of some interest to relate this particular conception of
organisational structure to the results of the early sentence recall
experiments by Miller, Mehler, and others. Rather than to construe
these results as reflecting the recall of "kernel sentences" with
attendant loss of various transformational "corrections" (cf. Mehler,

1963) - a conception which has not held up well under testing with
a larger variety of syntactic constructions - we can view them as
reflecting instead the following general principle: ease of recall
is directly related to the level of information structure involved.
Thus denotational information is remembered best (Who were the
participants to the action?) relational information next best (What
roles did these participants play with respect to the verb?), sentential information even less well (Was that a statement or a
question? Positive or negative?), and contextual information worst
of all (What was said earlier on in the conversation? What or who
was the main focus of attention?)

There is also one important point related to our conception of relational information which bears some clarification here. We construe the traditional "semantic" relational concepts, such as
"agent", "patient", "recipient", etc., to be part of the speaker's
general knowledge of the world, rather than as components of those
classes of information which are specifically encoded linguistically
(i.e., via either suprasegmental, lexical or syntactic
devices). In our view, an important part of the "beauty" of
communication is the relative simplicity which this achieves through
trading off a relatively few and straightforward formal linguistic
devices against the background of knowledge which speakers and
hearers already bring to the communicative situation on the basis
of their past experience. It is this vast store of general knowledge which allows for rather complex interpretations to be associated with utterances which provide only a modicum of essential,
language-transmitted information. This process can be illustrated
by means of a simple expression such as "It's cold". Information
about the "coldness" concept is linguistically encoded by the
speaker, though other relevant information is not, such as information regarding the topic of conversation (the weather), the locale
referred to (the same as the locale of the speech act) and the time
referred to (also the same as that of the speech act). Yet all
four "pieces" are clearly part of the hearer's interpretation of
this simple utterance, so much so, in fact, that this interpretation
would normally be maintained even in the face of counterfactual
events (say, very hot weather at the place and time of the speech
act) and the utterance thus construed as ironical! This is one
of the main advantages of viewing language in a "non-autonomous" way,
rather than as an isolated, independent "grammatical system": it
is not only more realistic, insightful and useful to do so, but
even more economical. Work which the "language" has to do in the
one framework, the language user can do in the other.

Let us now return for a quick overview of the remaining components
of Figure 1. Having partitioned his intended message into its
various appropriate "information" components, the speaker's main
task is next to choose from among the various linguistic devices
available to him, from those which he deems most suitable for com-

municating this information to the speaker.   What are these linguistic devices?   We construe them to be a set of learned <u>perform-
ance rules</u> which map the various components of the information
structure of an intended message onto certain features of the intended surface syntactic structure of sentences.[11]   A single illustration of this kind of rule will have to suffice here, but the
interested reader is referred to some recent papers by Prideaux
(1975, in press) for other examples and a proposed formalisation.
The particular example which I want to discuss here relates to that
familiar thorn in the side of most analysts, namely, the active
vs. passive distinction in English.

It is common knowledge that there is no single, simple syntactic
device available in English for encoding the semantic notion,
"agent of an action" (sometimes referred to as the "logical" or
"deep structure" subject).   In simple active sentences such as our
earlier example, "The boy hit the girl", the constituent which
denotes the agent is located in the surface "subject" position,
but in the "object" position in the corresponding passive, viz.,
"the girl was hit by the boy."   Notice, however, that in both of
these examples (as in English in general, it seems), there <u>is</u> a
single, simple device available for encoding the information
"grammatical subject" of a sentence.   Once the inner-bracketed Id
information has been encoded via appropriate linguistic devices
(e.g. the information related to the participants converted into
noun phrase (NP) constituents and the action information encoded
into a verb (V)) the information as to which NP is to be the
"grammatical subject" of the developing sentence can be indicated
by making use of the following simple rule: "Put that NP in front
of the V in its own sentence, with no other relational NP's inter-
vening."   This suggests to us that such semantic information as
"agent", "patient", etc. is not encoded <u>directly</u> into the surface
syntax of English, but that it is rather the relational information
categories "subject", "object", etc. which are (just as in the
highly inflected Indo-European languages).   "Subject" may thus be
a secondary encoding of <u>either</u> a topicalised agent <u>or</u> a topicalised
"patient" - or even a topicalised "instrument", as in Fillmore's
example: "the key opened the door" (1968, p.25)

Do any particular advantages accrue to the speaker in doing things
this way, rather than in mapping the primitive semantic relations
directly onto the surface syntax?   Indeed they do.   These advantages become clearer when other examples are considered.   Compare
the active vs. passive pair, "The boy threw the ball" and "the
ball was thrown by the boy."   In such sentences it is not necessary for the speaker to inform the hearer as to which constituent is
the "agent", for this is already obvious to all parties concerned
from their knowledge of how the world works: boys often throw
balls, but balls do not ordinarily throw boys.   It is apparently
on this basis, in fact, that young children are able to determine

the "agents" of sentences like these long before they have managed
to come to grips with such subtle Ic notions as "topic" and the
associated syntactic complexities of passive sentences (cf. McNeill,
1970, pp.122-4).   They simply use "plain old common sense".   And if
common sense will do the job so much of the time, it would be extremely wasteful for the speaker to retain a special, unique device
for directly encoding each of these primitive semantic notions.[12]
So, in our formulation, at least, what he encodes instead is the
information "subject" and "topic" (where subject = topic in the
"unmarked" or "default" case.   See Prideaux, 1975 for further discussion.)

This strategy does, however, introduce one new problem for the
hearer: if "agent" and "patient", for example, can both be topicalised via the "NP before main V" device, how is he, the hearer,
going to tell them apart in, for example, a so-called "reversible"
passive sentence, where both NP's represent animates and hence
potential agents (as in our earlier "The boy hit the girl" illustration)?   This, in our view, is precisely the main function of
the special "verb morphology" of passive sentences: the inflected
form of the verb be (and also, redundantly, the preposition by in
this particular case) serve to inform the hearer that in a sentence
such as "The girl was hit by the boy", the subject is to be interpreted as a "topicalised patient" rather than as the "agent", as
in the active, "default" case (cf. Reid, 1976).[13]

Setting aside for now any further discussion of the details of how
these mapping rules might work and interact, the result of their
implementation by a speaker is presumably a kind of labeled and
bracketed syntactic "tree", which is denoted in Figure 1 as the
linguistic structure (x) of the developing utterance.   The speaker's
last task, then, is to convert this string into an appropriate
physical utterance (y) by invoking various motor plans which we
need not go into here.   The result of this step is an acoustic
wave form, something which is capable of ready transmission from
the sound-producing apparatus of the speaker to the sound-receiving
apparatus of a hearer.

Assuming for the sake of this discussion that the acoustic signal
arrives at its destination more-or-less intact - and also assuming
that the peripheral-auditory system of the hearer is functioning
normally - we can now complete the picture fairly readily: the
hearer (provided that he is motivated to pay attention to the
speaker) perceives this acoustic wave form and on the basis of a
learned (but poorly understood) set of recognition rules he arrives
at a perceived linguistic structure (x') - hopefully one reasonably
close to that which the speaker had originally envisioned in producing the utterance.   At this point the hearer takes advantage of
his own store of general knowledge, motivations, attitudes, and
available linguistic skills and devices, on the basis of which he

attempts to reconstruct the intended information structure of the
utterance - and from that to infer some estimate (m') of the
speaker's intended message. Finally, he evaluates this construed
message (m"), and on this basis decides what to do next (e.g. keep
listening, answer a question, nod assent, call the speaker a liar,
walk away in a huff, etc.)

One last general observation about our view of language as a process
is the sharp distinction we make between the activities of the speaker and the activities of the hearer, a distinction which is completely obliterated in "autonomous" accounts of language as a thing,
since such accounts do not concern themselves with activities at
all.   The speaker's task is to decide which aspects of his intended
message must be encoded linguistically in order to "get the message
across" in some particular apperceived communicative situation and
then to choose a suitable set of linguistic devices to achieve this
end.   The hearer, on the other hand, has a very different problem:
he has little or no idea initially of what the speaker's intended
message is going to be and so he has to make a "best guess" based
on the information available to him, whether provided by his own
knowledge, perceptions and "mental set" or by his attempted reconstruction and interpretation of the (sometimes very meagre) linguistic information which is supplied to him by the speaker.

IV. Why Linguistics Failed

We trust that the foregoing has served to relate at least some of
our reasons for believing that the "autonomous" linguistic tradition
has led psychological inquiry astray in its attempts to understand
the phenomenon of linguistic behaviour or language use. In this
concluding section we shall try to understand why this happened -
indeed, why it was inevitable that it should happen.

Until very recently, psycholinguists have tended to take the products of linguistic research for granted and to construe their own
discipline as one whose goal was either to establish the "psychological reality" of the proposed grammatical constructs (cf. Greene,
1972; Fodor, et al., 1974) or else to show how the proposed
grammars might fit into a model of the language process (cf. Bever,
1968, p.482).   But what reason have we to believe that grammars
must - or even can - function in this way (cf. Osgood, 1968, p.506;
Bartsch & Vennemann, 1972, p.10; etc.), or, indeed, that the way in
which language knowledge is organised in the mind bears any relation
to the way in which it is organised in grammars?   Language is
neither learned nor used in isolation, so why should it be described
that way?

How, then, should language be described? It is important to recognise that this question cannot be answered in the abstract or in
general; one can "describe" in endless ways. Consequently, one

does not simply "describe" in science, but rather describes for some reason.   The form of a description is determined by the particular use to which that description is going to be put.   A biological organ (such as the heart) is described in the way it is (in terms of chambers, valves, etc.) because of the particular function which that organ is construed to perform within the body. Likewise, a grammar for some language which was constructed for pedagogical purposes might be very different in form and content from a grammar which was written, say, best to reflect the history of that language.   Other goals which have been proposed for grammars include those to "generate all and only the grammatical sentences of a language" (Chomsky, 1957), to capture "various linguistically significant generalisations" (Chomsky, 1965), and even to character- ise the syntactic basis for various kinds of linguistic "intuitions" regarding sentence relations, paraphrase, ambiguity, etc.   But each of these assumed goals begs one or more important questions:

(1)   Do native speakers agree as to which sentences "belong to" their language and which do not?   Can even the individual speaker decide on a clear-cut answer (or is this set, too, of the "fuzzy" variety; cf. Ross, 1972).   How important a consideration is "sentencehood" in ordinary language process- ing?[14]

(2)   On what basis does the analyst decide which generalisations are "linguistically significant" (cf. Derwing, 1973)?   Is the ordinary language learner even interested in formal issues of this kind (cf. Derwing, 1975)?

(3)   Why need we necessarily assume that the basis for linguistic intuitions is syntactic in origin, and hence must be character- ised by a kind of rule which serves to "derive" one syntactic structure from another?

On this last point, even Fodor et al (1974) readily admit that there has been a good deal of confounding of syntactic and semantic var- iables in most of the psycholinguistic experiments which have been concerned with the question of sentence similarity and relatedness. In a few studies, however, in which semantic "content" was better controlled, they find evidence of specific "syntactic effects" (p.272), namely, judgements of relatedness which reflect common syntactic patterns or presumed syntactic relationships (as between active and passive sentences, for example).   But why can't these relationships, too, be semantic in nature, (cf. Johnson-Laird, 1975)? It is not a "fact" but rather a supposition that "sentences which differ in their constituent structure may nevertheless be syntactic- ally related" (Fodor et al., 1974, p.226).

Some years ago, in an effort to force subjects to focus their atten- tion upon syntactic patterns rather than semantic content, we de- signed a "concept formation" experiment at Alberta in which three-dimensional target types involved one or the other voice (active vs. passive), mood (declarative vs. interrogative) and modality (affirmative vs. negative).   In order to discover his target type in this study (Baker et al., 1973), a subject had to abstract the pattern similarities of sentences such as "Wasn't the hunter track- ing the grizzly?" and "Hasn't the boy played my guitar?" (i.e. active, interrogative and negative in this particular example), and post-experimental interviews revealed that this was indeed the way in which our subjects approached their task.   Invariably, how- ever, it was the active vs. passive "voice" dimension which was the last one to be sorted out, regardless of the target type to which a subject had been assigned.[15]   Since the formal difference be- tween our active and passive sentences was much more highly "marked" than either of the other two dimensions, we can only interpret this finding to mean that our subjects were not actually responding to the syntactic patterns per se, but rather to the semantic signific- ance of these patterns.   Evidently, at least for sentences presented in isolation from their context, the difference in meaning between an active sentence and its corresponding passive is much less sal- ient (same lexical elements and semantic roles, but a different "focus" or "topic") than is the difference in meaning between statements and questions or affirmatives and negatives.[16]   There is thus a factor of "semantics of type" (i.e. the meaning of a particular syntactic pattern) which must also be taken into account in such experiments.   As a consequence, there appears to be no un- equivocal evidence whatsoever in support of the idea that the kind of syntactic relationships which are posited in formal, transform- ational grammars are required in order to account for native speaker judgments concerning "sentence relationships".

Another experiment was performed at Alberta in order to extend Miller's (1962) proposed test for the psychological validity of the notion of the grammatical "transformation" per se.   Using the same variety of sentence types as employed in the Miller & McKean (1964) study, subjects were required to "transform" sentences from one form to another, on cue from a visual display (Baker & Prideaux, 1975).   Whereas Miller & McKean required their subjects to perform only 12 of the transformational manipulations represent- ed in the sentence family under investigation, the subjects in the Alberta study were required to perform all 64 of the logically possible syntactic changes available (i.e. from every possible stimulus type to every possible response type).   The results were then compared with the predictions of various theoretical, perfor- ative models, (e.g. transform made directly from stimulus to res- ponse, from stimulus to response via a "kernel" structure, from "kernel" to response, etc.).   The latency data did not correlate highly with any of the models, but error frequency was strongly indicative of a straightforward "efficiency" strategy, in which the subject starts with his stimulus sentence and then proceeds to make the appropriate number of elementary syntactic changes

required in order to arrive at the desired response (r = .81).
This suggests that although subjects can be trained to "transform"
one syntactic pattern into another, they do so only in a purely
mechanical way - a clear indication of the artificial nature of
the task. The act of "transforming syntactic structures" is a very
dubious part of normal language behaviour.

Rather than belabour the point any further with an entire cata-
logue of negative evidence, let us try to put the matter back into
proper perspective. Would the linguist ever have seen any need
to posit such things as "deep syntactic structures" or "grammatical
transformations", had he not been motivated initially to achieve a
parsimonious description of the language product?   If the goal
he had chosen had rather been Wundt's goal, i.e. to describe
language processes, such notions would simply not have entered
into his thinking.[17]   His first question would have been, "What
are languages used for?", and his second, "How are these goals
accomplished?"   The kind of knowledge he would have sought to
describe would have been process knowledge, not product knowledge:
language competence is largely a matter of knowing how to, not
knowing what or that (cf. Derwing, 1973, pp.251-70).   The linguist's
fundamental error was that he started at the wrong place, and everything else went naturally downhill from there. He started by examining the tool, rather than by asking questions about the use to
which that tool was put.

If I were to pass to any of you a note containing the message,
"Your house is on fire", you would react to the meaning of that
message and to its significance for you: you would not be much
interested in its syntactic structure or the particular lexical
items used - any more than in the kind of material it was written
on, whether a pen or pencil was used, the colour of the ink (or
lead), or even - if you were bilingual - the particular language
in which it was written.   All tools, language devices included,
are means to ends; they may be essential to achieving those ends,
but once the goals are reached, the tools can be discarded (cf.
Deese 1970, pp. 46-7; Bransford & Franks, 1971, etc.).   If we
take as our goal the parsimonious description of sentences in
isolation from their use, we come up with all kinds of alternative
descriptive "grammars", some of which may contain quite a variety
of complex and abstract "syntactic structures" and "grammatical
operations" - and with this view that the role of psycholinguistic
research is to "find some use" for these grammars. If, on the
other hand, we take as our initial goal the description and understanding of the language process, we come to see man as a semantic
processor at base, and we tend to view syntax as just one type of
vehicle for delivering information which is required for this processor to work. Our task in this view, then, is

(1) to try to ascertain what <u>kinds of information</u> are conveyed by
     languages (as opposed to other means) and then,

(2) to try to ascertain what <u>specific linguistic devices</u> are em-
     ployed in various languages in order to achieve this.

The result, so far, for English is a tentative set of rules which
don't "move" constituents or "hop" affixes, but which rather <u>put</u>
constituents in certain positions for some purposes (e.g. an auxiliary verb is placed in the first or "most highlighted" position[18]
to ask a Yes/No question, or the WH constituent to ask a content-
question, etc.) or which <u>mark</u> certain elements for others (e.g.
tense is marked on the leftmost verb of a sentence). And now
Prideaux (in press) has also shown that such rules may not only
serve to describe language use, but may also be helpful in explaining the kinds of errors which children make in language
acquisition.

Call them whatever you like, autonomous grammars are still descriptions of the language <u>product</u> which say nothing at all about
the language <u>process</u>. Yet it is the language process - the language user's <u>competence to perform</u> - which is the object of ultimate interest in language study, and the value of product descriptions is only as good as the use to which such descriptions
can be put in this larger frame of reference. Individual linguists have nurtured certain formal skills, a patient penchant
for detail, and a familiarity with a wide diversity of language
types, and these are commodities which language study can scarcely
afford to do without. But autononous linguistics has failed as a
discipline. The reason it has failed is because it has consistently had too narrow a view of the problem of language, and because
it has reversed its priorities. It has tried to describe "language"
without prior consideration of what language is good for. Sooner
or later language study must be restored to its proper place within
psychology - and it is that discipline which will determine its
goals, prescribe its methods, and provide its general explanatory
principles. The idea that there can be an "autonomous science of
language" is doomed - for the simple reason that language does not
exist in a vacuum (cf. Rommetveit, 1974, p.22). Language is a
means to an end - and that end is achieved by putting linguistic
devices at the service of all the other cognitive mechanisms which
constitute the general mental make-up of man.

<u>Notes</u>

1. It should perhaps be pointed out that even this terminological
    revolution came about rather slowly, as a matter of fact. See
    Lyons (1970), Derwing (1973) and Steinberg (1975) for documented
    evidence that Chomsky's philosophical position was that of a
    straight Bloomfieldian formalist in his early years (Say to

around 1956, when Roger Brown may have been the first to put
the psychological bee in Chomsky's bonnet (cf. Brown, 1970,
p.17) - by which time of course, the development of the formal
apparatus of transformational-generative grammar (henceforth
TGG) was already well under way).

2. It could reasonably be argued at this point that very little
progress was actually achieved, in fact, during the structuralist era as far as the description of sentence-sized linguistic
units was concerned. But this again just makes TGG look more
and more like a perfectly natural extension of structural
grammar. Moreover, as Prideaux has pointed out, the filling
of this "syntax vacuum" had more than a little to do with the
popularity which TGG very rapidly achieved among linguists
(cf. Harris, 1970, pp.69-70).

3. McCawley (1974, p.178) seems to be the only reviewer of
Derwing (1973) to have perceived it as a critique not merely
of TGG per se, but rather of the entire tradition of autonomous
linguistics, of which TGG is merely the most recent and most
influential representative.

4. We are using the term TGG to refer to any brand of grammatical
theory which has had its ultimate origins in the particular
neo-Bloomfieldian tradition espoused by Chomsky. This includes everything from the original semantics-free formalisms
of Syntactic Structures through the "standard" theory of
Aspects, the so-called "generative semantics" of the Lakoff-
McCawley-Ross exis, the "case grammar" of Fillmore (1968), and
the more recent "relational grammar" of Perlmutter & Postal and
"trace theory" of Chomsky (1975), etc.

5. Professor Baker is largely responsible for the more original
aspects of the views outlined in this section, a definitive
statement of which is provided in Baker (in press).

6. We think that Miller was right when he said that, in science,
"at least half the battle is won when we start to ask the right
questions" - though, of course, we also think that he was mis-
taken in his view that "a description of language and a description of a language user must be kept distinct" (1965, p.18).
As argued above, we believe that it has been the linguistics-
inspired attempt to keep these two notions apart which has been
largely responsible for the deep-seated conceptual confusion
which today pervades virtually the entire field of psycholinguistics.

7. Please understand quite clearly that this diagram is in no
sense to be construed as a process model of what speakers and
hearers do. It is simply intended as a heuristic device to
list or indicate the major factors which we feel must be considered if we are to develop a proper perspective for the
analysis of communicative behaviour.

8. Our use of the term "information" should not be associated
    with the technical, mathematical usage of Shannon & Weaver
    (1949).

9. The active-passive pairs, as usual, present a somewhat more
    complicated picture, as illustrated below. See Baker et
    al. (1973) for a general discussion of the way in which the
    confounding of semantic and syntactic variables has thoroughly
    muddied the interpretation of most psycholinguistic experiments
    concerned with the notion of the "similarity" or "relatedness"
    of syntactic structures.

10. This is consistent with the observation of Fodor et al. (1974)
    that there is no unequivocal experimental evidence in support
    of the psychological reality of (transformational) "grammatical
    operations", but that there is support for various kinds of
    "structural relations" (pp. 510-12). In our proposed model,
    the relationships are preserved at various levels of information
    structure, but the transformations (e.g. rules which "move" or
    "delete" constituents, etc.) are dispensed with.

11. The epithet "surface" is actually redundant here, since we see
    no need to posit any more than one level of syntactic represent-
    ation for sentences.

12. This same principle of using a single linguistic device to
    serve multiple semantic functions is clearly operative in child
    language as Bloom (1970),in particular, has shown.

13. In other sentences, such as "the man felt sick", for example,
    it is now information provided by the verb which informs the
    hearer that the subject is not an "agent", but rather an
    "experiencer", etc. Note that Chomsky's famous eager/easy
    example (1964, p.66) can also be handled in a parallel way,
    since "eagerness" is a common property of persons, not actions,
    whereas for "easiness" the opposite is true.

14. Clearly the native speaker can not only comprehend "sentence
    fragments" in context, but all manner of ill-formed utterances,
    from the relatively minor distortions of child and immigrant
    language to the atrocities of a comedian's anacoluthic speech.
    It is difficult to see how the TGG-based sentence perception
    strategies of Fodor et al. (1974, pp.275-375) can account for
    this skill.

15. The same result was also found in Reid's aural replication of
    this experiment (1974).

16. Another, earlier, study (Prideaux and Baker, 1974) confirmed
    the finding of Clifton and Odom (1966) that the interrogative
    and negative dimensions are not orthogonal, as implied by formal
    grammatical descriptions, but rather interact strongly with one
    another.

17. Blumenthal is surely rewriting history when he says that "Chomsky did what Wundt would have wanted to do" (1970, p.198).

18. Andrew (1974) has shown that the prominence of a constituent is judged to increase as that constituent is placed farther forward in a sentence (cf. Wundt's original theory, as cited in Blumenthal, 1970, p.29), and that such syntactic devices as "passivization" (i.e. putting the semantic "object" in the grammatical "object" position) is one form of "focusing" phenomenon, just like contrastive stress. Millar(1976) shows that the same thing is true of various "clefting" devices, as well.

References

Andrew, C.M. (1974). An experimental approach to grammatical focus. Unpublished Ph.D. dissertation, University of Alberta.

Baker, W.J. (in press). An "information structure" view of language. In The Canadian Journal of Linguistics.

Baker, W.J. and Prideaux, G.D. (1975). Grammatical simplicity or performative efficiency? Ms. in Prideaux G.D., Derwing, B.L. and Baker, W.J. (eds.) Experimental Linguistics.

Baker, W.J., Prideaux, G.D. and Derwing, B.L. (1973). Grammatical Properties of sentences as a basis for concept formation. Journal of Psycholinguistic Research, 2, 201-220.

Bartsch, R. and Vennemann, T. (1972). Semantic Structures. Frankfurt/Main: Athenaum Verlag.

Bever, T.G. (1968). Associations to stimulus-response theories of language. In Dixon, T.R. and Horton, D.L. (eds.) Verbal behaviour and general behaviour theory. Englewood Cliffs, N.J.: Prentice-Hall.

Bloom, L. (1970). Language development: form and function in emerging grammars. Cambridge, Mass: MIT Press.

Bloomfield, L. (1914). An introduction to the study of language. New York: Holt.

Bloomfield, L. (1926). A set of postulates for the science of language. Language, 2, 153-164.

Bloomfield, L. (1933). Language. New York: Holt, Rinehart & & Winston.

Blumenthal, A.L. (1970). Language and psychology: historical aspects of psycholinguistics. New York: Wiley.

Botha, R.P. (1975). The justification of linguistic hypotheses. The Hague: Mouton.

Bransford, J.D. and Franks, J.J. (1971). Abstraction of linguistic ideas. Cognitive Psychology, 2, 331-350.

Brown, R. (1970). Psycholinguistics. New York: The Free Press

Chomsky, N. (1957). Syntactic Structures. The Hague: Mouton.

Chomsky, N. (1964). Current issues in linguistic theory. In Fodor, J.A. & Katz, J.J. (eds) The Structure of Language: readings in the philosophy of language. Englewood Cliffs, N.J.: Prentice-Hall.

Chomsky, N. (1965). Aspects of the theory of syntax. Cambridge, Mass.: MIT Press.

Chomsky, N. (1968). Language and Mind. New York: Harcourt, Brace and World.

Chomsky, N. (1975). Reflections on Language. New York: Pantheon Books.

Clifton, C. and Odom, P. (1966). Similarity relations among certain English sentence constructions. Psychological Monographs, 80, (Whole No. 613).

Cofer, C.N. (1968). Problems, issues and implications. In Dixon, T.R. & Horton, D.L. (eds.) Verbal Behavior and General Behavior Theory. Englewood Cliffs, N.J.: Prentice-Hall.

Davis, M. (1958). Computability and unsolvability. New York: McGraw-Hill.

Deese, J. (1970). Psycholinguistics. Boston; Allyn & Bacon.

Derwing, B.L. (1973). Transformational Grammar as a theory of language acquisition: a study in the empirical, conceptual and methodological foundations of contemporary linguistics. London: Cambridge University Press.

Derwing, B.L. (1974). Review of F.W. Householder, Linguistic Speculations. Language Sciences, No. 30, 25-32.

Derwing, B.L. (1975). Is the child really a "little linguist"? Revised version of a paper read at the Conference on Language Learning and Thought, Montreal, May, 1975. In MacNamara, J. (Ed.) Language Learning and Thought, to appear.

Derwing, B.L., Prideaux, G.D. and Baker, W.J. (to appear). Experimental linguistics in historical perspective. In Prideaux, G.D., Derwing, B.L. & Baker, W.J. (eds.) Experimental Linguistics.

Fillmore, C.J. (1968). The case for case. In Bach, E. & Harms, R.T. (eds.) Universals in Linguistic Theory. New York: Holt, Rinehart & Winston.

Firth, J.R. (1957). Papers in linguistics 1934-1951. London: Oxford University Press.

Fodor, J.A., Bever, T.G. & Garrett, M.F. (1974). The psychology of language: an introduction to psycholinguistics and generative grammar. New York: McGraw-Hill.

Greene, J. (1972). Psycholinguistics: Chomsky and psychology. Baltimore: Penguin Books.

Harris, P.R. (1970). On the interpretation of generative grammars. Unpublished M.Sc. thesis, University of Alberta.

Harris, Z. (1957). Co-occurrence and transformation in linguistic structure. Language, 33, 283-340.

Hymes, D. (1972). Models of the interaction of Language and social life. In Gumperz, J.J. & Hymes, D. (eds.) Directions in Sociolinguistics: the ethnography of communication. New York: Holt, Rinehart and Winston.

Hymes, D. (1974). Ways of speaking. In Bauman, R. & Sherzer, J. (eds.) Explorations in the ethnography of speaking. London: Cambridge University Press.

Jenkins, J.J. (1968). The challenge to psychological theorists. In Dixon, R.R. & Horton, D.L. (eds.) Verbal behaviour and general behaviour theory. Englewood Cliffs, N.J.: Prentice-Hall.

Jespersen, O. (1924). The philosophy of grammar. London: George Allen & Unwin.

Johnson-Laird, P.N. (1975). Is all that is not verse just prose? Review of J.A. Fodor, T.G. Bever & M.F. Garrett, The psychology of language. Nature, 255, 263-264.

Kling, J.S. and Riggs, L.A. (1971). (Eds.) Woodworth & Schlosberg's experimental psychology. (3rd edn.) New York: Holt, Rinehart and Winston.

Lakoff, G. (1971). On generative semantics. In Steinberg, D.D. & Jakobovits, L.A. (eds.) Semantics: an inter-disciplinary reader in philosophy, linguistics and psychology. London: Cambridge University Press.

Langendoen, D.T. (1969). The study of syntax. New York: Holt, Rinehart and Winston.

Linell, P. (1974). Problems of psychological reality in generative phonology: a critical assessment. Reports from Uppsala University Department of Linguistics, 4.

Lyons, J. (1968). Theoretical linguistics. London: Cambridge University Press.

Lyons, J. (1970). Chomsky. London: Fontana/Collins.

McCawley, J.R. (1974). Review of B.L. Derwing, Transformational grammar as a theory of language acquisition. The Canadian Journal of Linguistics, 19, 177-188.

McNeill, D. (1970). The acquisition of language: the study of developmental psycholinguistics. New York: Harper & Row.

Mehler, J. (1963). Some effects of grammatical transformations on the recall of English sentences. Journal of Verbal learning and Verbal Behaviour, 2, 346-351.

Millar, B. (1976). Clefting, contrastive stress and focus: an experimental study. Unpublished M.Sc. thesis, University of Alberta.

Miller, G.A. (1962). Some psychological studies of grammar. American Psychologist, 17, 748-762.

Miller, G.A. (1965). Some preliminaries to psycholinguistics. American Psychologist, 20, 15-20.

Miller, G.A. and McKean, K.A. (1964). A chronometric study of some relations between sentences. Quarterly Journal of Experimental Psychology, 16, 297-308.

Minsky, M. (1967). Computations: finite and infinite machines. Englewood Cliffs: Prentice-Hall.

Osgood, C.E. (1968). Toward a wedding of insufficiencies. In Dixon, T.R. & Horton, D.L. (eds.) Verbal behaviour and general behaviour theory. Englewood Cliffs, N.J.: Prentice-Hall.

Prideaux, G.D. (1971). On the notion "linguistically significant generalisation". Lingua, 26, 337-347

Prideaux, G.D. (1975). An information-structure approach to syntax. Paper presented at the University of Ottawa, March, 1975.

Prideaux, G.D. (in press). A functional analysis of English question acquisition: a response to Hurford. Journal of Child Language.

Prideaux, G.D. and Baker, W.J. (1974). A performative definition of sentence relatedness. Lingua, 34, 101-114.

Prideaux, G.D., Derwing, B.L. & Baker, W.J. (eds.) (to appear) Experimental Linguistics. Ms. under consideration by Cambridge University Press.

Reid, J.R. (1974). Sentence-type variables as aural concept formation dimensions. Journal of Psycholinguistic Research, 3, 233-245.

Reid, J.R. (1976). On the semantic implications of voice syntax in English. Unpublished Ph.D. dissertation, University of Alberta.

Robins, R.H. (1969). A short history of linguistics. London: Longmans.

Rommetveit, R. (1974). On message structure: a framework for the study of language and communication. London: Wiley.

Rosenbloom, P. (1950). The elements of mathematical logic. New York: Dover.

Ross, J.R. (1972). The category squish:endstation Hauptwort. In Peranteau, P.M., Levi, J.N. & Phares, G.C. (eds.) Papers from the eighth regional meeting of the Chicago Linguistic Society.

Sanders, G.A. (1970). On the natural domain of grammar. Linguistics, No. 63, 51-123.

Saussure, Ferdinand de. (1959). Course in general linguistics.
  Translated by Wade Baskin. New York: Philosophical Library.

Schank, R.C. & Wilks, Y. (1974). The goals of linguistic theory
  revisited. Lingua, 34, 301-326.

Shannon, C.E. and Weaver, W. (1949). The mathematical theory of
  communication. Urbana, Ill.: University of Illinois Press.

Slama-Cazucu, T. (1973). Introduction to psycholinguistics. The
  Hague: Paris.

Spencer, N.J. (1973). Differences between linguists and non-
  linguists in intuitions of grammaticality-acceptability. Journal
  of Psycholinguistic Research, 2, 83-98.

Steinberg, D.D. (1975). Chomsky: from formalism to mentalism
  and psychological validity. Glossa, 9, 218-252.

Uhlenbeck, E.M. (1963). An appraisal of transformation theory.
  Lingua, 12, 1-18.

Wall, R. (1972). Introduction to mathematical linguistics.
  Englewood Cliffs, N.J.: Prentice-Hall.

LANGUAGE ACQUISITION AS THE ADDITION OF VERBAL ROUTINES

D.A. Booth

University of Birmingham

INTRODUCTION

The theme of this paper takes me further than others at
this Conference along the line of underinterpretation
of language. Yet just this theoretical economy appears
to open a route for both experimental and computational
psycholinguistics to begin to create manageable explanations of the dauntingly complex phenomena of language,
plausibly at least in early speech comprehension and
production, but conceivably even in adult language also.

Two particular empirical contexts will be used to detail
most of the argument. However, some of the flavour of
my theme was detectable in my intention before I
actually began empirical work. I planned to identify
the adequate stimulus characteristics of "No, don't"
or "No, no" at the time that such utterances begin to
be effective at inhibiting the infant's current activity. Gesell (1934) dated this at 10 months. It seemed
a good problem - restricted in scope, but hopefully of
some importance because explicitly negative terms are
semantically formal and implicit negatives turn up in
connectives and comparatives. The idea was to track
acoustic stimulus control back to interestingly quasi-
biological roots in nonverbal phonology and in gestures
and facial expression, and to track forward the apparent
linguistic comprehension to "No more", "Not now", "It's
not hot" and the comprehension of negative propositions,
to overlap with the published work on negation mainly
in the productive mode. I hoped that the imposition of

some experimental control on naturalistic performance
might permit identification of the mental processes
actually occurring in the infant, rather than those
imputed by the mother or the linguist.

In the event, I was rapidly and massively diverted by
our initial observations into an attempt to approach the
general question of how language begins to get meaning!
Under the influence of philosophers who were working
out the implications of Wittgenstein's (1951) refutation
of his earlier Picture Theory of meaning, I had watched
from the sidelines during the later '60s as machine
translators, transformational grammarians and radical
behaviourists misformulated (or just plain missed) what
seemed to me "the" psychological problem of language -
linguistic meaning, not linguistic structure, and in
particular how language comes to have meaning for an
individual human being or nonhuman primate. What I have
to suggest is that we are still residually misoriented
to this problem. Indeed, I found myself still carrying
some residual lumber when I first tried to make sense of
the observations we had made of the earliest effects
that words as such had on infants.

EARLIEST VERBAL RECEPTION

We set ourselves the task of characterising the earliest
types of specific response to lexemes, and following
reception through to the earliest production of speech.

Method

We studied 20 infants, all but three cared for by mothers
who had had some postsecondary education or training.
The infant and mother were seen in the home once every
1-4 weeks (generally fortnightly) between 6-8 months and
12-13 months of age. Visits lasted 30-90 minutes during
a routine play period or mealtime. We found that even
very frequent visits to an attractively arranged experimental room did not compare with studies in the home for
sensitivity to the infant's abilities (see Lytton, 1971).
Use of specially constructed toys or tasks, even during
test in the home, was not effective in determining what
the infant was normally capable of either.

The mother was asked to give an account of any occasions
when the infant appeared to her to understand speech.
When spontaneous recall appeared to be exhausted, the
mother was prompted by the mention of incidents from
personal experience or from anecdote, and (later in the

study) from observations of other infants of similar
development. By about halfway through the study we had
a loosely coordinated repertoire of types of putative
linguistic reception incident, with which to make both
the quizzing of the mother and the testing of the infant
reasonably exhaustive. Under questioning and by involvement in testing, the mothers became more sensitive to
the type of event that was relevant; this further reduced the risk of omissions.

Some, but by no means most, mothers professed and indeed
showed a substantial change in their behaviour toward
the infant because of the investigator's interest in
potential receptive incidents, a sceptical interpretation of them, and yet often striking specificity in
the infant's response. Relying on such changes inspired
in caretaker behaviour, intervention to improve language
development might therefore be attempted as early as
6-12 months with possible profit, a suggestion we hope
to follow up.

Results

Illustrative incidents

Goodbye wave. During conversation with the mother, the
investigator turned to Martyn (10 months) and said: "Do
you say goodbye, Martyn?". Martyn turned to look at the
investigator, and his mother interjected: "No, but he
waves". Martyn immediately waved his hand in farewell
fashion. No signs of departure had been made, and no
gestural models or cures for imitation had been provided.

Martyn was familiar with departure situations in which
his mother said: "Wave goodbye, Martyn", and herself
waved, and Martyn himself had been waving in such situations. The incident can be regarded as generalisation
to morphemes which are in common between "Wave goodbye"
and either or both the utterance "he waves" by the mother
and the utterance "goodbye" by the investigator in the
above incident. This exemplifies the competence at
purely acoustic abstraction and interpretation which can
be demonstrated with speech stimuli in the so-called
"pre-linguistic" infant.

Visual orientation to usual location of mentioned object.
Martyn regularly had meals in his high chair with his
back to a mantelshelf, the only safe place in the
sitting room for a cup of milk. On one occasion at 10½
months, no drink had been placed on the mantelpiece at

the beginning of the meal as had been usual. Martyn had
finished eating and had started to grizzle. His mother
said: "Would you like a drink?", looking at him not the
mantelpiece. Martyn stopped grizzling and twisted round
the obtuse angle that was necessary to bring within view
the usual place on the mantelpiece for a drink to be
put. Mother said straight away: "I'll get you a drink
from the kitchen" and the infant remained content while
she did so.

Although it is conceivable that any remark from the
mother in this context would have elicited a look at
the usual place for a drink to be left, the infant's
demeanour was more like anticipation of an offer being
carried out than a distressed begging for a forgotten
drink once the mother had signalled her attention to the
child. Possibly the mention of a drink helped to
characterise the infant's distress state for him, and
elicited turning to the usual source of a drink in preparation for its ingestion. As to object permanence,
at least with respect to unconfused egocentric coordin-
ates (Harris, 1975), note that the observation was of
orientation towards a place out of sight from which
the object was at that time absent and indeed in which
it had not been seen for probably a day.

Locomotor search for a friend's toy. Andrew at 9 months
3 weeks had his friend James of the same age at his
place to play. James brings his teddy bear with him and
no doubt Andrew has prior experience of sorting out
whose teddy is whose, with accompanying maternal verbalisations! He routinely goes on request to get the specified one of his several familiar toys, his cup or his
plate. In the middle of Andrew's and James' play
together (which was in part of the room different from
the part where the two teddies had been left), the word
"teddy" was used for the first time on that occasion
when the investigator said to Andrew without any gestural or gaze deixis: "Where is James' teddy?" Andrew
grunted interrogatively and then crawled across the room
to the teddies and brought James' teddy to the investigator.

Perception of "teddy" and "James" (or discriminative
aspects of the lexemes) was presumably necessary to
this receptive performance, although it would have been
desirable to establish which teddy Andrew preferred to
retrieve without mention of its owner. The possessive
need not have been perceived as such, but merely as an
attractive or even as part of an unfractionable word

phrase "James'-teddy". Note that this also is a case of
search for a familiar object which is not in sight, a
performance not uncharacteristic of Andrew's age. There
is little reason to regard object permanence as a concept which is either possessed or not: rather, I would
argue, cognitive routines of a types which in sufficient
number enable the child to meet criteria of competence
in the concept gradually accumulate in the later part
of the first year.

General results

The occurrence of a response from the infant which was
specifically related to the intention of the eliciting
utterance was taken as possible evidence that the infant
possessed a receptive linguistic routine. When no non-
verbal cues had been provided by the speaker or any
other person present, and no recent or current situation
was likely to have potentiated the response, appropriately oriented behaviour was taken as definite evidence
for possession of a linguistic routine. On occasion,
some attempt was made to define the aspect of the adult
utterance that was perceived. However, as no systematic
study was made of the adequate stimuli, this issue will
not be treated in detail. Our impression was consistent
with other evidence that the sufficient lexical stimulus
was a single word within the utterance, or a very few
words in no particular order, except that responding
was made more likely by the word having salient position, stress or intonation.

Although not one of these infants produced an identifiable word before 10 months, most of them gave evidence
of verbal perception at 8-9 months. Table 1 gives the
proportion of the sample at each of three age ranges
who showed contextually uncued behaviour appropriate
to the eliciting utterance.

It should be noted that these receptive incidents
involved movements not oriented to objects as early as
they involved acts relating to objects. At 8 months,
the infant's gaze can be turned immediately to the
actual or habitual location of a familiar object by an
adult saying "Look, there's an X", "Would you like X?",
"Where's X?", "Find X" or making some such mention of
the object. At 9 months, the mentioned object out of
a pair was observed in some infants to be consistently
grasped selectively. By 11 months, some infants would
point to the object in response to a suitably interrogative verbal mention without other cues or a predis-

Table 1.   Frequency of linguistic reception and
           production at different ages

Category of response	Category of verbal stimulus	Infants evidencing comprehension (%)		
		At 8-9 months	At 10-11 months	At 13-15 months
Inhibition	No	25	34	-
Wave	Goodbye	0	16	63
Locate X	Where's X?	38	75	100
Place X	Put X	13	16	25
Other responses	(Various)	25	38	50
Speech production		0	25	38

Note: A locating response may be visual orientation,
selective reaching and grasping, or selective locomotor
search, depending on age.

posing context. Nevertheless, in parallel with such
incidents, from 7 months some infants would stand or at
least lift their body to "Up you come!" or "Up?", at 8
months the occasional infant would make on instruction
out of context a farewell gesture or an idiosyncratic
noise with the mouth, and by 11 months most were
observed to initiate procedures such as giving a kiss,
brushing the hair or playing patacake, in response
purely to verbal request. Empirically there seems no
basis for expecting linguistic deixis before linguistic
triggering of acts as implied by Macnamara (1974):
indeed, there are epistemological grounds for regarding
such a sequence to be impossible (cf. Wittgenstein's
(1951) analysis of ostensive definition). Further, the
"Original Word Game", which might appear some months
later than our observation, could be regarded as aug-
mentation of such receptive routines by another discrete
routine, or small set of productive routines, which -
far from being the central feature in language acquisition - is a late, sophisticated, isolated and possibly
unusual trick. Mere naming is language on a holiday:
the Original Word Game may seldom be played where psycholinguists or their interests have not intervened and
introduced a vocabularly-testing routine into the family.
The strikingly rapid expansion of vocabularly it permits
can only occur on a base of a substantial collection of
more critical linguistic routines which meet most of

Table 2. Types of putative verbal reception incident

Name and definition of a Schank "primitive"	Range of earliest ages (m,d) instance seen		Infants (%) observed with uncued specific instance up to 13,0
	Cued or non-specific reaction	Uncued specific reaction	
ATTEND: directing sense organ towards object	6,23	8,8 - 8,20	100
MOVE: movement of part of the body	6,14 - 8,0	9,4 - 11,13	100
SPEAK: producing sounds from the mouth	7,11 - 9,10	8,8 - 11,12	35
GRASP: taking object into the hand	9,0	9,18 - 11,20	95
PROPEL: application of force to object	6,0 - 7,21	7,25 - 9,19	85
PTRANS: physically transferring object to another location	7,0 - 11,13	9,19 - 11,22	40
ATRANS: transferring abstract relation of possession	-	?11,0	5
INGEST: taking object into the mouth	9,27 - 10,23	?10,11	10
MTRANS: transferring mental information between organisms	-	?11,2	5
DO: carrying out an unspecified action	7,11 - 9,10	8,22 - 9,22	55

the fundamental semantic problems.

The receptive incidents are classified in Table 2 according to the twelve action primitives which Schank (1975) claims are sufficient to programme the interpretation of a large class of natural language sentences.

The table gives the earliest ages at which either a
specific uncued response was seen or a nonspecific or
cued response was seen, and the percentage of infants
who were observed to show a type of routine in each
class. The ages are of course aggregates which do not
apply to an individual infant, who shows only a few
isolated types of routine and often an idiosyncratic
version of a type of routine. Schank's mental acts
MTRANS, CONC and MBUILD are absent, as would be expected. ATRANS, which would include transfer of ownership,
was not convincingly represented in our corpus (as
distinct from passing a named object to a named person).
Neither were there any definite cases up to 13 months
of responses involving INGEST or EXPEL, which was a
little surprising. It may be that the concept of containment attained towards the end of the first year is
not readily extended to placing into and out of the
mouth: visual monitoring is lost, and strong involunt-
ary reactions to intraoral stimuli may inhibit the
development of oral discrimination and intention.

The classifiability of our reception incidents under
Schank's action primitives must not be misinterpreted.
It does not amount to a semantic analysis. The mere
fact that an utterance elicits a movement or an action
is not evidence that the infant perceived the utterance
to refer to the act. The fact that an utterance elicits
orientation to an object is not sufficient criterion
that the utterance was percieved as naming the object,
or even as referring to it in any decently fullblown
sense of "reference". There is no need even to posit
that the utterance elicits a mediating motor or perceptual image. I see no problems in hypothesising of
the incidents observed up to 12-13 months that some
aspect of the lexical percept leads according to an
acquired rule directly to the observed organisation of
action or reaction. This could help to solve problems
of psychological explanation, not create them, because
the older child's referring, naming, meaning, and
possession of concepts are logical abstractions of the
achievements that are possible by use of a more complete collection of such acquired routines.

Slobin (1970) and Edwards (1974) offered case-grammat-
ical analyses of early speech in relation to concurrent
cognitive development. It might be urged that the
infant comprehended all the adult utterances in an
Actional case if no other, but the data do not justify
even such a modest grammar. The mental process could
simply be the generation of a particular behavioural

tendency by the perception of a lexical item which has
hitherto been presented in an appropriate contingency
to an action or object.

Thus the linguistic reception incident is not properly
called an instance of linguistic comprehension. The
incident by itself is evidence only that the infant
possesses a routine, i.e. ability to follow a rule
which specifies the emission of a particular act when
a particular verbal stimulus is presented in a particular physical and/or social context. Later in the
infant's development, the same routine may help him to
meet the criteria for some linguistic competence, e.g.
aid him to comprehend references to the act which was
simply elicited by the utterance in the earlier reception incident. At the time of the original performance,
such competence criteria were not approached, however,
and it would be misleading even to afford the courtesy
title of "linguistic comprehension incident".

PROCESSES YIELDING COMPETENCE

Benedict (1976) presented a technically much more
adequate corpus of both reception and production observations from 9 months. After most of our data had been
collected, we saw Huttenlocher's report of reception
incidents from 10½ months or older. From their later
starting point, both these studies gave results comparable to ours. Furthermore, Benedict acknowledges a
very early "S-R" phase in language reception. However,
she does not consider the power of flexible coordination
of situation-action rules to account for languages and
cognition through the second year of life. Harrison
(1972), in his philosophical "ground-clearing" for
experimental psycholinguistics, not only supposes that
the infant must start with "linguistically sterile
devices" - no more than tricks like those a dog does
to command. He also argues that it is the later
acquisition of a useable collection of such routines by
the human infant which is sufficient to generate
creative language. Newell & Simon (1972) have shown
how adults' performance in complex problem solving
tasks and their comments on their thinking can be
accounted for by computable information processing
analyses involving lists of "productions". A production
is a rule that an act be emitted if specified conditions
are met, e.g. a particular goal and a few percepts be
current. Such analyses would be even more appropriate
where the task structure and the steps in performance
are much more overt, as in Piagetian tasks in children.

Baylor & Gascon (1974) and Young (1975) have shown how
a relatively small "kit" of productions and additions
to it can account for performances in seriation tasks
and their successive stages. It seems to be that such
analyses should be most productive of all in naturally
occurring behaviour in infants and young children which
have now become so justifiably the focus of interest in
developmental psycholinguistics.

'IN', 'ON' and 'UNDER'

To illustrate how a collection of routines might coordinate under task demands to show cognitive and
linguistic competence, let us consider the task reported
by Clark (1974) and at this Conference by Hoogenraad,
Grieve, Baldwin and Campbell (1976), in which children
of about 18 months and older are instructed to place a
small toy on, in or under a large toy.

Language as a Toolkit

I have drafted what amounts to a rough and ready speculative analysis of the processes typical of performance in the On/In/Under task. The collection of routines
is written in the style of a productions "kit" in an
attempt to show the immediate relevance to experimental
psycholinguistics of a computational psycholinguistics
in the spirit of Harrison's (1972) exposition of a
Wittgensteinian view of language as a toolkit. The
collection below is not equivalent to a fully specified
productions programme. Indeed the collection has some
undesirable properties even in a paper-and-pencil check.
Nevertheless, it may serve to illustrate the explanatory
power of a collection of information-processing routines
which are set in little or no fixed hierarchy or other
network structure.

Selection of an active routine

Which routine is supposed to be operative at any instant
during performance of a task is determined by the number
of conditions in a routine which has all of its conditions satisfied. The routine with the greatest number is
the next one used, or in other words the most heavily
specified action is the one which is supposed to interest the child most. Young (1975) suggested that this
was a psychologically more realistic convention than
the ones under which production programming has often
operated. I believe that it would be psychologically
even more realistic to allow intensity factors also to

operate: salience, recency, degrees of confidence and
like variables should affect the selection of the
routine with the highest instantaneous priority. However,
in the present draft, perceptual conditions are simply
present or absent and a conventional goal stack is assumed, i.e. the satisfaction of the current motivational
condition or goal restores the motivation to that which
had previously been most recently current. Thus the
act which occurs in consequence of the operation of a
routine may return the system to its prior goal or it
may specify a new goal which can lead to the activation
of necessary subroutine strategies. The act put out by
a routine also may have an environmental effect which
changes perceptual conditions and also shifts activity
to another routine.

Routines are numbered below, and subcollections are
given titles, for convenience of discussion. Neither
numbers nor titles have any role in the selection of
routines during performance.

Subcollections relevant to the On/In/Under task might be
titled (numbered as follows:

Attending to adult's intentions (1-10)

Identifying a mentioned entity (50-59)

Grasping an object (70-79)

Manipulating a grasped object (80-89)

Relating a grasped object to a standing object (100-139)

Speaking about a current state of affairs (200-249)

Identifying a mentioned entity

Table 3 gives an abbreviation of four routines which com-
prise a subcollection permitting a variety of performances when the young child is faced with some objects
with or without being instructed which object to manipulate. These are the routines particularly relevant
to small objects and grasping a mentioned toy. They
all have a motivational condition of play or manipulation, which is not mentioned in the table.

'Sighted' in these routines is shorthand for the condition that an entity has just been visually located.
(In the older child, there could be an additional
routine with the condition that the entity appear to
be small enough to grasp.) 'Heard' means that an
unallocated lexeme has just been heard. Routine 54

Table 3   Some routines for identifying a graspable
           entity

	CONDITIONS	CONSEQUENCE
52	sighted	grasp
54	heard	identify
56	sighted & heard	identify
58	sighted & mentioned	grasp

could be activated again once a toy had been grasped
and other objects had been scanned for the grasped
object to relate to - by then the name of the grasped
toy will have been allocated, leaving only any name of
a standing object that was mentioned in instruction.
"Mentioned" in routine 58 means that the lexeme has
just been found to be allocatable to the entity in
sight.

"Identify" is an act that will succeed if the sighted
object is allocatable to the heard lexeme. If the
lexical entries attached to the perceptual analysis of
the object do not match any lexeme detected in the
instruction, this routine will fail and some other
routine may have the highest priority set of conditions
met, e.g. examining other objects.

"Grasp" will succeed if the object is small enough, but
fail if the object is too big. A goal of grasping under
the condition that the object is out of reach will
activate a routine which changes the current goal to
approaching the distant object by locomotion.

Even this very simply subcollection has some realistic
properties. If the system began dealing with objects
before verbal instruction had been issued (or when
instructed with unfamiliar words), it could ignore
instructions. If on the other hand, instructions have
been received before full involvement with the objects,
the system could pass over the first object sighted if
its mention was not identified and thus have a change
of selecting the correct object to grasp. Furthermore,
if a lexeme allocatable to a graspable object is
uttered first of two object names in the instructions,
this would prevent failure to grasp or diversion into
attempt to grasp the wrong object which would be pre-
cipitated by mention of a standing object first. That

is, apparent syntactic perception would in fact be a
consequence of action structure, not of the operation
of syntactic processing.

Relating grasped entity to standing entity

Table 4   Some routines for relating a grasped entity
          to a standing entity

	CONDITIONS	CONSEQUENCE
102	grasped & another sighted & that other mentioned	abutt
105	abutted	identify relatability
111	relatability identified	combine
113	heard	identify relatability
115	relatability identified & heard	identify relatability
117	relatability identified & that relatability mentioned	combine

Table 4 gives us another subcollection which will deal
with much of the processing involved in relating a
grasped object to a standing object, with or without
verbal instruction. The terms are self-explanatory or
have just been defined, except for "identify relata-
bility". This is conveniently brief jargon for an act
of examining a standing (large) object for possible
relationships into which the grasped object can be
entered. The range of relationships which can be
effectively noticed depends on the relational routines
which exist in the system. Support can be identified
if there is a routine in which perception of a horizon-
tal surface yields placement of an object in the grasp
onto the surface. Containment can be identified if
there is a routine for placing-in which has perceptual
conditions applicable to some aspect of the standing
object. These conditions and the placing act subroutines
might be present at, say, 8 months but in too restricted
a form to squeeze a doll into a cylinder barely wide
enough to take it. Coverage could be identified if a
routine could be applied which permitted placing-under
by some procedure. The relational perceptions would

also be subject to saliency constraints - familiarity,
recent practice, easy accessibility without displacement
or rotation of standing object.

Note that this collection also can respond successfully
to instruction without parsing the adult's utterance
and without processing via discrete semantic structures.

This anlysis does not invoke "meanings" and non-linguistic strategies which get in the way of meanings, as
Clark (1974) did. The whole system is nothing but a
collection of strategies, a subclass of which have
lexical conditions or consequences. These lexicalised
routines often get priority because they tend to be
more restrictive. Yet familiarity or other salience
factors could override priorities which would be normal
in a more mature system in which degrees of confidence
in lexical perception or lexical matching to object
labelling would be higher.

Such a subcollection should produce the effect described
by Rommetveit (1976) in which performance is less
effective when processing of instructions is forced to
occur before processing of the perceptual conditions of
the task. Only if both speech and object perception are
processed simultaneously into action is maximum
advantage taken of the specificity available in a given
collection of routines. That is, such processing is
highly homogeneous, without the inflexibility from which
hierarchical processing or even heterarchy (Winograd,
1972) suffers.

Other routines

Some routines which might contribute to speech production in the older child are specified in Table 5. If
confident perception and use of support and containment
combinations has developed and then a productive collection of routines for comment is acquired (e.g. routines
210 and 230) then appropriate utterance of "on" and "in"
should occur, although "under" may provide difficulties,
as observed, at a time when reception of "under" is
still subject to canonical bias.

The running through a routine in reality or in imagin-
ation (i.e. not stopping because actual motor acts would
fail or because perceptual conditions are not immediately
present) and articulating the lexical items which are
attached to percepts or act conditions will permit multiword utterances similar to those seen in the 1-year-old.

Table 5.   Some routines which would contribute to
           speech production

	CONDITIONS	CONSEQUENCE
200	Goal is to manipulate & a goal has just been satisfied	Change goal to commenting on a state of affairs
202	Goal is to manipulate & a goal has just failed & an adult is present	Change goal to adult co-operation in state of affairs
204	Goal is adult cooper-ation in state of affairs	Change goal to commenting on state of affairs and using imperative intonation
210	Goal is to comment on state of affairs	Change goal to uttering articulable label elicited
212	Goal is to comment on state of affairs & no label currently elicited	Change goal to imaginary running of routines for achieving state of affairs until label elicited
230	Goal is to utter an elicited label & per-ceptual taxonomic twig is labelled	Change goal to uttering that particular label
235	Goal is to utter a par-ticular label & pho-nological constraint operative	Articulate that label vocally, applying con-straint (e.g. imperative intonation)
240	Goal is to utter a par-ticular label & it is an entity label & have just uttered an entity label	Change goal to uttering an action or relation label

However, they will be generated without any syntactic
processing rules, and indeed there is no semantic processing distinct from the homogeneous processing of
perceptual analysis into action control. At about 2,
a syntactic routine such as 240 might be acquired: the
evidence for this would be the impossibility of eliciting "Soldier truck in" even when the child's attention

was led through the objects in succession before they
were brought together after "Soldier truck" had begun
to be uttered.

As in Tables 3 and 4, Table 5 contains routines which
can enter into several different tasks of the On/In/
Under variety, and indeed into totally different tasks,
particularly in the case of Table 5 routines. The
flexibility of these collections of routines suggests
that in principle this style of theory could provide an
economical account of the daunting fact that the human
mind is not a system but a potential for system organisation, forming its own system to deal with the task in
hand.

EXPERIMENTAL IMPLICATIONS

There are immediate consequences for experimental
design from this view that language is a collection of
lexicalised routines on top of a large number of unlex-
icalised cognitive routines. We need data on individual
cognitive strategies towards small and large objects
without instruction any more specific than "Put these
together", in the same children in the same session
before observing the strategies under instruction to
"Put in" or "Put the soldier in" or "Put it in the
truck". A control group would be needed to determine
whether uninstructed practice sets up perceptual or
response biasses, i.e. gives certain routines higher
salience than they would otherwise have had. Murray
(1976) compared performance under instruction in one
group with performance in another group without instruction, but the force of the present analysis is to
address data from the individual subject with the question - what cognitive routines are sufficient to account
for uninstructed performance and then what specifically
linguistic routines would be sufficient in interaction
with those nonlinguistic routines to generate the observed degree of comprehension of instructions?

GENERAL IMPLICATIONS

There is not space to detail the wider theoretical and
methodological ramifications of the theme of this paper,
but some major implications must at least be alluded to.

DEVELOPMENTAL PSYCHOLINGUISTIC THEORY

Syntax

One particular theoretical suggestion would be that
syntax may hardly exist psychologically in the 1-year-
old. Even the crude word sequence rules which Smith
(1970), Nelson (1972) and others have shown in comprehension and Bloom (1973) has shown for production may
be a product of nonlinguistic processing constraints,
not the operation of separable syntactic rules. Some
arguments in principle against structural overinterpre-
tation of early child speech have been well detailed
by Campbell (1974). Indeed, even though everyday adult
speech is fairly grammatical (Labov, 1970), its grammar
may be very simple compared with the complexity of the
concept of grammaticality. Sensorimotor action patterns
may not directly be conditions for acquisition of
syntax, but the source of an illusion of syntax. Syntactic routines may only be acquired and applied in
comprehension or production when it becomes necessary
to internalise adults' conventions to disambiguate
their utterances or permit them to disambiguate the
child's utterances.

Lexicon

A more programmatic theoretical suggestion is that we
should consider the mental lexicon to be distributed
around the processing of perception and action, and not
to be placed in its own dictionary box, which creates
unnecessary search problems.

Meaning and function

From the viewpoint of this paper, semantics and pragmatics are also distributed through the structurally
unfixed cognitive routines of individuals who have
similar linguistic routines. Meaning and function are
immanent in the routines of several users as they operate together, creating (as Rommetveit (1974) has it) an
'intersubjective reality'. To offer a semantic or
functional analysis of a speech corpus as a psycholinguistic theory is to confuse the logic of success with
the mechanisms which attain it. Ask how the words
came to be uttered or understood, not what they mean
or what function they fulfill.

As for example Hacking (1975) has recently emphasised,
when Wittgenstein said "Don't ask for the meaning, ask

for the use" he was not proposing a Use Theory of
Meaning to replace the Picture Theory. He was saying
that no problem was likely to be solved by directly
trying to construct any theory of meaning. Meaning is
there all the same - if it were not, there would be
nothing to explain. Yet the way to understanding what
meaning might be is to drop that formulation of the
problem and to work out what is actually going on when
language is in use.

"What is going on" is of course rule-following behaviour.
MacKay (1956, 1965) has aruged that propositional
knowledge can be represented in a collection of "conditional readinesses", i.e. tendencies to produce specifiable actions under specifiable conditions. De
Laguna (1927), Piaget (1962), Bruner (1975) and many
others have emphasised the action-bound origin of language. It is logically coherent to go much further and
root adult language and thinking in implicit action, as
a collection of performance rules or strategies - many
non-linguistic and fewer having either at least one
linguistic condition or consequence. Speech production
is not the translation of some pre-existing meaning
process into language. Speech reception is not the
translation through some meaning structure separate from
potential action. To presume that such must be the
case is to succumb once again to the fallacious Picture
Theory of Meaning. We should eliminate the tendency
that still persists to read the abstract structure of
competence into an alleged explanation of performance
processing.

The choice of one criterion of competence rather than
another should not be part of the psychological explanation of performance, although it might help arrange the
data summary tables. "Did the infant understand what
you say?", "Does the child have a concept of ...",
"What does that utterance mean, or what is the intention
behind it?" - such questions are not purely psychological. They involve philosophical issues as to the
questioner's concepts of understanding, the concept,
meaning and linguistic intention. The productive
scientific question is what particular information
processes are occurring.

Acquisition of linguistic competence

It struck me forcefully from our observations how isolated these linguistic routines are from each other
initially, at most the fragments of a potential system.

Yet with their accumulation in parallel with the increasing numbers and stricter specifiability of non-
linguistic routines, the linguistic routines become
increasingly coordinatable according to their internal
characteristics by the constraints of the linguistic,
social and physical environment to which they are adapted and by their possessor's perceptual and motivational states. When a set of well-adapted routines is
sufficiently large (and that need not mean a great
number), the individual will produce a variety of actions
in relation to various situations (including linguistic)
actions and/or linguistic situations) and this variety
may be sufficient to meet the criteria for linguistic
competence. The production "kit" for seriation consists of only about a dozen routines (Young, 1975).
Furthermore, improvements of performance through Piaget's
successive Stages, as well as the phenomena of decalage
horizontale, are accounted for by the successive
addition of one or two productions on top of the previous list, leaving no need to postulate a mysteriously
total reorganisation of the control structure in the
mind of the child.

Such an approach to cognition and its development would
require an account of how an individual routine or each
class of routine is in fact acquired. This can appear
a very difficult problem if the wrong paradigmatic
assumptions are made. The following two presuppositions
seem to me to reduce the size of the task of providing
an account of performance acquisition.

First, behaviour is not driven but selected, and represented contingencies do the selecting. For a start,
many types of representation of the environment have
response tendencies innately attached to them - stimuli
predicting pain and stimuli predicting nutrition cause
fear and appetite respectively. Given that response
tendencies can also be attached to situations as a
result of experience, there remains no danger of the
associative system being left 'lost in thought'.

Secondly, there is every reason to believe that brains
are constructed in a way which subserves the construction of a permanent representation of every contingency
or causal relation perceived in the way it is perceived
(Uttley, 1976). The simplest form of such registration
is classical conditioning (seen as expectancy formation,
not stimulus substitution). Acquisition of discriminative

stimulus control of operant behaviour could similarly be
the modelling of both discriminative stimuli and response-
produced stimuli as conditions for the occurrence of the
reinforcing stimulus. Response organisation and skill
might similarly be coordinated subcomponents discretely
acquired. Knapman (1975) has a computer system in which
a process can become a procedure and so acquire competence through experience.

Because human behaviour is obviously not merely a string
of simple overt respondents and operants, there is a
feeling that it does not get us very far to admit that
conditioning and reinforcement processes are important
in early development and a fair description of at least
some aspects of the acquisition of the earliest responses to language. However, if the acts can be overt
or covert and the interpretation of the stimulus controls
behaviour, not its physical presence, then we must consider the possibility that a flexibly selected string
of percept-act connections could indeed be the process
of human thought and behaviour (compare Newell, 1973).
When a cognitivist orientation finds this too behaviour-
istic and a behaviourist orientation finds it too men-
talistic, I dare to hope that the theme may be nearer
the truth than prejudgments in either camp.

METHODOLOGICAL ADVANTAGES

Analysis of performance as the use of a collection of
routines has advantages for empirical method which
have been illustrated by Young's (1975) work. The
requirements of clinical method and experimental method
can be reconciled, individuals' protocols and group
data are handled in similar terms, the idiographic/
nomothetic polarity is removed and individual differences
are accommodated within a process theory. A cycle of
observation, analysis and reobservation to select
between alternative models is applicable over a short
time interval within the same child. Data interpretation in this style is predictive theory.

This style of theory is also obviously programmable in
productions analysis form or equivalent AI techniques.
Therefore, it is a much needed bridge between experimental psycholinguistics and computational psychol-
linguistics. Furthermore, this particular style of AI
has two distinct advantages as theoretical psychology.

First, in being ideally suited to early developmental
studies where the subject's information processing is

so overt, the method promises to correct the major
weakness in the modelling of adult cognition - so many
different models are plausible. If we determine what
the mechanisms are when the system is small and overt,
and if the larger, covertly operating system is built
on this foundation without total reorganisation, then
the developmental theory will put the needed constraints
on the theory of adult cognition.

Secondly, the computer model's components have individually minimal power, and have no more power than
necessary to encapsulate behaviour at a given instant.
As Steedman and Johnson-Laird (1976) themselves say of
their computational psycholinguistic model, the processes
invoked (such as an augmented transition network for
sentence parsing) have embarrassingly great power.
Psychological realism is more likely to be preserved by
dispensing with processing that has not been proved to
exist or to be necessary.

Mature science is, I believe, characterised by its
inclusion of the style of theory that specifies a
system of processes which computably behaves as a whole
like the real system, but in addition the individual
processes and relationships between them are formally
equivalent to processes independently observed to be
operative in the real system. Such a systems analysis
is genuinely explanatory, and resolves the polarity
between reduction and holism. This style has recently
been found to be highly productive in the field of
motivation (Booth, 1976; Booth & Mather, 1977) and I am
a little taken aback to find myself trying to operate
in a similar style so soon after getting down to observations of early language.

REFERENCES

Baylor, G.W. & Gascon, J. (1974) An information processing theory of aspects of the development of weight
  seriation in children. Cognitive Psychology, 6, 1-40.

Benedict, H. (1977) Language comprehension in 9-15
  month old infants. Volume I.

Bloom, L. (1973) One Word at a Time. The Hague: Mouton.

Booth, D.A. (1976) Approaches to feeding control. In
  T. Silverstone (Ed.), Appetite and Food Intake, 417-78.
  Berlin: Abakon Verlagsgesellschaft.

Booth, D.A. & Mather, P. (1977) Prototype of a model of human appetite, growth and obesity. In D.A. Booth (Ed.), Hunger Models: Computable Theory of Feeding Control. London: Academic Press.

Bruner, J.S. (1975) From communication to language - a psychological perspective. Cognition, 3, 255-287.

Campbell, R.N. (1974) Propositions and early utterances. In Akten des 1. Salzburger Kolloquiums über Kindersprache, 247-259. Tübingen: Verlag Gunter Narr.

Clark, E.V. (1974) Non-linguistic strategies and the acquisition of word meanings. Cognition, 2, 161-182.

De Laguna, G. (1927) Speech: Its Function and Development. New Haven: Yale University Press.

Edwards, D. (1974) Sensory-motor intelligence and semantic relations in early child grammar. Cognition, 2, 395-434.

Gesell, A. (1934) Infant Behavior: Its Genesis and Growth. New York: McGraw-Hill.

Hacking, I. (1975) Why Does Language Matter to Philosophy? 184-187. London: Cambridge Univ. Press.

Harris, P.L. (1975) Development of search and object permanence during infancy. Psychological Bulletin, 82, 332-344.

Harrison, B. (1972) Meaning and Structure: An Essay in the Philosophy of Language. New York: Harper & Row.

Hoogenraad, R., Grieve, R., Baldwin, P., & Campbell, R.N. (1976) Comprehension as an interactive process. Volume I.

Huttenlocher, J. (1974) The origins of language comprehension. In R.L. Solso (Ed.), Theories in Cognitive Psychology. New York: Wiley.

Knapman, J. (1975) Some principles of artificial learning that have emerged from examples. Paper presented at IV International Joint Conference on Artificial Intelligence, Tbilisi. (Research Report No. 6, Department of Artificial Intelligence, University of Edinburgh.)

Labov, W. (1970) The study of language in its social context. Stud. gen., 23, 30-87

Lytton, H. (1971) Observation studies of parent-child interaction: a methodological review. Child Development, 42, 651-684.

MacKay, D.M. (1956) Towards in information-flow model of human behaviour. Br. J. Psychol., 47, 30-43.

MacKay, D.M. (1965) From mechanism to mind. In J.R. Smythies (Ed.), Brain and Mind, 163-200. London: Routledge and Kegan Paul.

Macnamara, J. (1972) Cognitive basis of language learning in infants. Psychol. Rev., 79, 1-13.

Murray, B. (1976) Cited in Hoogenraad, et al, 1977.

Nelson, K. (1973) Structure and strategy in learning to talk. Monog. Soc. Res. Child Dev., 38, No. 149.

Newell, A. (1973) Production systems: models of control structures. In W.G. Chase (Ed.), Visual Information Processing, 463-526. New York: Academic Press.

Newell, A. & Simon, H.A. (1972) Human Problem Solving. Englewood Cliffs: Prentice-Hall.

Piaget, J. (1962) Play, Dreams and Imitation in Children. New York: Norton.

Rommetveit, R. (1974) On Message Structure, London: Wiley.

Rommetveit, R. (1976) On the relationship between children's mastery of Piagetian cognitive operations and their semantic competence. Volume I.

Schank, R.C. (1975) Conceptual Information Processing. Amsterdam: North Holland.

Smith, C.S. (1970) An experimental approach to children's linguistic competence. In J.R. Hayes (Ed.), Cognition and the Development of Language. New York: Wiley.

Slobin, D.I. (1970) Universals of grammatical development in children. In G.B. Flores d'Arcais & W.N. Levelt (Eds.), Advances in Psycholinguistics. Amsterdam: North Holland.

Uttley, A.M. (1976) Neurophysiological predictions of a two-pathway informon theory of neural conditioning. Brain Research, 102, 55-70.

Winograd, T. (1972) Understanding Natural Language. Edinburgh: Edinburgh University Press.

Wittgenstein, L. (1951) Philosophical Investigations. Oxford: Blackwells.

Young, R.M. (1975) Seriation in Children: A Production-Systems Approach. Zurich: Birkhauser Verlag.

SYNTAX AND SEMANTICS FOR THE PSYCHOLINGUIST

J. MILLER

University of Edinburgh

I Introduction

It is unfortunate that many, perhaps most, psycholinguists are not aware of the diversity and richness of
linguistic theories that are at their disposal, and this
unawareness is not alleviated by the fact that those
psycholinguists who are aware of the range of theories
choose, or are obliged, to focus their attention on only
one of the theories, Chomsky's generative grammar.

That this rather harsh judgment on psycholinguists is
not without foundation is demonstrated by Brown's
remarks to the effect that linguistics always seems to
be nearer linguistic 'truth' than psychology is to
psychological 'truth'. Brown observes that this could
be an illusion fostered by the fact that in psychology
different schools flourish simultaneously, whereas in
linguistics different schools succeed one another, with
the result that at any one time there is only one school
in evidence (Brown, 1971:ii).

The purpose of this paper is to show that within Chomskyan generative grammar there are controversial topics
that deserve to be exposed, since their controversial
nature is of some relevance to psycholinguistics, and to
suggest other views of syntax and semantics that might
also be of interest to the psycholinguist.

II Psychological reality and the standard model

II.1 Status of transformations

The first controversial topic I want to consider, though very briefly, is the nature of transformations. Abstract automata can be understood as dynamic or static. On the dynamic interpretation the rules of a generative grammar create structure. They describe the operations that are carried out in the creation of deep structure and the steps by which one proceeds from the deep structure to the surface. On the static interpretation the rules do not create structure. The phrase structure rules merely check already-existing structures to see if they are well-formed and the transformations indicate that a structure of one general sort is related to a structure of another general sort.

The static interpretation has been explicitly adopted by Lakoff (1971, pp. 232-234) and the dynamic interpretation has been explicitly adopted by Postal (1972, fn. 19) and, apparently, by Chomsky (cf. Chomsky, 1972: fn. 19).

It is important to point out the two interpretations because there was a time when many experiments were conducted that were alleged to demonstrate that the operations described by the transformations as set out in various descriptions of English corresponded to the operations carried out by the native speakers of English as they interpreted sentences and responded to them. Fortunately, psycholinguists seem to be retreating from the view that the transformations postulated by the linguist correlate with the operations carried out in real time by the speaker. As Fodor, Bever and Garrett (henceforth FBG) put it: "We have seen, however, that when the logic of the matching experiments is investigated closely, the hypothesis that the operations that the subjects performed were grammatical transformations is actually disconfirmed by the data, since the rules the subjects were using did not observe the constraints on ordering that transformations obey". (FBG, 1974, p.241).

The alleged correlation between transformations and real-time operations is in any case most improbable, for a number of reasons. In the first place there is the sort of problem mentioned by Seuren (1969, pp.51-60) concerning the generation of sentences containing relative clauses. Seuren argued that unless there is an algorithm

for ensuring that an embedded sentence in a relative
clause structure contains a noun phrase identical with
a noun phrase in the matrix sentence both with respect
to syntactic structure and lexical items then a vast
number of structures will be generated that do not contain matching NPs and that will have to be rejected.
This argument is advanced independently by Friedman
(1969), who discovered that attempts to write a computer
program based on the (extended) standard model were
costly failures until an algorithm was developed that
controlled the generation of sentences from the first
line in the derivation onwards.

In the second place there is the problem that even within
the standard and extended standard models there is a
great deal of variation with respect to the formulation
of various transformations and their order in the cycle
and anyone who believes in a correlation between transformations and real-time operations carried out by the
native speaker has the ticklish task of deciding which
particular version of TG he would be best to adopt. So
ticklish, indeed, is the task that I suggest (though I
do not have space to develop the proposal in detail) that
transformations are the weak spot even in pure linguistic
theory. Linguists have intuitions about semantics which
can often be supported by the surface syntax and morphology. Taking the view that transformations simply state
that a relation exists between two structures, one
might postulate semantic structures and **surface syntactic**
structures and then state which semantic structures are
associated with which surface syntactic structures, and
having said that one might assume that there is no more
to be said, no mysterious intermediate structures to
be described.

To give an example, one might postulate a semantic
structure "John cause it: book move from John to Bill"
and state that this is related to John gave the book to
Bill or John gave Bill the book, without any reference
to, e.g., a transformation of Dative Shift. Of course,
it is open to any linguist to set up intermediate
structures, each one differing from the preceding one
in respect of one item, in order to show explicitly
which element in the deep structure (syntactic or semantic) is related to which element in his surface structure.
This, however, brings us to another difficulty, mentioned
by Braine (1974).

Braine observes that in doing generative phonology

linguists strive to provide the simplest and most general
lexical entry, even if the entry is very abstract and
can only be related to its surface realisations by a
complicated series of rules. Pursuing the analogy of
the brain as a computer, Braine points out that there is
another sort of economy which might be preferable, namely
ease in looking up addresses in the lexicon. This task
is made extremely difficult by the phonologists with
their abstract lexical entries but it would be made
almost absurdly simple by a description that simply
listed all the allomorphs in a set instead of deriving
them by rule from a single underlying item. The burden
of the argument is that, given the tremendous capacity
of the human brain for rote learning, there is good
reason to assume that native speakers of a language
store their morphophonemic knowledge in the form of
lists and not in the form of abstract lexical entries
with derivational rules.

It is possible to apply Braine's remarks not just to
phonology but also to syntax and semantics and to say
that there is no reason to assume that the native
speaker of any language stores his knowledge of his language with the simplicity, elegance and generality that
are the goal of the linguist.

II.2   Status of deep and surface structures

In any case, FBG adopt the position that it is the deep
and surface structures posited in the (extended) standard model that have some claim to psychological validity
and not the transformations. But even this position is
open to attack, since there is some difficulty in determining what the psychological reality might be and
since it is not clear whether the (extended) standard
model is entitled to any special credit.

Consider some of the examples presented by FBG. They
cite an experiment by Wanner which allegedly demonstrates
that we 'need to appeal to detailed properties of the
notation that the standard theory uses to represent
grammatical relations' (FBG, op. cit. p.261). When we
look at the actual sentences used by Wanner they turn
out to be

(1) The governor asked the detective to prevent drinking.

and

(2) The governor asked the detective to cease drinking.

When <u>detective</u> was used as a prompt, (2) was recalled
more often than (1). This fits with an analysis in
which <u>detective</u> occurs three times in the deep structure
of (2) but only twice in the deep structure of (1). But
the occurrences of <u>detective</u> will be the same in any
analysis that provides a satisfactory description of (2),
whether the linguist is using NPs and VPs or predicates
and terms.

In any case, the differences in recall might simply have
been caused by the fact that (1) is an odd sentence. I
have presented several educated native speakers of English
(not linguists) with a list of sentences including (1)
and each informant has expressed doubts about (1). Put
in a few modifiers - <u>The governor asked the detective</u>
<u>to prevent the illegal all-night drinking in unlicensed</u>
<u>clubs</u> - and the sentence becomes unexceptionable, but
as it stands it is odd to the point of being unacceptable.

Another example cited by FBG is an experiment by Blumenthal (FBG, <u>op.cit.</u> p.258). The crucial point in
Blumenthal's experiment is that subjects who had been
presented with sentences (3)

(3) <u>Gloves were made by tailors</u>.

and

(4) <u>Gloves were made by hand</u>.

responded much better to <u>tailors</u> as a prompt than to
<u>hand</u>. FBG suggest that <u>tailors</u> is a better prompt
either because it is in a higher position in the deep
structure and therefore closer to <u>gloves</u> or because (3)
derives from a deep structure containing one sentence,
whereas (4) derives from a deep structure containing
two sentences.

Quite apart from the possibility of also deriving (3)
from a deep structure containing two sentences, it is
possible that the crucial factor is that <u>tailors</u> is an
agent noun, 'agent' and 'patient' being central roles
in the sentence, whereas <u>hand</u> is a relatively minor
item. Moreover, there is the relation between (3) and
<u>Tailors made gloves</u>, but there is not a sentence *<u>Hand</u>
<u>made gloves</u>. Finally, given that the sentences describe
the making of gloves, there is semantic connection between
<u>tailors</u> and <u>gloves</u> that does not hold between <u>hand</u> and
<u>gloves</u> (this ties in with the remark about <u>tailors</u> being
an agent noun).

That is, the fact that <u>tailors</u> is the better prompt might arise not from the standard model analysis but from such factors as which items are semantically important and which sentence is related to a corresponding active sentence with the same lexical items.

III   Weakness in Chomskyan syntax

III.1   Syntax and generative grammar

The preceding discussion of syntactic structures and psychological reality has rested on the argument that the results of the experiments could be attributed to various causes, which are not considered by the authors of the experiments or by FBG, and that the results are therefore completely inconclusive. The next stage in the discussion is to show that the claim to psychological reality for the structures is exaggerated and premature, because the structures embody rather inadequate notions of syntactic structure.

Let me begin by drawing a distinction betwen the statements one might want to make about syntax, together with the notions incorporated in these statements, and the organisation of the statements into a coherent system. It is obvious that the major contribution of generative grammar has been its investigation of such systems of statements. The first part of the this paper has argued that it is this part of linguistic theory, the organisation of the grammar, that is not associated with any psychological reality. On the other hand, it is reasonable to suppose that the linguist's analysis of the surface syntax of a language has some connection with the unconscious knowledge possessed by the native speaker of the language, for instance, the possible constituents in noun phrases and prepositional phrases.

But it is precisely in this area that Chomskyan generative grammar is weakest. It is descended from the item-and-arrangement (IA) analysis of syntactic structures and it might seem on a casual reading of Bloomfield, Harris, Hockett and Chomsky that all four make use of the same constituent structure analysis. In the following paragraphs I will attempt to show that in fact there are important and unmentioned differences between the analyses of the first three and Chomsky's and that the IA view suffers from certain fundamental weaknesses that have been inherited by Chomskyan generative grammar.

III.2 Endocentric and exocentric constructions

The central notions of the earlier IA approach were those
of endocentric and exocentric constructions, the head of
a construction and its modifiers, privilege of occurrence
(or, distribution, expansion, substitution) and closure.
The notions of head and modifier, which need no explanation, are crucial to the concepts of endocentric and
exocentric constructions. An endocentric construction
was defined as one that contained a head constituent,
which constituent was the head being determined thus.
If two forms x from class X and y from class Y made up
a sequence whose distribution was the same as the general distribution of class Y, then Y was held to be the
head of the construction or sequence. If the distribution of xy was the same as the distribution of members
of X, then X was held to be the head of the sequence or
construction. (It is important to remember that what
was at stake was not the distribution of particular
lexical items but the distribution of the general class
of nouns or adjectives or whatever). If the distribution
of xy was not the same as the distribution of the members of X or the members of Y the sequence or construction was held to have no head constituent and to be
exocentric.

The notion of closure was usually invoked with respect
to endocentric constructions but it was also relevant
to exocentric constructions. It came into discussions
of syntax whenever endocentric constructions were thought
of as consisting of a constituent to which some other
constituent had been added to yield a sequence whose
distribution was the same as that of the first constituent, e.g. a sequence of adjective and noun in English,
the adjective being added to the noun. It could then
be asked whether a third constituent could be added to
the sequence without changing the latter's distribution.
For instance, the sequence this exotic food has the distribution of the class of nouns and to it can be added
all to yield all this exotic food, which sequence has
still the same general distribution as the class of nouns
in that it can occur after prepositions, after a main
verb, before a main verb or an auxiliary. But the
sequence is now closed because no constituent can be
added to the left of all to yield a sequence that has
the same general distribution as the class of nouns.
(For the sake of simplicity I will ignore the possibility
of adding to the right of food sequences like for sale
in that delicatessen my wife likes so much.)

Bloomfield, Harris and Hockett all considered noun
phrases like <u>all this exotic food</u> or <u>unpleasant surpri-
ses</u> to be endocentric. Bloomfield does not discuss
sequences like <u>prepared some food</u> (Bloomfield, 1935) but
Hockett does, and these he regards as exocentric, along
with prepositional phrases. It is worthwhile pointing
out that sequences consisting of a verb and an object
noun do share one environment with a single verb, namely
N - (# stands for 'clause boundary'), but the important
point is that sequences that are treated as endocentric
usually share several environments with the head constituent.

It is a bit more difficult to find out whether Harris
regarded sequences like <u>prepared some food</u> as endocentric
or exocentric. The trouble is that the formalism developed in his 1946 paper is designed to capture several
things. The symbols in his notation consist of letters
and raised numbers, e.g. N^1, N^4, V^4. The letters stand
for general classes - N for noun, V for Verb - and indicate the head of a construction. The numbers are connected with the notion of closure, though Harris talks not
of one constituent being added to another but of constituents being inserted into frames. I have argued
elsewhere (Miller, 1973) that the raised numbers can
be interpreted as levels on which different frames are
available. With respect to noun phrases, on the lowest
level, denoted by the raised number 1, the available
frames consist of central constituents like adjectives.
The next level up has frames consisting of determiners
like <u>the</u>, <u>this</u> and the highest level has frames consisting of phrases like <u>all</u> and <u>some of</u>. The highest
level is denoted by the raised number 4, and once that
level is reached the frames available for the noun phrase
to be inserted into are all verb sequences. Since the
insertion of a noun into any particular frame might be
optional, N^4 might be simply <u>boys</u> or it might be <u>some
of those nasty rough boys</u>.

The general formula for a sentence is N^4V^4, which can
be interpreted as saying that to the right of the subject
noun in English there is a constituent labelled V^4, and
from the exposition in Harris' paper we know that this
V^4 can consist either of a verb or of a verb and a noun.
The occurrence of the symbol V certainly gives the impression that Harris considered the V to be the head of the
construction, but Robinson (1970a, p.269) has indicated
a passage from a later paper (Harris, 1962) that seems
to show that Harris thought of verb-object noun construc-

tions as being exocentric. Certainly, once one looks
beyond the simplest sort of active declarative sentence
the distributional evidence does not support the treatment
of such sequences as endocentric.

A comparison of Harris' formulae with those of the (extended) standard model reveals that the latter do not capture the head-modifier distinction in any obvious
fashion. This defect has been remedied by Chomsky
(Chomsky, 1970) but his suggestions do not appear to
have been generally adopted, though it is clear that
only a grammar that does reflect the head-modifier
distinction can have any claim to psychological validity.

The topic of endocentric and exocentric constructions
is not yet exhausted, because in Chomskyan generative
grammar verb phrases and prepositional phrases are not
regarded as exocentric but as endocentric, having the
verb and the preposition as their respective heads.
Since this runs counter to Hockett's position and to
one of Harris' positions, and since such an analysis
is not supported by distributional evidence, one is
entitled to ask for this view to be justified.

No argument for this view is to be found in the writings
of Chomsky or his colleagues but justification is found
in a paper by Robinson (1970b). Instead of looking at
the privilege of occurrence of, e.g., verb-object noun
sequences she looks at which constituent is the characteristic one in the construction. With respect to verb
phrases the answer is, of course, that the verb is the
characteristic constituent, and with respect to prepositional phrases the answer is that the preposition is
the characteristic constituent. This is why some linguists regard all constructions as endocentric, even
verb phrases and prepositional phrases.

IV   Constituents and relations

IV.1   Prepositions

There are two important points to be made now concerning
syntactic structure. The first, and minor one for
present purposes, is the analysis of prepositions. It
is plain from sentences like Into the room came a tiny
old lady that a preposition and the noun phrase that
follows it function like a single unit in some respects.
It is equally plain that the function of the prepositions
is to relate nouns with nouns or verbs with nouns, and
it is this relational function of prepositions that the

IA analysis and Chomskyan generative grammar fail to
capture. Using a dependency model, Robinson is able
to represent what she calls the "Janus-like nature" of
prepositions and also to show that prepositions and the
noun phrase that follows them do constitute a single
constituent.

The question that arises out of these considerations is
this. Given that the relational function of prepositions
is indisputable, how much heed should be paid to the
claims to psychological reality of a syntactic description that does not give any place at all to the relational
function? Not very much, I think would be the answer.

IV.2   Arguments for the centrality of the verb

The second point I want to make about syntactic structure
has to do with the status of the verb in the sentence,
in particular with the view that the verb is the central
constituent on which all the other constituents are
dependent. There are several ways in which this view
can be arrived at but the argument I will set forth
first is an extension of the characteristic constituent
criterion described above. Defining the sentence as a
unit postulated by the linguist in order to account for
the dependencies between constituents - for example,
concord and government - and having asked what the characteristic constituents of noun phrases and prepositional
phrases are, it is possible to ask what the characteristic constituent of a sentence is.

Before answering that question, I would like to draw a
distinction between nominative sentences and what, for
convenience, can be called predicative sentences. The
former are sentences like A beautiful sunlit morning or
Night, without verbs or predicative adjectives, and the
latter are sentences like The car is grey or My brother
bought a Renault. If the question about the characteristic constituent is confined to predicative sentences,
the characteristic constituent turns out to be the
verb. In My brother bought a Renault, bought is the
head constituent of the sentence, and in The car is
grey the head is is. (Note that there are languages
in which a single verb form constitutes a complete
predicative sentence.)

To avoid unnecessary controversy, let me say that what
is being analyzed is the surface syntax. There have been
proposals (Lakoff, 1970 and Lyons, 1966) that surface verbs

and adjectives should be associated with a single category in the deep structure, Lakoff calling the single
deep category a 'verb' and Lyons calling it a 'predica-
tor'. Lyons, arguing that the copula in English and
other languages be treated as a dummy verb that is
inserted into certain sentences at the level of surface
structure, adduces in support of his analysis the fact
that there are languages in which a distinction between
verbs and adjectives in the surface structure is difficult if not impossible to draw and the further fact that
even in languages in which the distinction can be drawn
there may be no copula in surface sentences containing
predicative adjectives. The languages in which no distinction can be drawn present no problem, since the
characteristic constituent will always be a verb (that
being the label that is normally used in grammars of
such languages) and the languages for which the distinction can be drawn but which may have predicative adjectives occurring without a copula do not present any
real problem either. The crucial point is that it is
usually the non-past copular forms that can be omitted
but not the past or 'future' tense forms and that an
alternative auxiliary verb may be used to supplement the
missing copular forms (as is the case in Modern Russian).

For these reasons I will continue to treat the verb,
including the copula, as the characteristic, and consequently the head, constituent in predicative sentences
and I will assume that in some languages some forms of
the copula can be omitted from the surface structure.
The question of the deep syntax of sentences with predicative adjectives does not arise for me because I am
assuming that the linguist's task is to analyse the
surface syntax and morphology and to postulate semantic
structures without setting up a level of deep syntax.

As I said earlier, the view that the verb is the head
constituent of a predicative sentence can be arrived at
in various ways. One way is via a consideration of
what modifies what in a sentence. Consider, for example,
the English sentence Peter sold the bike to Michael.
It is part of the European grammatical tradition that
the phrases the bike and to Michael modify sold. It
is mentioned by Hjelmslev in his Principes de Grammaire
Générale that early mediaeval grammatical theory, like
the Arab one, also held the subject noun to be a modifier of the verb, but that this notion became submerged
in later grammatical theory. The credit for reviving
this analysis of the relationship between the verb in a

sentence and its accompanying noun phrases and prepositional phrases must go to Fillmore. One of the important
observations that he made in his early work on case
grammar is that the subject noun in a sentence may enjoy
special status in descriptions of the theme-rheme system
but that there is no reason why it should enjoy special
status in descriptions of constituent structure, since
nouns that turn up in subject position occur in construction with the same set of lexical items but preceded
by a preposition, as in Peter sold the bike to Michael
and The bike was sold to Michael by Peter. Even object
nouns can be preceded by a preposition in certain constructions, e.g. the selling of the bike to Michael by
Peter (was done without his father knowing). Fillmore
concludes that in descriptions of constituent structure
there is no good reason to distinguish between 'pure'
relations (in English expressed without a preposition)
and 'labelled' relations (expressed in English by prepositions), since all noun-verb relations can be regarded
as labelled, the lack of preposition being associated
with neutral topic position. All noun phrases and
prepositional phrases, that is to say, can be regarded
as 'adverbs', as modifiers of the verb.

The centrality of the verb is supported by Benveniste
(1946) in his treatment of the category of person in
the verb. He draws attention to the curious fact that
in many languages that have personal affixes there are
special affixes for first and second person, the forms
denoting the speaker and the hearer, but that the third
person form lacks a personal affix. He suggests that
it is only the first and second person forms that
deserve the label 'personal', since the 'third-person'
form, while indicating that reference is being made to
someone or something, does not specify the person or
thing. (There is usually no doubt as to who the speaker
and hearer are with respect to a particular speech act).
Normally it is only 'third-person' verb forms that are
accompanied by full noun phrases and Benveniste argues
that in many languages the noun phrases are added only
if the reference of the verb is not clear. He glosses
the Latin sentence Volat avis as 'It is flying - the
bird, that is'. (Interestingly enough, there is a
description of a Mexican Indian language in which noun
phrases are analysed as 'tagmemes of detail' - cf. Elson
and Pickett, 1965, p.72).

Finally, it is worthwhile mentioning that Jackendoff
(1974), in his extension of the bar-X notation introduced by Chomsky (1970), is obliged to adopt what looks

like a verb-dependency analysis. He wants to develop
a notation that will express cross-category generalisations, and one of the generalisations is that in The
enemy's destruction of the city and The enemy destroyed
the city enemy is the subject noun. In the former
example it is the subject of the noun phrases, in the
latter - the subject of the sentence. To express these
two facts in one notation Jackendoff has to replace S
by a double-bar V, which, if the conventions are adhered
to, is equivalent to regarding the subject noun as the
specifier of the V and the object noun as its complement.
This, in turn, is equivalent to saying that both the
subject and the object noun modify the verb.

The essential point in this account of verb-dependency
theory is that linguists working independently have, for
different reasons, taken the verb to be the central constituent in a predicative sentence. This, taken in conjunction with Fillmore's remarks (Fillmore, 1968) on
the importance of the verb in semantic interpretation
and Bever's findings that the perception and decoding
of sentences appear to depend on the verbs (Bever, 1970)
signals the potential of the verb-dependency approach
as a source of insights for the psycholinguist.

IV.3 Arguments for VPs

As the adoption of the verb-dependency analysis of
surface syntactic structure entails the abandoning of
the verb phrase, it is worthwhile examining briefly the
arguments that have been advanced for this constituent.
One major argument has been that do so replaces either
a verb or a verb and an object noun and that if a constituent labelled 'verb phrase' is set up, then the
transformation introducing do so can be supplied with a
relatively simple structure index and structural change
indicating that it applies to a VP. Another argument
is that a VP constituent simplifies the coordinate deletion transformation postulated by Koutsoudas (1971).

The real problem is that if other evidence is lacking
we have to weigh the relative simplicity in the statement of transformations against the importance of expressing those aspects of syntactic structure that are captured by the verb-dependency model. I will choose the
latter because that is what I consider important, but
someone who is attracted by notational elegance above
all else will be attracted to the former. Fortunately,
there is other evidence.

We should note that the do so transformation is not as
convincing as it once seemed, since Bolinger (1970) has
put together some examples from English from which it
can be concluded that there should not be any single
entry for do so but rather separate entries for do and
so. Bolinger puts forward this suggestion because he
has seen a semantic contrast between do it and do so,
and he would like to set out the restrictions on the
choice of so as opposed to it in the lexical entry for
so. If Bolinger is correct, do replaces the main verb
and so replaces whatever follows the main verb in the
sentence, and one argument for the unitary nature of
verb phrases vanishes.

Dougherty (1971) attempts to write phrase-structure
rules that generate conjoined constituents directly in
the base. The interesting feature of this attempt, for
present purposes, is that he can generate conjoined noun
phrases and conjoined adjective phrases and even, I
suppose, conjoined prepositional phrases. The one construction he cannot generate is that consisting of conjoined verb phrases, because, of the verb phrases that
are conjoined, one might be produced by the passive
transformation and the other by a raising transformation
e.g. John was sold the house by Bill and is likely to
regret it. Of course, in a grammar without transformations this difficulty will not arise. Incidentally,
Dougherty makes it look as though he generates conjoined
VPs directly in the base. In a sense, he does, but one
of the VPs is empty and is filled at a later stage in
the derivation with the material dominated by VP in an
S conjoined with the S dominating the empty VP. Conjunction, in a word, provides evidence for other constituents
than VP but not for VP.

What are being contrasted is a view of constituent structure which sees the object noun as more closely connected
with the verb than the subject noun and a view which sees
the verb as surrounded by nouns that, with respect to
constituent structure as opposed to thematic structure,
are all of equal status. The latter view is supported
by the passive transformation, whose job after all is
to permute nouns, and it is supported by the phenomena
associated with concord and government. One function of
the latter two is to link the verb in a clause with both
its subject noun and its object noun. Notice that
although it is the object noun that is sometimes taken
to be especially closely connected with the verb it is
in fact the subject noun with which the verb agrees in
number.

In IE languages and in languages of other families
agreement and government involve morphs of person,
number and case. It has already been mentioned that the
verb can be regarded as governing the case of all its
accompanying nouns, even the nominative case of the
subject. It is also a simple matter in a transformational grammar to assign features to the verb that then
determine the number on the subject noun. From a purely
syntactic point of view this is a perfectly reasonable
solution. The reason why it is usually the subject noun
that is made to determine the number on the verb is
semantic: number is a property of objects and nouns
denote objects.

The category of person has already been discussed, but
I would like to draw attention to the treatment of person by Hockett (1958) and Bloomfield (op. cit). Hockett
treats agreement in person not as government or concord
but as a special type of syntactic linkage called
'cross-reference'. Hockett seems to be implying
that neither the verb nor the noun can be viewed as the
governing constituent, and Bloomfield seems to embrace
both the traditional and the verb-dependency approaches
when he says "A latin finite verb such as cantat 'he
(she, it) sings' includes substitutive mention of an
actor. It is joined in cross-reference with a substantive expression that makes specific mention of the
actor, as in puella cantat '(The) girl she-sings'".
(Bloomfield, op. cit. p. 193).

V Semantics and psycholinguistics

Almost as a postcript I would like to squeeze in some
comments on semantics, the perfunctory nature of this
section being justified by the existence of a general
introduction (Anderson, 1971) to the semantic theory
that I think is of great potential interest to the
psycholinguist, namely localism. Although FBG confine
themselves to the need to associate syntactic structures
with speech acts and to determine the entailments of
sentences, there are other topics that deserve to be
explored. I suggest that by allowing his intuitions
about semantic structure to be guided by surface syntax
and morphology the linguist can produce semantic analyses
that can be correlated with the analyses of the psycholinguist, particularly the psycholinguist studying the
acquisition of cognitive and linguistic skills.

The essence of localism is that all relations between
objects are regarded as being based on the notions of

location in space and movement from one point to another.
The localist does not take that proposition as self-
evident but appeals to the fact that over language after
language the morphs that express temporal locality,
causality, agency, benefactive relations, aspect, etc.
also express location or movement in space. (For aspect
see Anderson (1973), Miller (1972, 1974a, 1974b),
Bolinger (1971); for case rules see Anderson (1971),
Miller (1975); for what is essentially a localist analysis
of complement structures see Jackendoff (1976). Cf.
also the localist nature of Gruber's work (Gruber, 1965,
1967)).

Slobin (1972), following Piaget, talks of a constant
rate and order of development of semantic notions across
languages, and of cognitive development preceding linguistic development. Localism provides the linguistic
underpinning of these hypotheses and roots linguistic
and cognitive development firmly in experience, thus
reducing the role played by alleged innate knowledge
or whatever.

Let me give one brief and programmatic example of what
I mean by the statement that surface syntax and morphology provide an insight into semantic structure. Consider comparative constructions like John is bigger than
Bill. Big is a scalar adjective, and it is usual in
in discussions of such adjectives to talk about a scale
of bigness on which John and Bill are placed, John being
ahead of or above Bill on the scale. A survey of comparative constructions in various languages reveals some
interesting facts (some of which are mentioned by
Campbell and Wales (1969), who do not draw the conclusions I am going to).

In the Finno-Ugric languages the noun corresponding to
Bill has attached to it a postposition that elsewhere
expresses 'movement out of/away from'. (An exception
might seem to be Finnish itself, in which the partitive
case is said to occur in the comparative construction,
but according to the grammar the original   meaning of
the partitive and one it still has, is 'movement from').

In Tonga, a Bantu language, John is better than Bill is
expressed by a sentence that can be glossed literally
as "John be good pass Bill", i.e. "John is on the scale
of bigness and moves away from Bill on that scale".
The relevant examples, taken from Hopgood (1953),
are the following:

Nywebo mu-la-inda ku Kasompa   "You (emphatic) you - Tense
                                 - pass to Kasompa", i.e.
                                 "You passed (on your way)
                                 to Kasompa".

Ku-Masuku nku-botu kwiinda ku Chipembi   "To - Masuku it-
                                 is-good to-pass to Chipembi"
                                 i.e. "Masuku is better
                                 than Chipembi".

kwiinda = ku + inda.

In Indo-European we find some interesting and relevant
data. In classical Greek the noun corresponding to
judge in a sentence like the general is wiser than the
judge would be in the genitive case, which case can be
analysed as expressing 'movement from' as its basic
meaning. In Modern Greek the comparative construction has
apo ('from'). In Latin we find the ablative of comparison and in a descendent of Latin such as French
there is a reflex of the Latin in such phrases as
plus de vingt soldats. Old Church Slavonic has a comparative construction involving the genitive case, which
can be analysed as expressing 'movement from', and
Modern Bulgarian, which has lost its case inflexions,
uses the preposition ot ('from') in the comparative.

The essential fact is that a great many languages have
comparative constructions that contain a form - verb,
postposition, case-ending, preposition - that elsewhere
expresses 'movement from'. This suggests that the
semantic description of comparative constructions should
make reference to an object X being located on a scale
such that X possesses the property P and to X moving
away from another object Y on that scale.

This does not enable us to predict that such a structure
is appropriate to every language, for there are apparently
languages in which the comparative construction contains
an allative, i.e. a form that expresses 'movement to'.
These constructions, however, can be analysed in terms
of structures like John is big compared with Bill,
comparison being treated as based on movement towards.

VI Conclusion

In conclusion I wish to assert unequivocally that I do
not agree with FBG's statement that '... this work on
the psychological reality of syntactic structures will
ultimately be concerned...more and more with choosing
between candidate analyses where the linguistic

arguments are equivocal' (FBG, op.cit p.512). It
should be clear from the opening section of this paper
that I do not believe that any amount of experimentation
in the psychology laboratory will provide grounds for
choosing between linguistic analyses, at least not
beyond the grossest of features, such as the postulation
of noun phrases.

I do believe, however, that the linguist's analysis
should provide the psycholinguist with hints as to possible avenues of exploration and that the psycholinguist's
analysis should provide a certain amount of corroboration
for the linguist's description. For instance, noun
phrases are so obvious as constituents and turn up in
so many grammatical theories and in the descriptions of
so many languages that it would be peculiar if the
psycholinguist failed to find any evidence at all that
the linguist's noun phrase corresponded to a syntactic
unit employed by the native speakers of a language. On the
other hand, linguists would not abandon noun phrases if
psycholinguists found no evidence for them and would
doubtless ignore experimental evidence that apparently
demonstrated that native speakers of, say, English do
not employ a syntactic unit corresponding to the noun
phrase. Which goes to show, regrettably, that the link
between linguistics and the psychology of language is
rather more tenuous and indirect than has been supposed
in some quarters.

References

Anderson, J.M. (1971). The Grammar of Case. Cambridge:
    Cambridge University Press.

Anderson, J.M. (1973). An Essay Concerning Aspect. The
    Hague: Mouton.

Benveniste, E. (1946). Structure des relations de per-
    sonne dans le verbe. Bulletin de la Société de Lin-
    guistique, XLIII. Reprinted in E. Benveniste (1966)
    Problèmes de Linguistique Générale. Paris: Gallimard.

Bever, T.G. (1970). The cognitive basis for linguistic
    structures. In J. Hayes (ed.), Cognition and the
    Development of Language. New York: Wiley.

Bloomfield, L. (1935). Language. London:Allen & Unwin

Bolinger, D. (1970). The meaning of do so. Linguistic
    Inquiry, 1, 140-144.

Bolinger, D. (1971). The nominal in the progressive.
    Linguistic Inquiry, 1, 246-250.

Braine, M.D.S. (1974). On what might constitue learnable phonology. Language, 50, 270-299.

Brown, R. (1971). Psycholinguistics. New York:Free Press.

Campbell, R.N. and Wales, R. (1969). Comparative structures in English. Journal of Linguistics, 5, 215-251.

Chomsky, N.A. (1970). Remarks on nominalisation. In R.A. Jacobs and P.S. Rosenbaum (eds.) Readings in English Transformational Syntax. Waltham, Mass.: Ginn.

Chomsky, N.A. (1972). Some empirical issues in the theory of transformational grammar. In Peters (op. cit.).

Dougherty, R.C. (1971). A grammar of coordinate conjoined structures II. Language, 46, 298-339.

Elson, B., and Pickett, V. (1965). An Introduction to Morphology and Syntax. Santa Ana: Summer Institute of Linguistics.

Fillmore, C.J. (1968). Lexical entries for verbs. Foundations of Language, 4, 373-393.

Fodor, J.A., Bever, T.G., and Garrett, M.F. (1974). The Psychology of Language. New York: McGraw Hill.

Friedman, J. (1969). Directed random generation of sentences. Computational Linguistics, 12, 40-46.

Gruber, J.S. (1965). Studies in Lexical Relations. Ph. D. Thesis, M.I.T.

Gruber, J.S. (1967). Look and See. Language, 43, 937-947.

Harris, Z.S. (1946). From morpheme to utterance. Language, 22, 161-183.

Harris, Z.S. (1962). String analysis of Sentence Structure. The Hague:Mouton.

Hockett, C.F. (1958). A course in Modern Linguistics. New York: Macmillan.

Hopgood, C.R. (1953). A Practical Introduction to Tonga. London: Longmans, Green & Co.

Jackendoff, R. (1974). Introduction to the Bar-X Convention. Indiana University Linguistics Club.

Jackendoff, R. (1976). Towards an explanatory semantic representation. Linguistic Inquiry, 7, 89-150.

Koutsoudas, A. (1971). Gapping, conjunction reduction and coordinate deletion. Foundations of Language, 7, 337-386.

Lakoff, G. (1970). Irregularity in Syntax. New York: Holt Rinehart and Winston.

Lakoff, G. (1971). On generative semantics. In D.D. Steinberg and L.A. Jakobovitz (eds.) Semantics. Cambridge:Cambridge University Press.

Lyons, J. (1966). Towards a 'notional' theory of the 'parts of speech'. Journal of Linguistics, 2, 209-236.

Miller, J. (1972). Towards a generative semantic account of aspect in Russian. Journal of Linguistics,8, 217-236.

Miller, J. (1973). A note on so-called 'discovery pro- cedures'. Foundations of Language, 10, 123-139.

Miller, J. (1974a). Further remarks on aspect. Russian Linguistics, 1, 53-58.

Miller, J. (1974b). 'Future Tense' in Russian. Russian Linguistics, 1, 255-270.

Miller, J. (1975). A localist account of the dative case in Russian. In Brecht, R.D. and Chvany, C.V. (eds.) Slavic Transformational Syntax, 244-261. Ann Arbor: University of Michigan.

Peters, S. (ed.)(1972). Goals of Linguistic Theory. Englewood Cliffs, N.J.: Prentice-Hall.

Postal, P. (1972). The best theory. In Peters (op. cit).

Robinson, J.J. (1970a) Dependency structures and trans- formational rules. Language, 46, 259-285.

Robinson, J.J. (1970b). Case, category and configuration. Journal of Linguistics, 6, 57-80.

Seuren, P.A.M. (1969). Operators and Nucleus. Cambridge: Cambridge University Press.

Slobin, D.I. (1972). Cognitive prerequisites for the development of language. In D.I. Slobin and C.A. Ferguson (eds.) Studies of Child Language Development. New York: Holt, Rinehart and Winston.

SYNTAX AND SEMANTICS OF RELATIVE CLAUSES

RENATE BARTSCH

University of Amsterdam

It is a widely accepted assumption that the difference
between restrictive and non-restrictive relative clause
constructions should be expressed in their syntactic
categorical structure as well as in their semantics.
This view is held, e.g., by Rodman (1972). There, a syntactic function combines a common noun and a sentence to
form a complex common noun in the case of restrictive
relative clauses, and, in the case of non-restrictive
relative clause construction, a syntactic function combines a term (noun phrase with determiner or a proper
name) with a sentence to form a complex term. Rodman
formulates this assumption in a Montague grammar in a
way that rests on the transformational approaches to
relative clause constructions. He thus has to formulate
conditions that say under which circumstances which
elements of a sentence can be relativized, namely, within a relative clause no other element can be relativized
and in the case of non-restrictive constructions, a
further restriction has to be imposed to the effect that
coreference between the governing term and the relativized
element has to be assumed. The formulation of restrictions like these can be avoided, if we do not assume
that relative clauses are constructed from sentences by
relativizing one of their elements and moving it towards
the front of the sentence, but rather make quite a different assumption to begin with, namely: relative clauses
are not built from sentences but from predicates. With
this assumption we follow Frege's (1892) analysis of
restrictive relative clauses and not the tradition of
transformational grammar.

In what follows, this approach will be elaborated for
restrictive as well as for non-restrictive relative
clauses. It then will be discussed whether it is really
wise to give two relative clause constructions in syntax,
namely as noun modifiers and as term modifiers for
restrictive and non-restrictive interpretations, respectively, or whether we should not rather follow Thompson
(1971) in just assuming one syntactic construction, but
different strategies of interpretation, depending on
certain pragmatic factors present in the context. If
we follow that line the term-modifying construction is
the only candidate for a single syntactic structure of
relative clause constructions, because the construction
"proper name + relative clause" can never be incorporated in a noun modifying construction. This is not
possible, because a proper name is a primitive term
that cannot be analysed into a "Determiner + Common noun"
-construction. Although the rules of semantic interpretation for the term-modifying construction have some draw-
backs with regard to the elegance of their form, there
is further evidence in favour of assuming that this syntactic construction applies also to restrictive relative
clauses - evidence arising from the problem of interpretation of the so-called Bach-Peters sentences. This
problem can be solved in the proposed framework.

Proposal I

In proposal I, I assume that the relative pronoun is the
morphological realisation of two different operators,
REL_n and REL_t. REL_n maps a pair consisting of a noun and
a one-place predicate onto a noun. Of course, the noun
and the one-place predicate can be complex, i.e. the
result of a syntactic construction. In this way we have
the noun-modifying construction, generally assumed as
the restrictive relative clause construction. REL_t maps
a pair consisting of a term and a one-place predicate
onto a term. This is the term-modifying construction,
which is widely assumed to be the non-restrictive clause
construction. We have the following syntactical and
semantical rules:

Syntactical rules:

S1: Adnominal relative clause

   If α is a noun, β the relative operator REL_n
   and γ a one-place predicate, then $\beta(\alpha,\gamma)$ is a
   noun; $\beta(\gamma)$ is an adnominal relative clause,
   and $\beta(\alpha,\gamma) = (\beta(\alpha))(\gamma)$.

S2: Adterminal relative clause

If α is a term, β the relative operator REL_t and γ a one-place predicate, then $\beta(\alpha,\gamma)$ is a term; $\beta(\gamma)$ is an adterminal relative clause, and $\beta(\alpha,\gamma) = (\beta(\gamma))(\alpha)$.

Remark: By means of S1 and S2 we get the categorial structures of relative clause constructions. To get the surface clauses, $\beta \cdot REL$ has to be realized as a relative pronoun agreeing in gender and number with the governing noun or term, respectively, and according to the morphological case realisation of the λ-bound position of the one-place verb γ. The serialisation of the relative clause $\beta(\gamma)$ is according to the general rules of subordinate clause serialisation, i.e. in German the natural

Operator (Operand) $\rightarrow$ Operator + Operand.

Categorical constructions, by convention, are always such that an operator stands in front of its operand. They are thus not to be understood as linearly ordered but rather as hierarchically ordered, with a unidirectional representation of the hierarchies when we write them down on paper (compare Bartsch 1976).

Semantical rules:

T1: Adnominal relative clauses (restrictive)

If $\lambda x\alpha'(x)$ and $\lambda x\gamma'(x)$ are the translations of α and γ from S1, then the translation β' of β is a mapping with the following characterisation:

$\beta'(\lambda x\alpha'(x), \lambda x\gamma'(x)) = \lambda x(\alpha'(x) \ \& \ \gamma'(x))$.

The translation of the adnominal relative clause $\beta(\gamma)$ then is $(\beta'(\lambda x\gamma'(x))))$ with:
$(\beta'(\lambda x\gamma'(x)))(\lambda x\alpha'(x)) = \lambda x(\alpha'(x) \ \& \ \gamma'(x))$.

T2.1: Adterminal relative clauses (non-restrictive)

If $\lambda P P(a)$ is the translation of the proper name α and $\lambda x\gamma'(x)$ the translation of γ from S2, then the translation β' of β is a mapping with the following characterisation: $\beta'(\lambda P P(a), \lambda x\gamma'(x)) = \lambda P(\gamma'(a) \ \& \ P(a))$.

The translation of the adterminal relative clause $\beta(\gamma)$ then is $(\beta'(\lambda x\gamma'(x)))$ with:
$(\beta'(\lambda x\gamma'(x)))(\lambda P P(a)) = \lambda P(\gamma'(a) \ \& \ P(a))$.

Since we also have the non-restrictive constructions with the specific indefinite article, T2 needs to be extended to cover these cases. But before doing so we need to describe how definite descriptions are interpreted. As long as we simply adhere to the Russellian analysis of

definite descriptions and translate (basically following
Montague's "Proper Treatment" (Montague 1974) and Thoma-
son (1974) and Ballmer (1975)) the by $\lambda Q \lambda P(\exists x)(Q(x)$ &
$(\forall y)(Q(y) \rightarrow x = y)$ & $P(x))$, we could incorporate into T2
the following extension:

T2.2: If $\alpha = ({}^{\mu}{}_{D<def>}({}^{\kappa}{}_N){}_{T<def>})$, then $\beta'(\lambda P(\exists x)(\kappa'(x)$

     & $(\forall y)(\kappa'(y) \rightarrow x = y)$ & $P(x)), \lambda x \gamma'(x)) = \lambda P(\exists x)$

     $(\kappa'(x)$ & $(\forall)(\kappa'(y) \rightarrow x = y)$ & $\gamma'(x)$ & $P(x))$.

But even this is unnecessary. We could just treat the
non-restrictive relative clauses with definite terms by
S1 and T1. No mistake can arise this way: if $\kappa'(x)$ is
satisfied for exactly one individual then, if also $\gamma'(x)$
is satisfied for the same individual following T2', then
the definite description with $\lambda x(\kappa'(x)$ & $\gamma'(x))$, constructed according to S1/T1, is satisfied. The difference
between restrictive and non-restrictive would be recog-
nisable by the context, but would have no effect on the
semantics. That an analysis, solely based on Russell's
theory of definite descriptions, is inadequate as an
analysis of the use of the definite article in natural
language has long been recognised and, actually, never
was intended by Russell to be adequate in this way.
Russell's theory does not cover contextual use of the
definite article and has to be modified quite a bit to
be applicable even for the semantics of the definite
article in contextual use. We want to be able to give
the semantics of examples like the following:

A man and a woman sat at a table in a restaurant, having
a drink.
Suddenly, a man appeared at the entrance. The man, who
briefly surveyed the scene, took his gun and shot the
woman.

We want to interpret the sentence such, that the man
refers back to the man who entered, and not to the man
mentioned first. On the other hand, in the following
text, we want to refer to the man mentioned first:

A man and a woman sat at a table in a restaurant, having
a drink. Suddenly, a man appeared at the entrance. The
man who sat at the table, took his gun and shot at the
stranger.

Obviously, in the second example, the relative clause
plays a role in identifying the man meant out of the
two mentioned, and in the first example it does not.
We, therefore, need an analysis of definite terms that
allows us to account for these facts. To do this, we

need a theory of interpretation that, among other things, provides the connection between a definite term and a previously used indefinite term. The interpretation is based on two kinds of reference act formulated by means of reference instructions. This concept of reference instruction was introduced in Ballmer (1972). The interpretation of an indefinite term "a(n)α" in a non-negative context is accompanied by an instruction to perform an act of referent-introduction and the interpretation of a definite term "the α" relies on a reference act performed according to an instruction to search for a suitable referent. The referent-introduction and the referent search is done in an universe of discourse constructed in the course of interpretation of a text (compare also Ballmer, 1975). This universe consists partly of representations of entities which the speaker assumes to be commonly known to the partners of the discourse, and partly   of representations of entities introduced by the discourse. This realm of entity-representations is structured for each partner according to his knowledge about the entities represented and gets further changed by the discourse. We can look at this realm of entity-representations with its structure as a model for the set of propositions assumed to be true by the discourse partners. The discourse partners deviate from each other in their models but in order for the discourse to be felicitous,a correspondence of parts of the realm of entities and their structural properties (i.e. properties and relations satisfied by them and represented in the models) is necessary. For the purpose of reference introduction we need to narrow down the concept of a discourse universe to the concept of a <u>relevant universe of discourse</u>.

A relevant universe of discourse (for referent-introduction) is determined as follows: if there is an intensional verb to which "a(n)α" is applied

1.   the relevant universe of discourse is the universe of those worlds that are specified by the intensional verb and its subject term, if the term "a(n)α" is applied with regard to a place of the verb marked as intensional.

Otherwise (i.e. in all <u>de re</u> constructions) we have (2).

2.   the relevant universe of discourse is the universe of the speaker-hearer discourse, i.e. the basic universe of discourse of assumed common speaker-hearer knowledge, including information from the previous text.

The indefinite article can be interpreted in a specific or non-specific reading with regard to the speaker's way of referring (cf. Bartsch, 1976). The non-specific reading is primary. The specific reading is only possible if indicated by the context.

The indefinite article:

specific: a(n) translates into $\lambda Q \lambda P(Q(\nu)$ & $P(\nu))$, with ν being a newly introduced constant.

non-specific: a(n) translates into $\lambda Q \lambda P(\exists x)(Q(x)$ & $P(x))$.

Referent-introduction:

R1: If a(n) is not in the scope of a negation or preceded by a term of the form "every..." or "no..." then add, after the translation that specifies the variables Q and P, a new individual name with the specifications of Q and P to the relevant universe of discourse.

Remark: If the new name is ν then we have to add ν to the relevant universe of discourse with Spec $[Q]$ (ν) and Spec $[P]$ (ν), where Spec $[Q]$ and Spec $[P]$ are the specifications of Q and P, respectively.

The definite article (specific):

the translates into $\lambda Q \lambda P(Q(\nu)$ & $P(\nu))$ with ν determined according the following reference instruction:

Reference instruction:
for: the α.

1. Take the last individual name introduced or used with the interpretation of the last sentence such that the property α', i.e. the predicate specifying Q, is true of its referent. If there is no such name, do (2).

2. Take the second last individual name introduced or used with the interpretation of the second last sentence such that the property α' is true of its referent. If there is no such name, do (3).

3. Decide whether it holds in the knowledge induced by previous text information and presupposed speaker-hearer knowledge that: $(\exists x)(\alpha'(x)$ & $(\forall y)(\alpha'(y) \rightarrow x = y))$. If that is the case use the introduced individual name ν with $\alpha'(\nu)$, or, if there is no such name, introduce a new name ν with $\alpha'(\nu)$. If it is not the case, do (4).

4. Decide whether in the presupposed speaker-hearer
    knowledge that does not belong to the previous text
    information it holds that $(\exists x)(\alpha'(x)$ & $(\forall y)(\alpha'(y)$
    $\rightarrow x = y))$. If that is the case use the introduced
    individual name y with $\alpha'(y)$, or, if there is no
    such name, introduce a new name y with $\alpha'(y)$.

There exists the following restriction of use for the
definite articles. The α can only be used if
$(\exists x)(\alpha'(x)$ & $(\forall y)(\alpha'(y) \rightarrow x = y))$ holds in one of the
realms of information described in the strategy (1) -
(4) above. The restriction gives rise to the assumption
of $(\exists x)(\alpha'(x)$ & $(\forall y)(\alpha'(y) \rightarrow x = y))$ for one of the
realms of information described in the strategy (1) -
(4).

After this excursus on indefinite and definite articles
we are able to formulate the interpretation of non-
restrictive relative clause constructions for definite
terms and specific indefinite terms in the same way as
for proper names. We add to T2:

T2.2:   If the α or a α translates into $\lambda P(\alpha'(y)$ & $P(v))$
and γ into $\overline{\lambda x \gamma}'(x)$, then β translates into the mapping β'
with: $\beta'(\lambda P(\alpha'(v)$ & $P(v)), \lambda x \gamma'(x)) = \lambda P(\alpha'(y)$ & $\gamma'(v)$ &
$P(v))$. And, accordingly, the relative clause $\beta(\gamma)$ trans-
lates into $(\beta'(\lambda x \gamma'(x)))$ with: $(\beta'(\lambda x \gamma'(x)))(\lambda P(\alpha'(v)$
& $P(v))) = \lambda P(\alpha'(v)$ & $\gamma'(v)$ & $P(v))$.

The restrictive relative clauses are taken care of by
S1/T1, by which a complex $\lambda x \beta'(x)$ is constructed that
goes into the syntax and semantics of term constructions,
following rules S_T and T_T.

S_T: If α is a determiner and β is a noun, then $\alpha(\beta)$ is
        a term.

T_T: If α' is the translation of α and β' the translation
        of β from S_T, then the translation of $\alpha(\beta)$ is:
        $\alpha'(\beta')$.

The translations of non-specific determiners, thus far,
are formulated, following Montague (1974) and Ballmer
(1975):

   $\underline{a}$: $\lambda Q \lambda P(\exists x)(Q(x)$ & $P(x))$

   $\underline{every}$: $\lambda Q \lambda P(\forall x))Q(x) \rightarrow P(x))$

   $\underline{no}$: $\lambda Q \lambda P(\forall x)(Q(x) \rightarrow -P(x))$

and, according to the Russellian analysis:

the : $\lambda Q \lambda P (\exists x)(Q(x) \& (\forall y)(Q(y) \rightarrow x = y) \& P(x))$.

This treatment of determiners will be revised in the course of this paper.

To be able to handle some interesting types of relative clause construction we need to incorporate an interpretation strategy for pronouns. The interesting cases are constructions that cannot be handled in a Montague Grammar so far. For example, the sentence (1)

(1) Every man who buys a car insures it.

cannot be handled in Montague's PTQ with coreference for a car and it, because it remains unbound by the quantifier introduced with the translation of a car. Another interesting group of constructions that have not been handled successfully in a Montague Grammar are the so-called Bach-Peters sentences. The solution by introduction of double term constructions, as proposed by Rohrer (1975), has as a drawback that the method needs to be generalised if we want to have a unified way of interpreting terms. It is fairly ad hoc if we treat term interpretation in sentences that are not Bach-Peters sentences differently than in Bach-Peters sentences. And since problems are imaginable where we need triple term introduction or even n-tuple term introduction,we end up at a generalisation of Rohrer's method that is very costly to handle, if at all. Furthermore, not all the interesting types of Bach-Peters sentences have been analysed in Rohrer's framework. Before getting into details of the problem of Bach-Peters sentences,a treatment of pronouns has to be sketched that makes possible pronominal coreference with terms outside and inside a sentence, and that, specifically, serves to handle the interpretation of sentences like sentence (1) and the so-called Bach-Peters sentences. Like the interpretation strategy for definite terms the interpretation strategy for pronouns has the status of a tentative proposal and certainly needs revision in the light of special empirical studies on the subject matter. But this provisional account of pronouns should at least be sufficient to handle the following illuminating linguistic facts.

A provisional treatment of pronouns

The sketch of pronominal reference given in this paper will at least take care of the following linguistic facts, which are contextual properties of pronouns.

(A) Within relative clause constructions, for every
term a pronoun can occur   in the following text
that is coreferential with it.

(1) A man who loves the woman who loves him ...
     No man who loves the woman who loves him ...
 Every man who loves the woman who loves him ...
    The man who loves the woman who loves him ...
     John who loves the woman who loves him ...

But it seems doubtful whether terms "every ..." or
"no ..." in subordinate relative clauses can have a
coreferential pronoun in superordinate clauses, e.g.,

(2) A man who loves every woman who loves him dates
     her.

(B) Within conditional and adverbial clause constructions coreference between a term in the subordinate
clause and a pronoun in the main clause is possible
only for terms with the indefinite or definite article
and proper names in preceding subordinate clauses, and
the indefinite term may not be in the scope of a
negation operator or follow a term of the form "every..."
or "no...".

(2)   * If every participant is invited to the party he
         has to buy some liquor.

     * Because every participant is invited to the party
       he has to buy some liquor.

     * Although every participant is invited to the
       party he has to buy some liquor.

     * If no participant is invited to the party he has
       to buy some liquor.

     * Because it is not the case that a participant is
       invited to the party he has to buy some liquor.

     * If every man dates a woman, she is happy.

But: If a participant is invited to the party he has
       to buy some liquor.

       If a participant is not invited to the party he
       may come nevertheless.

       Because John is invited to the party he has to
       buy some liquor.

(C) If the full term is in the main clause and the pronoun in the subordinate clause in conditional or adverbial construction coreference is possible for all terms,

regardless of whether the subordinate clause precedes
or follows the main clause.

(3)   Every participant has to buy some liquor if he is
      invited.

      No participant needs to buy liquor if he is invited.

      No participant needs to buy liquor because he is
      invited.

      Every participant has to buy liquor although he is
      invited.

      Although he is invited every participant has to
      buy liquor.

(D)   Coreference extending over the sentence boundary
in parallel situations with those described under (B):
Coreference is possible between pronouns and proper names,
pronouns and definite terms, and between pronouns and
indefinite terms if these are not in the scope of a
negation operator.

(4)   * Every participant is invited. He has to buy
         some liquor.

      * No participant is invited. He has to buy some
         liquor.

      * It is not the case that a professor is invited
         to the party. He has to buy the liquor.

         A professor is invited to the party. He has to
         buy the liquor.

         A professor is not invited to the party. There-
         fore, he does not need to buy liquor.

         The professor is invited to the party. He has
         to buy some liquor.

         John is invited to the party. He has to buy some
         some liquor.

      * Every man loves a woman. Therefore she is happy.
But specific:

         Every man loves a woman. She is mother Mary.

(E)   If the preceding indefinite term appears under the
scope of an intensional operator or fills in an intensional position of a verb, coreference is restricted to
extensional readings if the pronoun is not under the
scope of the same intensional operator or another

intensional operator such that the set of possible
worlds specified by the first is included in those of
the second.

(5)   John seeks <u>a ball</u>. It is red.   (extensional
                                                reading)

     * John seeks <u>a wife</u>. <u>She</u> is tall   (intensional
                                                reading)

But:

        John seeks <u>a wife</u>. <u>She</u> must be tall.

        John believes there is <u>a red-eyed unicorn</u>. He
        seeks <u>it</u>.

(F)   Furthermore a theory of pronominal reference must
be such that the so-called Bach-Peters sentences can be
handled.

(6) EVERY MAN who finds <u>a car</u> that HE likes buys <u>it</u>.

     A MAN who finds <u>the woman</u> who loves HIM dates <u>her</u>.

     THE MAN who loves <u>her</u> dates <u>the woman</u> who loves HIM.

The pronouns in the sentences under (6) have to be interpreted in such a way that they are coreferential with
the corresponding term. But, furthermore, the interpretation procedure must also permit the interpretation
with no coreference within the sentence, but with coreference with a term outside the sentence as is
required in the texts under (7).

(7) John has <u>a fine old car</u>. Every man who wants the
     car he likes will buy <u>it</u>.

     John has <u>a nice girlfriend</u>. Every man who chases
     a woman who he likes dates <u>her</u>.

To account for the linguistic facts (A)-(F), as far as
pronominal reference is concerned, we develop an interpretation procedure that does the following: Pronouns
in subordinate clauses get translated by means of variables that have to be bound by the same quantifier that
is used in the translation of the corresponding terms
in the main clause. If the term is a proper name the
pronoun is translated by means of a constant. With
specific indefinite and definite terms both procedures
are possible, the procedure <u>via</u> a variable or <u>via</u> a
constant.

Pronouns in the main clauses behave differently as to
whether they are in relative clause constructions or in
adverbial and conditional clause constructions. In
relative clause constructions they may be coreferential
with an indefinite term in a relative clause, and then
translate <u>via</u> the variable or constant used in the
translation of that term. In adverbial and conditional
clause constructions they can only be translated by a
constant used in the translation of the preceding term.
In texts, pronouns with coreference that extends over
the sentence boundary can only be translated by means
of constants used or introduced with the translation of
the preceding term.

To be able to formulate a translation procedure, which
is taken to be the interpretation procedure of natural
language sentences as far as the translation into an
interpreted logical language goes, we first have to
modify our previous procedure of translation for terms.
This is done by using the method used in logical deduc-
tions, which is known as existential instantiation,
universal instantiation, existential generalisation and
universal generalisation. To avoid difficulties, we introduce a method of blocking of variables which says that
a variable blocked with the translation of a term may
not be used in the translation of other terms in the
course of interpretation of a sentence (clause construction), except in the translation of those pronouns that
are coreferential with the term for which the variable
was blocked.

The general interpretation procedure, or more precisely,
"translation procedure", IN, for terms that translate
with a quantifier consists   of two parts: an instantiating translation and a generalisation. Pronominal
reference, as far as it happens via variables, occurs
within the instantiating procedure and then takes part
in the generalisation and thus is captured inside the
quantifier scope. By this means, the problem of dang-
ling variables in Montague grammar is avoided.

<u>IN</u>: In translating non-specific terms of the form
    "<u>a(n)</u>...", "<u>every</u>...". "<u>no</u>...", "<u>the</u>..." block a
    new variable ν with the translation (<u>instantiating</u>
    <u>translation</u>):

    "<u>a(n)</u>   ...": $\lambda Q \lambda P(Q(\nu) \ \& \ P(\nu))$
    "<u>every</u> ...": $\lambda Q \lambda P(Q(\nu) \rightarrow P(\nu))$
    "<u>no</u>   ...": $\lambda Q \lambda P(Q(\nu) \rightarrow -P(\nu))$
    "<u>the</u>   ...": [2] $\lambda Q \lambda P(Q(\nu) \ \& \ (\forall y)(Q(y) \rightarrow y = \nu) \ \& \ P(\nu))$.

After finishing the translation of a sentence on the
basis of this preliminary translation of terms the
final translation is achieved by existential generalisation and universal generalisation according to the
following rule. The order in which these operations
are applied follows the order of the terms within the
sentence beginning from the inside, i.e. first we have
generalisation regarding the innermost term (i.e. the
last term) and last we have generalisation regarding
the outermost term (i.e. the first term).

Generalising interpretation:

(1)
Rule of the order of application:
The order of application of quantifiers reflects the
linear order of the terms in the sentence translated.

(2.1)
If a term A is preceded by an uneven number of head
terms of the form "every..." or "no..." in embeddings
of relative clause constructions of which the instantiation translation has been given, we get for A the
following generalisation:[3] If A = "a(n)..." or A =
"the..." its generalisation is achieved by binding the
respective variables used in the instantiating translation of the term A by the universal quantifier ($\forall$.).

If A = "every..." or A = "no..." its generalisation is
achieved by binding the respective variables used in
the instantiating translation of the terms A by the
existential quantifier ($\exists$.).

(2.2)
Otherwise, we get the following generalisation:

If A = "a(n)..." or A = "the..." its generalisation is
achieved by binding the respective variables used in
the instantiating translation of the term A by the
existential quantifier ($\exists$.).

If A = "every..." or A = "no..." its generalisation is
achieved by binding the respective variables used in
the instantiating translation of the term A by the
universal quantifier ($\forall$.).

(3)
After the generalisation is completed, variable blocking
is cancelled.

The procedure IN is based on the idea that surface order
reflects the hierarchical order of deeper structure.
It is a kind of reverse "global constraint", but in a
more severe form than presented in Lakoff (1971). It pre-
vents the sentence <u>Every man loves a woman</u> from being
interpreted as ambiguous between "For every man there
is a woman such that he loves her" and "there is a
woman such that every man loves her". Only the first
interpretation is a direct interpretation of the sentence discussed. The second interpretation is stronger
then the first. It can be inferred from the first in
conjunction with sufficient contextual information as
in "Every man loves a woman. She is mother Mary". Here
we need to assume the specific interpretation for "a
woman", because otherwise we would not have an antecedent term for the pronoun <u>she</u>. From the translation
of this text we can infer the translation of "There is
a woman such that every man loves her". But this does
not mean that "every man loves a woman" is ambiguous
in the direct way assumed by Lakoff (1971) and, even,
Montague (1974).[4] If we want to express the second
reading, assumed by these authors, without a specifying
context we have to use passive voice: "A woman is loved
by every man" or the rather clumsy sentence "there is
a woman such that every man loves her". Note, that in
using the described translation procedure we avoid the
undesirable multiplicity of readings, that arises by
Montague's rule S14/T14 in Montague (1974).

Restriction on I<u>N</u>:

In the translation of subordinate clauses other than
relative clauses, generalisation has to be completed
for all non-pronominal terms before the clauses get combined with their main clauses.

The restriction excludes the unacceptable cases under
B and D. The acceptable ones, like <u>If A MAN buys A CAR</u>
<u>HE insures IT</u>, are taken care of by using the new con-
stant introduced with the translation of <u>a car</u> in the
translation of the pronoun. The rules of pronominal
reference are formulated so   that they permit that.[5]
Of course, a satisfactory theory of interpretation can
only be achieved if the study of the interpretation of
subordinate clause constructions is completed. So far,
the restriction on (IN) is just an <u>ad hoc</u> measure to
avoid obvious mistakes.

These preliminaries suffice to establish a theory of
pronominal reference that can account for the linguistic

facts mentioned above. We restrict ourselves to non-reflexive pronouns.

The interpretation of non-reflexive pronouns: he, she, it (him, her are related to he, she, respectively, by a morphological rule).

IP: REFERENCE-INSTRUCTION:

1. (a) Search within the clause construction for the nearest preceding term with gender-number agreement with the pronoun such that term and the pronoun do not bind places of the same verb.[6]
(b) If there is no such term search for the nearest following term with gender-number agreement in the superordinate clause within the clause construction such that that term and the pronoun do not bind places of the same verb.
(c) If there is no such term search for the second nearest preceding term with gender-number agreement, and so on.
(d) If there is no such term in the clause construction search for the next preceding term with gender-number agreement in the preceding text.

2. If a term has been found according to (1) then translate the pronoun by $\lambda^{PP}(\nu)$ with ν being the variable blocked in the course of translation of the term or with ν being the constant used or being introduced with the translation of the term.[7] If there is no such variable or constant follow procedure (1) for the next subroutine among (a), (b), (c), (d).

3. If the interpretation achieved via the translation according to (1) and (2) is improbable or destroys text coherence - then follow procedure (1) and (2) for next subroutine among (a), (b), (c), (d).

The procedure formulated in IP very likely will turn out to be deficient in a further study of pronominal patterns. But it seems to suffice for the present purpose. We need two other syntactic rules to be able to construct sentences, term construction (S3) and sentence construction (S4). The surface structures are related to the categorial structures, constructed according to S3 and S4, by realisation rules, including: serialisation rules, lexicalisation rules and morphological and morphophonemic rules. These will be neglected in the present fragment.

S3: If α if a D (determiner) and β is an N (noun), then $\alpha(\beta)$ is a T (term).
T3: If α' is the translation of α as a D and β' is the translation of β as an N, then $\alpha'(\beta')$ is the trans. of $\alpha(\beta)$.

Pronouns are a subcategory of the category of terms.
Their syntactic role is therefore described by the
syntactic rules for terms. In our present fragment we
shall use rule S3/T3 only.[8]

S4: If β is a V^n (n-place verb) with the set of term-
places K, $i \in K$, and α a T (term) then $(\alpha, i)(\beta)$ is a
V^{n-1} with the set of term places $K-\{i\}$. For this
we write $(\alpha_{T<i>}(\beta_{V^n})_{V^{n-1}})$.

The corresponding translation rule is:

T4: If α' is the translation of α as a T and $\lambda x_1, \ldots,$
$x_n \, \beta'(x_1, \ldots, x_n)$ is the translation of β as a V^n,
then the translation of $(\alpha, i)(\beta)$ is:
$\lambda x_1, \ldots, x_{i-1} \, x_{i+1}, \ldots, x_n(\alpha'(\lambda x_i(\beta'(x_1, \ldots, x_n))))$.

Note that this rule corresponds to rule S14/T14 in
Montague 1974c, which is the basis of construction for
de re readings. In the fragment presented in this
paper we only get de re readings; but there exists a
modification of this fragment with regard to S4 and T4
in part III of Bartsch/Ehrich/Lenerz (1976), which accounts
for de dicto readings, too.

Before discussing Bach-Peters sentences in detail, we
should be aware of the following property of the procedure for pronominal interpretation presented here:
This procedure relies on the analysis of the head term
of a relative clause construction as a single constituent. That is, it assumes that analysis (1) is taken
and not (2) as is usually done for restrictive clauses.

(1)   (2)

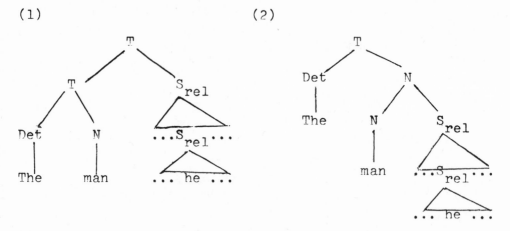

The reason for this is that a pronoun in a relative

clause that is embedded in a higher relative clause
may be translated by means of the variable used in the
translation of the determiner of the head noun (of the
higher relative clause) which by number and gender
agreement establishes that the head term is a candidate
for a preceding coreferential term. If we assume The
man not to be a constituent, as in (2), we need to
define a special surface structure relation between the
determiner and the head noun that has no other function
than to serve as a link between the establishment of
number-gender agreement between head noun and pronoun
on the one hand   , and the variable blocked in the
translation of the determiner that goes with the head
noun and that is needed for the translation of the
pronoun on the other. Since it does not seem advisable
to define, in grammar, two different kinds of constituents, those according to categorical structure and
those according to the needs of extablishing agreement
and coreference, we suggest that   structure (1), i.e.
term-modification, should be used as a single syntactic
structure for all relative clause constructions,
whether restrictive or non-restrictive.

There are also independent reasons for assuming a single
categorical structure for all relative clause constructions, as also suggested by Thompson (1971):

1)   There is only one surface structure for relative
     clause constructions, as far as word order and
     morphology are concerned.

2)   Because of (1), the question arises by means of what
     information one could decide, before entering the
     interpretation process, which categorical structure
     to take, the term-modifying one or the noun-
     modifying one.

3)   The problem mentioned in (2) can be decided for
     head determiners a, every, no, in favour of always
     taking the noun-modifying construction; and for
     proper names as head terms the only possible one
     is the term-modifying construction. But for head
     terms with a this would not be quite correct because
     non-restrictive relative clause constructions are
     possible, e.g., "a priest, who of course never
     cheats, shall be the accountant". And also for
     constructions with the, we would have to rely on
     intonation in deciding which categorical structure
     to assume. This is not a counter argument per se,
     but there are reasons to link information supplied
     by sentence intonation with contextual features

such as presuppositions and focus (compare Bartsch,
1976). With regard to the restrictive/non-
restrictive distinction, intonation can give information about whether the head term or the whole
term including the relative clause will be used
in the reference instruction that has to be followed
to find the individual a definite term refers to.
This means that a minor modification of our rule
of reference instruction for definite terms is in
order.

4) If we decide to favour a single syntactic relative
clause construction for all relative clauses we
cannot opt for the one in which the relative clause
is a noun-modifier, i.e. according to structure (2),
because relative clauses may occur with proper names
or pronouns as head terms. Thus, only structure
(1), i.e. the relative clause as term-modifier, can
be a candidate for single syntactic relative clause
construction.

Although our treatment of Bach-Peters sentences can
also be used with respect to structures with relative
clauses as noun-modifiers[9] we decide to abandon this
construction in favour of the term-modifier construction
as the only syntactic relative clause construction.
This decision is based, first, on the desirability to
have a head term as a constituent for our retrieval
of the variable or constant to be used in the translation of pronouns, and second, on the independent considerations mentioned under (1)-(4). The treatment of
all relative clause constructions as term modifying
ones amounts to cancelling S1/T1 and to supplying a new
formulation of T2.2.

T2.2: If $(\delta_D(\alpha_N)_T)$ translates into $\lambda P(\ldots\alpha'(\nu)\ldots)$,
       with ν as a variable or constant and γ translates
       into $\lambda_x \gamma'(x)$, then the relative clause $\beta(\gamma)$
       translates into a mapping to the effect that
       $\alpha'(.)$ is replaced by $\alpha'(.)$ & $\gamma'(.)$ in all its
       occurrences. I.e. $\beta'(\lambda P(\ldots\alpha'(\nu)\ldots), \lambda_x\gamma'(x))$
       $= \lambda P(\ldots \alpha'(\nu)$ & $\gamma'(\nu) \ldots)$, and, accordingly,
       $(\beta'(\lambda_x\gamma'(x)))(\lambda P(\ldots\alpha'(\nu)\ldots)) = \lambda P(\ldots\alpha'(\nu)$ &
       $\gamma'(\nu) \ldots)$.

Note, that if ν is a constant the reading is in fact
the non-restrictive one; if ν is a variable the reading
may be restrictive, depending on whether there are, in
fact, more individuals with the property α' than with
the properties α' and γ'. If δ is the definite determiner then it may be that the uniqueness condition is

not fulfilled with α' but rather with α' and γ' in conjunction. Note, that $\alpha(.)$ has also to be replaced by $\alpha(.)$ & $\gamma(.)$ in the uniqueness condition. It is possible that the translation of "the α" has no designation, namely if the uniqueness condition $(\forall y\ (\alpha'(y) \to y = y)$ is not fulfilled, but "the α that γ" has a designation, if the definite description with a restrictive relative clause is used correctly.

Examples

(1) <u>Every man who buys a car insures it</u>.

This sentence cannot be interpreted in Montague's (1974) fragment with coreference between <u>it</u> and <u>a car</u>. In the present fragment both interpretations are possible, with and without coreference within the sentence. In this example the interpretation with coreference is the most probable.

<u>A</u>: <u>Instantiating translation</u>:

<u>every man</u>:	$\lambda P(\underline{man}'(x) \to P(x))$; x blocked! [T3, IN]
<u>a car</u>:	$\lambda P(\underline{car}'(y)$ & $P(y))$;y blocked! [T3, IN]
<u>it</u>:	$\lambda PP(y)$ [IP]
<u>buys a car</u>:	$\lambda v(car'(y)$ & $\underline{buy}'(v,y))$ [T4, lexicon]
<u>every man who buys a car</u>:	$\lambda P(\underline{man}'(x)$ & $car(y)$ & $\underline{buy}'(x,y) \to P(x))$ [T2.2]
<u>insures it</u>:	λv <u>insure</u>$'(v,y)$ [T4]
<u>every man who buys a car insures it</u>:	$man'(x)$ & $\underline{car}'(y)$ & $\underline{buy}'(x,y) \to \underline{insure}'(x,\underline{y})$

<u>B</u>: <u>Generalising translation (complete translation)</u>:

$(\forall x)(\forall y)(\underline{man}'(x)$ & $\underline{car}'$ (y) & $\underline{buy}'(x,y) \to \underline{insure}'(x,y))$

Note that <u>a car</u> was preceded by one head term of the form "every..." and thus got generalised by $(\forall.)$.

(2) <u>Every man who finds a woman who loves him dates her</u>.

<u>A</u>: <u>Instantiating translation</u>

<u>every man</u>:	$\lambda P(\underline{man}'(x) \to P(x))$; x blocked. [T3,IN]

a woman:	$\lambda P(\underline{woman}'(y) \,\&\, P(y))$; y blocked	$[T3,IN]$
him:	$\lambda PP(x)$	$[IP]$
her:	$\lambda PP(y)$	$[IP]$
love(s):	$\lambda vz \,\underline{love}'(v,z)$	$[lexicon]$
love(s) him:	$\lambda v(\lambda PP(x))(\lambda z \,\underline{love}'(v,z))$ $= \lambda v(\lambda z \,\underline{love}'(v,z))(x))$ $= \lambda v \,\underline{love}'(v,x)$	$[T4]$
a woman who loves him:	$(REL(\lambda v \,\underline{love}'(v,x)))(\lambda P(\underline{woman}'(y) \,\&\, P(y))$ $= \lambda P(\underline{woman}'(y) \,\&\, \underline{love}'(y,x) \,\&\, P(y))$ $[T2.2]$	
find(s):	$\lambda vz \,\underline{find}'(v,z)$	$[lexicon]$
finds a woman who loves him:	$\lambda v((\lambda P(\underline{woman}'(y) \,\&\, \underline{loves}'(y,x) \,\&$ $P(y))(\lambda z \,\underline{find}'(v,z)) = \lambda v((\underline{woman}'(y) \,\&$ $\underline{love}'(y,x) \,\&\, \underline{find}'(v,y))$ $[T4]$	
every man who finds a woman who loves him:	$(REL(\lambda v(\underline{woman}'(y) \,\&\, \underline{love}'(y,x) \,\&$ $[T2.2]$ $\underline{find}'(v,y))(\lambda P(\underline{man}'(x) \rightarrow P(x)))$ $= \lambda P(\underline{man}'(x) \,\&\, \underline{woman}'(y) \,\&\, \underline{love}'(y,x) \,\&$ $\underline{find}'(x,y) \rightarrow P(x))$	
date(s):	$\lambda vz \,\underline{date}'(v,z)$	$[lexicon]$
date(s) her:	$\lambda v(\lambda PP(y))(\lambda z \,\underline{date}'(v,z))$ $= \lambda v \,\underline{date}'(v,y)$	$[T4]$
every man who finds a woman who loves him dates her:	$(\lambda P(\underline{man}'(x) \,\&\, \underline{woman}'(y) \,\&\, \underline{love}'(y,x)$ $[T4]$ $\&\, \underline{find}'(x,y) \rightarrow P(x)))(\lambda v \,\underline{date}'(v,y))$ $= \underline{man}'(x) \,\&\, \underline{woman}'(y) \,\&\, \underline{love}'(y,x)$ $\&\, \underline{find}'(x,y) \rightarrow \underline{date}'(x,y)$	

B: Generalising translation (complete translation):

1. regarding a woman (with preceding "every ..."):

$(\forall y)(\underline{man}'(x) \,\&\, \underline{woman}'(y) \,\&\, \underline{love}'(y,x) \,\&$
$\underline{find}'(x,y) \rightarrow \underline{date}'(x,y))$

2. regarding every man:

$(\forall x)(\forall y)(\underline{man}'(x) \,\&\, \underline{woman}'(y) \,\&\, \underline{love}'(y,x) \,\&$
$\underline{find}'(x,y) \rightarrow \underline{date}'(x,y))$ $[IN, (1)-(3)]$

With this, the translation is completed and the interpretation can be provided according to model theory for predicate logic.

Compare the following text with the interpretation of (2).

(3) John knows a nice girl. Every man who loves every
     woman that pays attention to him dates her.

Here her is not interpreted as coreferential within
the sentence but rather as coreferential with a nice
girl from the preceding text. Otherwise we would get
an incoherent text. We obtain as the translation of
the first sentence $(\exists x)(know'(h,x)$ & $girl'(x)$ & $nice'(x))$
with referent introduction: a with $know'(h,a)$ & $girl'(a)$
& $nice'(a)$. Then the pronoun her of the second sentence
can be translated by $\lambda PP(a)$. Thus we get as the appropriate interpretation: $(\forall x)(\exists y)(man'(x)$ & $(woman'(y)$ &
pays attention to$'(y,x) \rightarrow love'(x,y)) \rightarrow date(x,a))$. This
is equivalent to $(\forall x)(\forall y)(man'(x)$ & $(woman'(y)$ & pays
attention to$'(y,x)) \rightarrow love'(x,y)) \rightarrow date'(x,a))$ and to
$(\forall x)((man'(x)$ & $(\forall y)(woman'(y)$ & pays attention to$'$
$(y,x) \rightarrow love'(x,y))) \rightarrow date'(x,a))$

The following example has a pronoun preceding its coreferring term in the main clause. Rule (IP) takes
care of this.

(4) A man who finds her dates a woman who loves him.

Following the rules of our fragment we get the translation:

$(\exists x)(\exists y)(man'(x)$ & $woman'(y)$ & $love'(y,x)$ & $find'(x,y)$
& $date'(x,y))$

Compare the following text with the interpretation of
(4):

(5) Kathryn is a well-known psychologist. A man who
     listens to HER dates a woman who loves him.

Because of text coherence we have to follow the procedure (IP) until we get Kathryn as the coreferring
term with the pronoun her. If $\lambda PP(k)$ is the translation of Kathryn we get $\lambda PP(k)$ as the translation
of the pronoun her.

In the following example we get the particular interpretation and not the universal interpretation, which
would be more probable.

(6) The man who finds a woman who loves him dates her.

$(\exists x)(\exists y)(man'(x)$ & $woman'(y)$ & $love'(y,x)$ & $find'(x,y)$
& $(\forall z)(man'(z)$ & $woman'(y)$ & $love'(y,z)$ & $find'(z,y)$
$\rightarrow z = x)$ & $date'(x,y))$

The general interpretation that takes the man as a term
denoting an universal concept cannot be derived in this
fragment.

But we get another particular interpretation which is
the non-restrictive interpretation of

(7) The man, who finds a woman who loves him, dates her.

This interpretation is based on the translation of the
man which refers to a contextually introduced referent;
let it be b. Then we get:
$(\exists x)(man'(b)$ & woman$'(x)$ & love$'(x,b)$ & find$'(b,x)$ &
date$'(b,x))$.
Note that our fragment permits the translation of sentences like

(8) Every man who loves every woman is happy.

It is $(\forall x)(\exists y)(man'(x)$ & (woman$'(y) \rightarrow$ love$'(x,y)) \rightarrow$
happy$'(x))$. This is equivalent to $(\forall x)(man'(x)$ &
$(\forall y)(woman'(y) \rightarrow$ love$'(x,y)) \rightarrow$ happy$'(x))$. But it also
admits the derivation of a translation with coreference
between the underlined terms for sentences as

(9) Every man who finds every woman who loves him dates
     her.

and

(10) A man who finds every woman who loves him dates
     her.

The translations obtained are:

(9') $(\forall x)(\exists y)(man(x)$ & (woman(y) & love$(y,x) \rightarrow$
        find $(x,y)) \rightarrow$ date $(x,y))$
and

(10') $(\exists x)(\forall y)(man(x)$ & (woman(y) & love$(y,x) \rightarrow$
        find $(x,y))$ & date$(x,y))$,

respectively.

This might not be desirable because sentences (9) and
(10) seem, at least to me, unacceptable with coreference
between the underlined terms. If we have a close look
at the truth conditions of these translations we realise
that (9) would be true if for every man there is at
least one woman who loves him but he did not find her
(this makes the second part of the conjunction in the
antecedent of the main conditional false and therefore
the whole conditional true). Sentence (10) would be
true if there would be a man, but no y would be a woman
[or no y would love this man] and, he would date y.

We might want to block the interpretation of these
sentences as Bach-Peters sentences, because of their
likely unacceptability in the coreference reading.

The blocking of the translation of these sentences
could be achieved by placing a restriction on (IP) to
the effect that in a subordinate clause, i.e. including
relative clauses, no term of the form "every..." or
"no..." can be coreferential term for a pronoun in the
main clause. This has already been excluded for subordinate clauses other than relative clauses by a restriction on (IN). The restriction on (IP) would make
sentences (9) and (10) have only the interpretation in
which every man and her are not coreferential.

The treatment of Bach-Peters sentences given in this
paper suffers from the fact that some rules and restrictions have features that that give it an ad hoc appearance.
This is true of the restrictions on referent introduction, the restriction placed on the rule (IN) and
the restriction on (IP) mentioned above, and the generalisation procedure for instantiating translations
that made the choice of the quantifier dependent on
whether there is an uneven number of preceding head
terms (of relative clause constructions)of the form
"every..." or "no...". It seems to me that in the
course of the development of a more comprehensivetheory
of pronouns and subordinate clause constructions some
of these features could be covered by rules that are
more general.

Notes

1. This can be continued for the third last sentence
   and possibly further backwards. Further empirical
   studies are necessary to decide what factors determine the limit of searching backwards. It should
   be noticed that the procedure has to be extended:
   if according to the procedure formulated here, no
   individual name can be found then the procedure has
   to be repeated with the weaker requirement "the
   predicate specifying Q may be true of the
   individual" instead of "is true".

2. Note that "the..." can also be translated by means
   of a constant by the procedure introduced previously.
   This translation is appropriate for non-restrictive
   relative clauses.

3. The necessity of formulating the disjunction
   that complicates part (2) of this rule has been

brought to my attention by Zeno Swijtink and Theo
Janssen. Without this complication a sentence like
"Every man who finds a woman who loves him is happy"
would receive the wrong interpretation according to
$(\forall x)(\exists y)(man'(x)$ & $woman'(y)$ & $love'(y,x)$ & $find'(x,y) \rightarrow$
$happy'(x))$ instead of the right interpretation accord-
ing to $(\forall x)((\exists y)(man'(x)$ & $woman'(y)$ & $love'(y,x)$ & $find'$
$(x,y)) \rightarrow happy'(x))$ which is equivalent to the one we
get in the present paper, namely $(\forall x)(\forall y)(man'(x)$ &
$woman'(y)$ & $love'(y,x)$ & $find'(x,y) \rightarrow happy'(x))$ [compare
note 5]. Note further, that a coordination of head
terms will count as one head term; but the treatment
of constructions with coordinated terms goes beyond
the present treatment.

4. The assumption of the specific reading for "a woman"
   by the hearer is a matter of cooperation with the
   speaker by the hearer. The speaker uses the pronoun
   in an unexpected way. Therefore the hearer has to reconsider his interpretation of the preceding sentence
   and to assume that the speaker has somebody in mind
   that he is referring to by the term "a woman". The
   specific readings are always only secondary readings
   which the hearer can get at by means of context or
   situational information, as, for example, with: "Look,
   a boat is capsising", where the hearer might answer:
   "No, it is just heeling over." Certain verbs, especially verba sentiendi, e.g. see, hear, are an indicator that the hearer might assume a specific reading
   with regard to the speaker's universe of discourse.
   These are the only situations when a hearer may interpret a relative clause with "a(n) ..." as a head term
   as non-restrictive.

5. Note, that for conditional sentences the restriction
   on IN corresponds nicely with the fact that $(\forall x)(...$
   $x...) \rightarrow p$ is not equivalent to $(\forall x)((...x...) \rightarrow p)$ and
   that $(\exists x)(...x...) \rightarrow p$ is not equivalent to $(\exists x)((...$
   $x...) \rightarrow p)$. For a sentence like "If every man buys a
   car industry flourishes" we have to get the transla-
   tion $((\forall y)((\exists x)(man'(y) \rightarrow car'(x)$ & $buys'(y,x)))) \rightarrow$
   $flourish'(i)$. The scope of the universal quantifier
   may not be extended beyond the main conditional without changing truth conditions. Therefore, we are not
   able to provide a translation for the following sentence with every man and he being coreferential:

   If EVERY MAN buys a car HE profits from it.

   And, luckily, we do not need to supply a translation
   for this sentence because it is unacceptable with
   coreference.

6. Otherwise the pronoun would have to be a reflexive
    pronoun. It was brought to my attention by Martin
    Stokhof that it is not possible to establish on
    the surface level that this restriction holds for
    examples like: Every man who beats him, where every
    man should not be selected as a coreferential pre-
    ceding term for him. The information whether two
    terms are binding places of a single verb has to be
    formulated on the level of the translation of the
    sentence. On the surface level we might manage
    with the following restriction:

    A pronoun occurring in a relative clause as binding
    one of the places of the verb of that relative
    clause cannot have the head term of the relative
    clause construction as its preceding term with
    coreferentiality.

7. According to the restriction placed on IN, the terms
    "every...", "no...", and "a(n)..." under negation
    or preceded by a term of the form "every..." or
    "no...", are excluded as preceding coreferential
    terms in subordinate clauses other than relative
    clauses. They are excluded because the blocked
    variable is no longer retrievable, since the trans-
    lation of those subordinated clauses has been com-
    pleted before they got interpreted into the trans-
    lation of the clause construction. Thus we prevent
    the non-acceptable examples under (B) and (D). By
    way of cooperation, the hearer might assume "a(n).."
    to be intended in a specific use, if he does not
    find another term that   could be coreferential
    with the pronoun to be interpreted.

8. This rule and others are motivated for languages
    with a morphological case system and fairly free
    order of terms, like German, in Bartsch (1976). In
    English, restrictions are placed on the order of
    application of terms in the different case-positions,
    if we assume that linear word order reflects the
    hierarchy of structures formulated according to
    rules of categorical syntax.

9. This stand was taken by Theo Janssen in a criticism
    of this paper.

References

Ballmer, T. (1972). Gründe für eine formale Pragmatik.
In K. Hyldegaard-Jensen (ed.) Linguistik 1971.
Referate des. 6. linguistischen Kolloquiums 11.-14
August, 1971 in Kopenhagen. Frankfurt: Athenäum.

Ballmer, T. (1975). Sprachrekonstruktionssysteme.
Scripter Verlag: Kronberg.

Bartsch, R. (1976). The role of categorical syntax in
grammatical theory. In A. Kasher (ed.) Language in
Focus: foundations, methods and systems. Essays in
memory of I. Bar-Hillel. Dordrecht: D. Reidel.

Bartsch, R., Ehrich, V., and Lenerz, J.(1976). Ein-
führung in die Syntax. Kronberg:Scripter Verlag.
(forthcoming).

Frege, G. (1892). Sinn und Bedeutung. In Zeitschrift
fur Philosophie und philosophische Kritik. Nf 100;
25-50. Reprinted (1962) in Frege, G. Funktion,
Begriffe, Bedeutung. Göttingen:Vandenhoek & Ruprecht.

Lakoff, G. (1971). On generative semantics. In D.
Steinberg and L. Jakobovits (eds). Semantics. An
inter-disciplinary reader in philosophy, linguistics
and psychology. Cambridge, Mass:Cambridge University
Press.

Montague, R. (1974). The proper treatment of quantifi-
cation in ordinary English. In Thomason, 1974.

Rodman, R. (1972). The proper treatment of relative
clauses in a Montague Grammar. Papers in Montague
Grammar. Occasional Papers in Linguistics,2. UCLA.
Department of Linguistics.

Rohrer, Chr.(1975). Double terms and the Bach-Peters
paradox (preliminary version). Manuscript, University
of Stuttgart, October, 1975.

Thomason, R. (1974). Formal Philosophy. Selected papers
of Richard Montague (with an introduction by R.
Thomason) New Haven and London: Yale University Press.

Thompson, S.A.(1971). The deep structures of relative
clauses. In C.J. Fillmore and D.T. Langendoen (eds).
Studies in Linguistic Semantics. New York: Holt,
Rinehart and Winston. 79-96.

FOREGROUND AND BACKGROUND INFORMATION IN REASONING

LEO   G. M.   NOORDMAN

University of Groningen

Every meaningful message contains new information, or
at least information meant by the speaker to be new for
the hearer, embedded in old information. Every message
aims at communicating something new. If I were to say
now two and two is four, that would be meaningless,
because that sentence contains only old information and
nothing new is communicated. On the other hand, a sentence that contains only new information e.g. suddenly
he got extremely scared cannot be understood, it cannot
be assimilated to previous knowledge. The quite normal
sentence I met your brother yesterday contains as new
information that a certain meeting has taken place at a
certain moment. There is also old information: knowledge shared by the speaker and hearer which is presupposed to be true. The presupposed information in this
case is that the hearer has a brother. A linguist who
has recently studied verbal communication and semantics
in the framework of old and new information is Chafe
(1970, 1972). Related distinctions are topic and comment, given and new, theme and rheme (Halliday, 1967,
1970) presupposition and focus (Chomsky, 1971).

The psychological implications of new and old information have especially been studied by Clark (1973; Clark
and Haviland, 1976). He has developed a theory about
communication and comprehension in terms of the given-
new contract elaborating upon Grice's cooperative principle (1967). According to Clark, the speaker is cooperative if he communicates information he thinks the listener already knows as given information and information

he thinks the listener doesn't yet know as new information. He can do this by using a certain syntactic
construction or a certain stress pattern. The listener
divides incoming information into given and new information, he then searches memory for a unique antecedent
that matches the given information; the new information
is integrated into memory by attaching it to that unique
antecedent.

Related to the notions of given and new information are
the notions of presupposition and assertion. The
assertion of a sentence is the information that is
explicitly communicated; a presupposition is what is
assumed to be true by the speaker and hearer. The standard test for the claim that a certain proposition is a
presupposition of a sentence is to see whether it is
preserved under negation (Fraser, 1972; Horn, 1969;
Lakoff, 1971). The negation in the sentence I did not
meet your brother yesterday does not affect the old information that you have a brother. This notion of presup-
psotion is called logical presupposition. The notion
of presupposition is sometimes defined with respect to
the beliefs of the hearer and the speaker. A proposition
is considered as a presupposition of a sentence, if it
is not affected by the negation of the sentence as it
is understood by the hearer and the speaker. It is in
this way that Miller (1969) uses the notion of presupposition. His discussion is based on the implications of
a negation in a predicate nominal sentence, e.g. Leslie
is not a mother. It is an empirical question to determine the implications of a negation, as Miller says.
He suggests that the sentence X is not a mother implies
that X is a woman and denies that X is a parent; woman
then is the presupposition of mother, because it is not
affected by the negation; parent, which is negated with
respect to X, is the assertion.

It should be clear that this notion of presupposition
differs from the notion of logical presupposition given
above. The sentence Leslie is not a mother but a father
is not meaningless as is the sentence I met your brother
yesterday but you don't have a brother. The sentence
Leslie is not a mother does not logically exclude the
possibility that Leslie is a man. What is claimed
however, is that the speaker and the hearer of the sentence Leslie is not a mother presuppose or believe that
Leslie is a woman.

The point to be made is that only some information in a

sentence constitutes the intended message, whereas other
information is tacitly assumed to be true. The sentences
X is not a painter and X is a painter explicitly communicate something about the profession of X. That is the
foreground information. Not explicitly communicated is
e.g. that X is living, human. That is background information which is assumed to be true in both sentences.
That background information is not easily available
for a negation: X is not a painter but a stone sounds at
least strange. Similarly, if the analysis of Miller is
correct, the sentence Leslie is not a mother when used
in a context that necessitates to interpret Leslie as a
man, will cause the hearer some difficulty.

The topic of the present paper is new and old information,
explicitly communicated and presupposed information or
foreground and background information. More specifically
the present study will demonstrate the role of foreground
and background information in simple reasoning tasks.
Not all information is equally likely to be background
information. Some factors will be discussed that determine what information is likely to be foreground information and what information is likely to be background
information.

Presupposed or Background Information

In the first experiments to be reported, very simple
problem solving tasks have been used, each of them
involving a negative sentence. The task is such that in
order to arrive at the conclusion, a certain proposition
has to be negated. It is assumed that the solution of
these tasks is straightforward if a subject can negate
that proposition. If many subjects are not able to solve
the task, this will show that that proposition is wrongly
presupposed to be true. This existence of presuppositions
is inferred from the difficulty of the task. A task is
considered to be difficult when it can hardly be solved
or when it takes a relatively long time to solve it.

Experiment 1

The first experiment has been done in collaboration with
Levelt. The starting point of this experiment was the
following well known problem: "Two Indians, a tall one
and a small one were sitting at a fence; the smaller one
was the son of the taller one; the taller one was not the
father of the smaller one. How is this possible?" Most
people have great difficulty in solving this problem.
They don't manage to deny the masculinity of the tall

Indian. The negation in <u>the tall one is not the father
of the small one</u> did not affect the feature <u>male</u>. This
indicates that <u>male</u> is a presupposition of <u>father</u> which
agrees with Miller's analysis.

It could be argued however, that the difficulty of the
problem resides not only in the presuppositional structure of the kinship terms but also in some other factors.
Indians are normally assumed to be male Indians. This
factor can be eliminated by constructing the problem in
the form: <u>A is the son of B</u>, <u>B is not the father of A</u>.
Another factor attributing to the difficulty of the problem might be the presence of kinship terms of the same
sex in both sentences. When kinship terms contrasting
on the gender component are used in the sentences, it is
less likely that the gender is background information
and consequently more likely that the gender will be
affected by the negation. If this is correct, the following problem should be much easier: <u>A is the daughter
of B, B is not the father of A</u>. It is assumed that the
context determines what information is background and
what foreground information. Similar remarks can be
made with respect to the problems <u>A is the daughter of
B, B is not the mother of A</u> and <u>A is the son of B, B is
not the mother of A</u>.

These four problems constituted the items of the first
experiment. Because of the riddle character of the
items, each subject was given only one item. Each item
was presented to seventeen subjects. An item followed
by the question <u>how is this possible</u> was presented
acoustically. The time was measured from the end of the
presentation of the question until the subject gave the
answer. If the subject did not give the answer within
120 seconds he was considered unable to solve the problem
and the experimenter stopped the session.

The experimental data are the latencies and the number
of errors. The latencies have been analysed with the
Mann-Whitney-U test. To subjects who gave an incorrect
answer or no answer at all, the highest possible rank
was attributed in these tests. The median latency and
the number of errors for each item are represented in
Table 1. The median latency for item 1 is not reported.
Because that item has not been solved by more than one
half of the subjects, the median cannot be computed in
a non-arbitrary way.

Table 1. Items, median latencies (sec) and number of
          subjects*who gave the correct answer (Expt. 1).

Item	Median Latency	Number Correct
1. A is the son of B B is not the father of A	+	2
2. A is the daughter of B B is not the father of A	68	9
3. A is the daughter of B B is not the mother of A	26	12
4. A is the son of B B is not the mother of A	11	17

* n = 17 per item

+ for absence of median, see text.

The results confirm the predictions with respect to the
hardness of the problems. Many subjects could not solve
the problem they were given, and if they could solve
the problem, it took a relatively long time. It can be
concluded that the subjects have difficulty in negating
the sex component. The gender is a presupposition or
background information. The analysis of the errors confirms this conclusion: stepfather and stepmother are
frequent reactions.

Items with kinship terms of different sex are overall
easier than items with kinship terms of the same sex
($z = 2.77$, $p < .01$). So in the former items where the
kinship terms contrast on the gender component, the
gender is less likely to be background information and
more available for negation than in the same-sex items.

There is an additional significant result. The items
with B is not the father of A are more difficult than
the items with B is not the mother of A ($z = 6.22$, $p < .001$)
Apparently, it is easier to conclude from B is not the
mother that B is the father than from B is not the
father that B is the mother. One could say that the
masculinity of father is more strongly presupposed or
background information than the femininity of mother.
This asymmetry will be discussed later on.

The gender of the subjects did not affect the latencies;
male and female subjects reacted in the same way.

The conclusion so far is that the gender is the background information of the kinship terms father and
mother in these problems. It is plausible then that the
foreground information is parenthood. A negation affects
the foreground information. So it is expected that the
meaning component parenthood is most likely to be affected by a negation. This prediction has been tested in
two problems, where both the gender component and the
parenthood component can be negated.

Experiment II

The first problem is: A is the father of B, A is the
grandfather of C, B is not the father of C, what could
B be of C? There are three possible answers: mother,
uncle and aunt. Mother differs from father with respect
to the gender; uncle differs from father with respect to
parenthood; aunt differs from father with respect to
both meaning components. If parenthood is the foreground
information and the sex the background information, the
most frequent answer will be uncle. Similar predictions
can be made with respect to the problem: A is the mother
of B, A is the grandmother of C, B is not the mother of
C, what could B be of C?

The experiment has been run by students of Levelt. Each
problem has been presented to 40 subjects. The experiment was balanced with respect to the gender of the
experimenter and of the subjects. These variables,
however, had no effect.

The results of the father-grandfather problem confirm
the predictions. Twenty-eight subjects gave a correct
answer; the answer uncle was given 20 and mother 8 times.
Moreover, 11 out of the 12 erroneous answers were male
terms. The conclusion is that the sex component is the
background information and parenthood foreground information. An additional result is that no subject gave
the answer aunt. Apparently, the negation affects only
one meaning component.

The results of the mother-grandmother problem also confirm the predictions. The answer aunt is given by 12
subjects and father by 15 subjects. But 12 out of the
13 erroneous answers were female terms. The results are
less dramatic than those of the first problem. This is
in agreement with the observed asymmetry between father
and mother in the previous experiment.

Again no subject gave the answer uncle which differs on

two meaning components from <u>mother</u>.

The fact that in producing the answer only one feature
is changed, is in complete agreement with the rule of
minimal contrast formulated by Clark (1970) for word
association data. That rule says: change the sign of
only one feature, preferably the last feature in the
list. The most frequent association to <u>man</u> is <u>woman</u>
where only the sex is changed, then <u>boy</u> implying a
change in the feature adult, and then <u>girl</u>, implying a
change in both features. This rule of minimal contrast
has also been found by Levelt, Schroeder and Hoenkamp
in the domain of motion verbs (in this volume).

Differences in background information

The results so far demonstrate that the gender is background information of the kinship terms. A second result
was that the masculinity of <u>father</u> is more background
information than the femininity of <u>mother</u>. In other words
<u>father</u> and <u>mother</u> are asymmetric with respect to back-
ground information. An interesting question is whether
there are kinship terms that are more symmetric than
<u>father</u> and <u>mother</u>. According to Lyons (1968), that is
the case for <u>brother</u> and <u>sister</u>: "The fact that there
is no superordinate term for the two complementaries
<u>brother</u> and <u>sister</u>, is prima facie evidence that the
opposition between the two terms is semantically more
important than what they have in common." Consequently,
the asymmetry found for <u>father</u> and <u>mother</u> should not be
found for <u>brother</u> and <u>sister</u>. On the other hand, the
results for items with <u>son</u> and <u>daughter</u> should exhibit
the same asymmetry as those for <u>father</u> and <u>mother</u>.

Experiment III

The items are represented in Table 2. Condition I contains the <u>not son</u> and <u>not daughter</u> items, e.g. <u>A is the
father of B, B is not the son of A</u>. Condition II contains the <u>not brother</u> and <u>not sister</u> items, e.g. <u>A is
the brother of B, B is not the brother of A</u>. Each problem in condition I has been given to ten subjects; each
problem in condition II to twelve subjects. Each subject was given one item. The design was balanced with
respect to the gender of the subjects.

Assuming again that the gender is the background information of these kinship terms, it is predicted that subjects will need a relatively long time to solve these
problems, if they can be solved at all. The other

prediction is that the B is not the son items are more difficult than the B is not the daughter items and that there is no difference between B is not the brother items and B is not the sister items.

Table 2.   Items, median latencies (sec) and number of correct answers (Expt. III).

Item	Median Latency	Number Correct
Condition I (n=10 per item)		
1. A is the father of B B is not the son of A	+	3
2. A is the mother of B B is not the son of A	13	7
3. A is the mother of B B is not the daughter of A	9	7
4. A is the father of B B is not the daughter of A	10	9
Condition II (n=12 per item)		
1. A is the brother of B B is not the brother of A	+	6
2. A is the sister of B B is not the brother of A	+	6
3. A is the sister of B B is not the sister of A	34	8
4. A is the brother of B B is not the sister of A	+	4

+   for absence of median, see text.

The results are presented in Table 2. Again no medians have been reported for items that have not been solved by at least one half of the subjects. The latencies have been analysed with the Mann-Whitney U test. The sex of the subjects did not have a differential effect on the latencies.

The results are in agreement with those of Experiment I: subjects have difficulty in denying the gender. Many

subjects could not solve the problem they were given.
The gender is the background information in these problems.

The latencies for the items B is not the son are longer
than for the items B is not the daughter (z = 1.71,
p < .05 one tail). The masculinity of son is more background information than the femininity of daughter. On
the other hand, the items B is not the sister were as
difficult as the items B is not the brother as was expected (z = 0.46, p = .65). The gender is equally background information in brother and sister.

The asymmetry between father and mother and between son
and daughter has been interpreted in the sense that the
masculinity of these male terms is more background information than the femininity of these female terms. Male
is background information just as the features e.g.
human, living, concrete. These features are sometimes
called higher order presuppositions; they are presuppositions of presuppositions: man presupposes human,
human presupposes living, living presupposes concrete.
According to linguists, higher order presuppositions
are less likely to be affected by a negation than first
order presuppositions. The sentence this child is not
married is less strange than the sentence this stone is
not married. By attributing not married to stone not
only the presupposed information of the feature adult,
but also of the features living and human are denied.
Another example comes from Lakoff (1972):

1. few men have stopped beating their wives.

2. some men have stopped beating their wives.

3. some men have beaten their wives.

Sentence 1 presupposes sentence 2; 2 is the first order
presupposition of 1. Sentence 2 presupposes sentence
3; 3 is the second order presupposition of 1. A first
order presupposition can be suspended, as is demonstrated
in 4:

4. few men have stopped beating their wives,
    if any at all have.

A higher order presupposition cannot be suspended: sentence 5 is not acceptable:

5. few men have stopped beating their wives,
    if any have ever beaten them at all.

In line with this analysis one could say that in these

problems the masculinity of <u>father</u> behaves like a presupposition of a higher order than the femininity of
<u>mother</u>. The negation is less likely to affect the fea-
ture male of <u>father</u> than the feature female of <u>mother</u>.
The same remark can be made with respect to the asymmetry between <u>son</u> and <u>daughter</u>.

Many word pairs in natural language have such an asymmetry. Examples are marked-unmarked word pairs like <u>good</u>
and <u>bad</u>, <u>long</u> and <u>short</u>, and positive and negative prepositions like <u>above</u> and <u>below</u>. Similar psychological
phenomena have been found for these categories of concepts as for the kinship terms <u>father</u> and <u>mother</u>. In a
recall task (Clark & Card, 1969) <u>bad</u> is recalled as
<u>good</u> more frequently than that <u>good</u> is recalled as <u>bad</u>.
The transformation of <u>isn't above</u> into <u>below</u> took more
time than the transformation <u>isn't below</u> to <u>above</u> (Young
& Chase, 1971). With respect to marked and unmarked
words e.g. <u>good</u> and <u>bad</u>, Clark (1975) distinguishes between the nominal sense of <u>good</u> as in <u>how good was the</u>
<u>film</u> and the contrastive sense of <u>good</u> as in <u>the film</u>
<u>was good</u>. A marked word has only the contrastive sense.
These senses are represented by Clark in the following
feature notation: <u>good</u> in the nominal sense has the
components (+ evaluative (polar)); in the contrastive
sense it has the components (+ evaluative (+polar)),
<u>bad</u> has the components (+ evaluative (- polar)). The
asymmetry consists in the fact that the nominal sense
which exists for only the unmarked concept, is simpler
because it is not specified for polarity. For the positive and negative prepositions Clark proposes the following feature notation: <u>above</u> is coded as (+ vertical
(+ polar)); <u>below</u> as (+ vertical (- polar)). The
asymmetry is now explained by postulating that the
coding + is more easily obtained than the coding -.
This explanation is also offered as an alternative
explanation of the asymmetry between marked and unmarked
words.

An intriguing question is where this asymmetry comes
from. Why is masculinity more background information
than femininity in the previous experiments; why can
only <u>good</u> be used in a nominal sense and not <u>bad</u>. There
are certainly biological, educational and cultural factors involved. Clark (1973a) has discussed spatial
words in this respect. This problem however, will not
be discussed at present. The question to be studied
now is how the asymmetry has to be described. Is it due
to the lexical coding of the concepts themselves or is

it due to the use of the concepts. In other words, can
the asymmetry sufficiently be described in terms of differential lexical coding of the concepts or does the
asymmetry depend on factors such as foreground and background. The problem to be studied at present is whether
the asymmetry between male and female kinship terms can
be influenced by manipulating the foreground and background information.

Variability of Foreground and Background Information

In terms of background and foreground information, the
asymmetry consists in the fact that a particular meaning
component e.g. the gender, is more background information in one word than in another word. In that case,
the asymmetry should disappear if that component is
stressed as the focus of the communication. In terms of
lexical coding however, the asymmetry consists in the
fact that the positive value of the meaning component
can more easily be represented than the negative value.
This asymmetry will not disappear if that meaning component is stressed.

In the following experiments the subject is given some
information about a person X. He then has to make a
decision with respect to X. In one case the criterion
on which he has to decide is the generation; in the
other case it is the gender.

Experiment IV

In this experiment, the task is such that the attention
of the subject is directed to the generation. The
problems were of the following form: A is the father of
B, can B be the son of A. Items have been constructed
with father, mother, son and daughter. The nature of
the task stresses in fact the generation component as
the criterion on which the subject has to make his
decision. The answer is yes if and only if the kinship
term in the second sentence is of another generation
than the kinship term in the first sentence. It is clear
that the gender component is quite irrelevant for the
decision: if A is the father of B, B can be the son, but
also the daughter of A. According to the foreground-
background theory, it is easier to decide that B is
male than that B is female. The reason is that the masculinity in father and son is more background information
than the femininity in mother and daughter. Subjects
will be more inclined to assume that B is male than that

B is female. The theory of lexical coding predicts the
same results. In this case, the reason is that the
positive value on the gender feature as such can more
easily be represented than the negative value.

The items were presented visually. The sentences were
presented the one after the other. The latency was
measured from the onset of the presentation of the second
sentence until the subject pressed a button. The experiment was balanced with respect to the number of items
requiring the answer yes and no, with respect to the
gender of the subjects, and with respect to the location
of the yes and no button. There were 24 subjects, each
of them being given all the problems.

Table 3. Items and mean latencies (msec.) (Expt. IV).

	Item	Mean Latency
1 A is the father	of B; B is the son	of A	1634
2 A is the father	of B; B is the daughter	of A	1811
3 A is the mother	of B; B is the son	of A	1641
4 A is the mother	of B; B is the daughter	of A	1773
5 A is the son	of B; B is the father	of A	1311
6 A is the son	of B; B is the mother	of A	1819
7 A is the daughter	of B; B is the father	of A	1465
8 A is the daughter	of B; B is the mother	of A	1557

The results, shown in Table 3, confirm the predictions.
The latency for the items with a male term in the second
sentence was smaller than the latency for the items with
a female term in the second sentence (z = 3.46, p < .001;
Wilcoxon matched-pairs signed-ranks.)

Experiment V

In this experiment the attention of ths subjects was
directed to the sex of the kinship terms. The problems
were of the following form: A is the father of B, is A
male? Items have been constructed with the terms father,
mother, son and daughter for the first sentence and
with the concepts male and female for the second sentence.
According to the theory of lexical coding, the latency
for the male items should be shorter than for the female

items for the same reason as in the previous experiment.
According to the theory of foreground-background information, the latencies should not differ, because the gender is in both cases equally in the foreground. The experimental procedure was the same as in the previous experiment. The number of subjects was twelve.

Table 4. Items and mean latencies (msec.) (Expt. V).

Item	Mean Latency
1. A is the father of B, A is male	1135
2. A is the mother of B, A is female	1253
3. A is the son of B, A is male	1238
4. A is the daughter of B, A is female	1173

The results shown in Table 4 support the foreground-
background theory. There is no difference in the latencies between the items with A is male and A is female
(z = 0.21; p = .83; Wilcoxon matched pairs signed-ranks).

Several other experiments with marked and unmarked word
pairs lead to the same conclusion (Noordman, 1976).
If a subject knows that he has to answer the question,
e.g. which one is the highest or which one is the lowest
and if the required information is not stored in a way
congruent with one of these questions, there is no difference in RT between the marked and the unmarked
question.

The conclusion is that the asymmetry between male and
female kinship terms as well as between marked and unmarked adjectives depends on the background-foreground distinction, the asymmetry can best be described in terms
of background and foreground information.

Conclusion

The results of the present experiments lead to the
conclusion that the meaning components of the kinship
terms differ with respect to foreground and background
information. Foreground information is explicitly communicated information; background information is not the
proper content of the message; it is assumed to be true.
In general the generation is the foreground information
of the problems being studied and gender is the background information. For whatever reason however, the

masculinity of many male kinship terms is more background information than the femininity of the corresponding female terms. When processing a negation, one
searches the meaning component the negation refers to,
starting with the information that is most in the
foreground. This explains first the difficulty of the
items in the negation experiments and second the asymmetry between the tasks with male and female kinship
terms. That masculinity is more background information
and more likely presupposed than femininity accounts
also for the results of the latter experiments where a
decision had to be made with respect to the gender or
the generation. Explaining the asymmetry between male
and female kinship terms as well as between **unmarked and**
marked adjectives in terms of background and foreground
information has the advantage that one can easily
account for the effects of contextual factors, linguistic or psychological, on reasoning tasks with these
terms.

References

Chafe, W.L. (1970). Meaning and the structure of language.
  Chicago: University of Chicago Press.

Chafe, W.L. (1972). Discourse structure and human know-
  ledge. In Carroll, J.B. and Freedle, R.O. (eds.)
  Language Comprehension and the Acquisition of Knowledge.
  Washington: Winston & Sons.

Chomsky, N. (1971). Deep structure, surface structure
  and semantic interpretation. In Jakobovits, L.A. and
  Steinberg, D.D. (eds.) Semantics: An Interdisciplinary
  Reader in Philosophy, Linguistics and Psychology.
  Cambridge: Cambridge University Press.

Clark, H.H. (1970). World associations and linguistic
  theory. In Lyons, J (ed.) New Horizons in Linguistics.
  Harmondsworth: Penguin Books.

Clark, H.H. (1973a). Space, time,semantics and the child.
  In Moore, T.E. (ed.) Cognitive Development and the
  Acquisition of Language. New York: Academic Press.

Clark, H.H. (1973b). Comprehension and the given-new
  contract. Paper presented at the conference on "The
  role of grammar in interdisciplinary linguistic
  research". University of Bielefeld.

Clark, H.H. (1975). Semantics and comprehension. In
  Sebeok, T.A. (ed.) Current trends in Linguistics, Vol 12,
  The Hague: Mouton.

Clark, H.H. and Card, S.K. (1969). Role of semantics in remembering comparative sentences. Journal of Experimental Psychology, 82, 545-553.

Clark, H.H. and Haviland, S.E. (1976). Comprehension and the given-new contract. In Freedle, R. (ed.) Discourse Production and Comprehension. Hillside, N.J.: Lawrence Erlbaum Associates.

Fraser, B. (1971). An analysis of 'even' in English. In Fillmore, C.J. and Langendoen, D.T. (eds.) Studies in Linguistic Semantics. New York: Holt, Rinehart and Winston.

Grice, H.P. (1967). The logic of conversation. William James Lectures, Harvard University.

Halliday, M.A.K. (1967). Notes on transitivity and theme in English: II. Journal of Linguistics, 3, 199-244.

Halliday, M.A.K. (1970). Language structure and language function. In Lyons, J. (ed.) New Horizons in Linguistics. Harmondsworth: Penguin Books.

Horn, L.R. (1969). A presuppositional analysis of "only" and "even". In Binnick, R.J., Davidson, A., Green, G.M., and Morgan, J.L. (eds.) Papers from the fifth regional meeting of the Chicago Linguistic Society. Chicago: University of Chicago.

Lakoff, G. (1972). Linguistics and natural logic. In Davidson, D., and Harman, G. (eds.), Semantics of natural language. Dordrecht: Reidel.

Lyons, J. (1968). Introduction to Theoretical Linguistics. Cambridge: Cambridge University Press.

Miller, G.A. (1969). A psychological method to investigate verbal concepts. Journal of Mathematical Psychology, 6, 169-191.

Noordman, L.G.M. (1976). Reasoning with comparative concepts. Heymans Bulletin 76-HB-234 EX. University of Groningen.

Young, R. and Chase, W.G. (1971). Additive stages in the comparison of sentences and pictures. Paper presented at the Midwestern Psychological Association Meetings. Chicago (cited in Clark, 1975).

ANAPHORA: A PROBLEM IN TEXT COMPREHENSION

SIMON GARROD and ANTHONY SANFORD

University of Glasgow

A charactersitic feature of text is that there are many less entities talked about than there are noun phrases referring to them. The writer may introduce a certain person or object in one sentence and refer to it, perhaps on several different occasions, in subsequent sentences. An example of this is afforded by the following pair of sentences:

(1) A bus came trundling down the hill.

(2) The vehicle narrowly missed a pedestrian.

The object "bus" is introduced in sentence (1) and subsequently referred to as "the vehicle" in sentence (2). The phrase "the vehicle" is thus an anaphor of the phrase "a bus".

From the point of view of the reader this kind of anaphoric reference presents certain problems. The first such problem is simply one of deciding what entity in the previous text a particular anaphoric noun phrase refers to. This presumably involves identifying the anaphoric phrase with its associated antecedent. For instance in the example given above the reader will need to identify the phrase "a bus" as the antecedent to the phrase "the vehicle". In order to help him to do this there are a variety of cues in the text itself which may be either syntactic or semantic. For example, that the reader is dealing with an anaphoric phrase is indicated by syntactic cues such as the presence of the definite

article. That its antecedent is the phrase "a bus"
is derived from the semantic relationship between the
terms bus and vehicle. A second problem which arises
is one of representation, of how the reader represents
semantically the information which appears in both
sentences once the identification has taken place.

In this paper we will be using the paradigm of anaphora
as a way of illustrating some much more general aspects
of text comprehension. We will start by showing how
the identification process can be isolated experimentally as a component in understanding text. We will
then follow up some of the representational problems
raised by the results of this experiment and finally
we will suggest that the problem of identifying the
entities mentioned in a text and the problem of representing them in memory can be solved by the same process.

Identification as a component in text comprehension

In order to isolate identification as a component of
text comprehension one is able to capitalise on the
fact that linking sentences like (1) and (2) above
involves making an implicit class membership judgement
of the sort "a bus is a vehicle" and it is relatively
easy to manipulate the time it takes a person to retrieve
or utilise such information. For instance there have
been many studies showing that when a subject is asked
to make an overt class membership judgement the time
taken will depend upon the conjoint frequency[1] of the
items to be evaluated (e.g. Wilkins, 1971; Rosch,1973).

One would expect therefore that identifying an antecedent containing a high conjoint frequency noun would be
easier than identifying one containing a low conjoint
frequency noun and that this should have an effect on
the overall time taken to understand the sentence containing the anaphor. This point can easily be examined
using sentence pairs of the following sort:

(3) A $\begin{pmatrix} \text{Bus} & \text{(HCF)} \\ \underline{\text{Tank}} & \text{(LCF)} \end{pmatrix}$ came trundling down the hill

(4) The <u>vehicle</u> nearly flattened a pedestrian.

What we need to be able to show is that sentence (4)
takes less time to understand when it is preceded by a
sentence containing the HCF item "bus" than when preceded by a sentence containing the LCF item "tank".

In order to get some measure of comprehension time we
simply asked the subjects to read sentence pairs of
this sort in a task where they had to answer questions
about them. The subjects read the sentences and questions in succession, pushing a key whenever they were
ready to move on to the next sentence or question in the
sequence. This enabled us to get a measure of the time
the subject took to read each sentence in the sequence
and the time taken to answer the question.

In generating the materials we choose 16 categories
with high and low conjoint frequency instances which
had been shown to give a reliable conjoint frequency
effect in previous class membership evaluation studies
(Sanford & Garrod, 1975). 16 pairs of sentences were
then constructed around these items, so that each of the
16 pairs could contain either an HCF exemplar in sentence (1) or an LCF exemplar, thus giving 32 pairs of
sentences in all. Apart from conjoint frequency a further variable was also manipulated, this was the order
of the instance and the category in the sentence pair.
For half the subjects the category always appeared in
the first sentence as below:

(5) A vehicle came trundling down the hill.

(6) The (bus) nearly flattened the pedestrian.
         (tank)
whereas for the other half the more normal instance
first order was used (as in (3) and (4) above).

The crucial measure that is of interest is the time they
took to read the second sentence in each pair, since it
is comprehension of this sentence which involves "identification". These reading times are shown in Table
1 for both the instance first and the category first
conditions and both levels of conjoint frequency.

Table 1. Mean reading times (secs.) for the second
          sentence, Experiment 1.

Condition	Conjoint Frequency		
	Hi	Lo	
Instance First	1.321	1.402	1.361
			Min F' (1,25)
			= 4.5 p < 0.05
Category First	1.554	1.623	1.588
	Min F' (1,127) = 7.601 p < 0.025		

Analysis of variance on these data confirm all the trends revealed
in the Table, both conjoint frequency and instance versus category
first conditions are reliably different by min F'.

The presence of the conjoint frequency effect for the reading time
of the second sentence indicates that the process of identification
enters directly into our comprehension of anaphoric sentences.
Manipulating the difficulty of making the identification affects
the overall comprehension time for the sentence. The category
versus instance first difference raises other issues. Since there
was no interaction[2] between the two effects, the category - instance
effect is probably not attributable to any special difficulty associated with identifying an instance given a category or vice versa,
but rather with some more general feature of how the meaning of the
sentence is being represented. For instance, one simple explanation
of this difference could be in terms of the informativeness of the
second sequence in both cases. To learn that the bus or tank referred to in one sentence is a vehicle is not very informative, we
know this already, but to learn that a previously mentioned vehicle
is a tank or a bus is informative. The substantial difference in
comprehension time could then be attributed to the processing
associated with accounting for this extra information. Since the
extra information that we are talking about is in terms of the text
as a whole it is worth while to look at some of the ways in which
we might represent the information from several connected sentences.

Representing textual information

One requirement in representing textual information is that the
reader be able to identify the various major entities mentioned in
the passage. To this end he will need to have a representation
associated with each entity in which can be stored information identifying that individual from among all the other objects or people
mentioned. On encountering subsequent references the reader will
be able to relate the other information in the sentence to the
appropriate representation in memory. In this way, he can build up
an overall memory structure which revolves around the key objects
or people in the passage, part of the structure containing information identifying each of these and part containing the events in
which they are involved.

One way in which such a system might operate has been suggested by
Minsky (1975) in his article on Frame analysis. Minsky proposes
that the key words and ideas in the text evoke substantial thematic
or scenario structures which he refers to as Frame systems. Each
Frame within the system can be represented as an hierarchically
arranged set of questions to be raised about the particular entity
or situation which evoked it. As the reader encounters a new
sentence in the text the contents are assigned to the appropriate
terminals on the Frames already evoked. These assignments can then

themselves take the form of sub-Frames elaborating the Frames already generated. As each question in a Frame is given a tentative answer the corresponding sub-Frames are attached and the questions they ask become active in turn. In this way the terminals associated with any particular Frame accumulate indicators and descriptors which both expect and key further assignments. A given sentence in the text will serve both to answer questions raised earlier and perhaps pose new ones itself.

If we assume that each entity in the text evokes such a Frame, the problem of representing the information which identifies that individual is solved in terms of the markers on the terminals of the Frame which specify what can be assigned. So for instance if a terminal has acquired a "male" marker it will reject "female" assignments. Identification would thus involve deciding to which terminal in the Frame system a particular piece of information can be assigned. The conjoint frequency effect demonstrated earlier would reflect the time taken to check the anaphor against the various terminal markers and the category-versus-instance first effect would reflect the extra time taken to attach the new sub-Frame associated with the instance to the Frame already allocated to the antecedent category. This will be necessary in order to change the descriptors and indicators marking the Frame terminal.

One outcome of such a system is that each entity mentioned in a passage will have a unitary representation in memory. That is, if some entity 'X' is referred to on several different occasions as "a man", "the plumber", "Mr. Smith" etc. these facts about 'X' will all be incorporated into a single description of 'X' and any events which 'X' was involved in will be represented as having happened to the individual described rather than to the particular reference made within that sentence in the text. For example, when one reads a passage such as the following:

"A man came walking out of the house. Mrs. Jones recognised him as her plumber, and since her water tank had just burst rushed over to meet him. On seeing her, Mr. Smith the plumber fled down the nearest alley in order to evade his irate customer."

A Frame will be evoked by the phrase "A man" in the first sentence which as the reader encounters the subsequent sentences will acquire the descriptors 'plumber' and 'Mr. Smith'. When the reader has completed the passage he will have represented the information in the first sentence as "Mr. Smith, the plumber, came walking out of the house" although this information was not available at the time that he read this sentence.

An experiment will serve to illustrate this point. A group of subjects were given a passage to read in which several different anaphoric references were made to the same individual. On completing

the passage they were then required to indicate who had been involved in the various events mentioned in the passage. They did
this by answering a series of WH-questions which had been generated by inserting the interrogative pronoun into one of the sentences from the text in place of the original reference to the
individual.   For example, if the subject had been given the
passage shown above, he would be asked questions such as "Who
ran out of the house?", "Who fled down the nearest alley.", etc.
In this way it was possible to discover how the reader had represented the information from that particular sentence in the
text.

For twenty of the subjects the passage was written in such a way
that each subsequent reference was more specific than the previous
ones as in the passage shown above. For the other twenty this was
reversed. Each subject answered 11 questions so giving 440 responses in all. If we ignore the 5% incorrect answers there were
only 34% which matched the original reference to that individual
in the text.   Of the remaining answers 75% gave more information
than was given in the original sentence and nearly half of these
gave more information than was available to the reader at the time
of reading that sentence.

These results reinforce the observation that a reader builds up a
representation of the passage which is independent of the original
sentences. Beyond this, however, they show that there is a tendency to identify the individuals encountered in the most informative way possible. In terms of the Frames analysis, we can
assume that the reader is accessing the Frame associated with that
individual and reading off the descriptors accumulated as a result
of reading the whole passage, which will reflect not just the one
reference but all identifying information available.

Identification as a function of interpreting text versus identification as a function of interpreting a sentence

The Frames analysis offers a solution to the problem of identifying the various people or objects mentioned in a passage at the
same time as suggesting how information about them can be represented as the reader interprets the text as a whole. Each entity mentioned is associated with a Frame retrieved from memory.   The
Frame raises questions which, in a sense, direct the way in which
the reader interprets subsequent sentences in the passage. The
reader's main goal is at all times to try to use the transient
outputs of sentence analysis to answer these questions and so
build up and elaborate on the Frames already present.

Such an account can in part be contrasted with an alternative
model proposed by Haviland and Clark (1974) which was designed to
explain how people integrate the information from the sentences

that they are currently interpreting with that already in memory.
They suggest that as a reader interprets a sentence he first breaks
it down into its syntactically defined "Given - New"[3] structure, and
then he attempts to add the new information to memory. They argue
that the way in which the reader is able to identify the "Given"
information is by looking for the presence of particular syntactic
constructions such as restrictive relative clauses, cleft constructions, definite noun phrases and so on. Once the "Given" information
has been identified in this way the reader searches his memory for
a matching antecedent and on finding it adds the new information to
what is already stored there.

The Clark model contrasts with what we have been referring to as
the Frames analysis in terms of what it is that instigates the
identification process. In the case of the Frames analysis
identification comes about as a result of attempting to assign
sub-Frames to the terminals on the Frames already evoked. It is
in a sense instigated by the representation that the reader has
already built up. In the "Given - New" model the identification
search is instigated as a result of the reader's preliminary interpretation of the sentence. It is motivated not only by his representation of the previous sentences but by a syntactic analysis
of a sentence currently under interpretation. The reader's use
of the context is so to speak passive here: it merely filters out
certain interpretations of the current sentence.

It is to a certain extent possible to make differential predictions
from the two approaches in the context of the anaphora paradigm
that we have been talking about. For instance, consider how a
reader would deal with a sentence pair of the following sort according to the "Given - New" account:

(7)   A $\begin{pmatrix} \text{Tank} \\ \text{Bus} \end{pmatrix}$ came trundling down the hill.

(8)   It nearly smashed into some vehicles.

In this case as the reader interprets sentence (8) he will first
discover "It" can be taken as "Given" and the rest of the sentence
as "New". So no semantic comparison between "Bus" in sentence
(7) and "Vehicles" in sentence (8) would be entertained. In terms
of the first experiment mentioned above we would not expect to see
a conjoint frequency effect here for reading sentence (8). With
the Frames analysis however this need not be so. After interpreting
sentence (7) one can assume that a Frame has been evoked for the
bus or tank whose terminals will expect and key further assignments.
On encountering sentence (8) both the noun phrases "It" and "Some
vehicles" would be entertained as possible pointers for assignments
to this Frame,[4] since they would both be keyed by the terminal
markers. We would therefore expect to see the presence of a conjoint frequency effect for the reading of sentence (8).

In order to test this prediction experimentally three sets of
materials were generated in which the presence of semantic and
syntactic cues for anaphora could be manipulated.   Examples from
each set are shown in Figure 1.

In set A both the syntactic and semantic conditions for anaphora
are present.   The noun phrase containing "vehicle" is definite
and the semantic relationship between "bus" or "tank" and "vehicle"
is appropriate.   In set B the syntactic information would preclude
identification, since the noun phrase is not definite.   The third
set C merely acts as the control comparison since here the sentence
is identical to that in set B except for the addition of a modifier
in the "vehicle" noun phrase, which would preclude identification
on semantic grounds, (in terms of the Frame analysis "horsedrawn
vehicle" would be rejected by both a "bus" or "tank" Frame terminal).

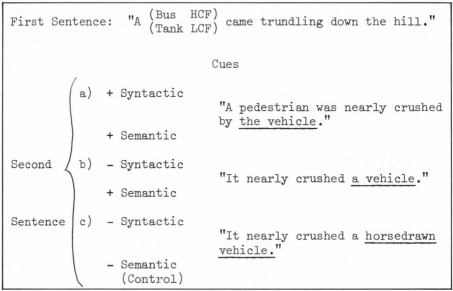

Figure 1   Examples of materials used in Experiment 3.

All of the materials used in the first experiment were modified in
this fashion so producing three sets of sentence pairs.   Three
groups of 12 subjects each were presented with set A, set B, or
set C materials and their reading times were measured as in
Experiment 1.   The second sentence reading times are shown in
Table 2 with the results of individual analysis of variance indicated.[5]   For set A there was a substantial and highly reliable
conjoint frequency effect as expected (Min F' $(1,25)$ = 6.95,
$p < 0.025$).   For set B we see a smaller effect of 46msecs., reliable by materials but just short by subjects (Materials
F $(1,16)$ = 4.70, $p < 0.025$, Subjects F $(1,11)$ = 2.55, $p < 0.05$), nine

out of the twelve subject showed evidence of the effect). In
set C there is no indication of any effect (all F's<1.).

Table 2 Mean reading times (secs.) for the second sentence,
         Experiment 3.

	Hi	Lo	Difference (msecs.)
Condition A	1.383	1.480	97
Condition B	1.348	1.394	46
Condition C	1.324	1.338	14

These results go against the prediction from the "Given - New"
model since even in syntactically unfavourable environments there
is some evidence that the reader is comparing antecedent items
with potential anaphors. This is quite consistent with the account
derived from Minsky's Frame analysis where the impetus for text
interpretation comes from the Frames already evoked, rather than
the sentence currently under interpretation.

Summary and general discussion

In this paper we have taken some problems associated with anaphoric
reference as a point of departure for looking at test comprehension.
The two major questions that were considered were first how the
reader is able to identify the entity that is being referred to in
an anaphoric phrase and second how the reader represents in memory
the information associated with this entity.

In attempting to answer these questions we discussed a very general
model for interpreting text proposed by Minsky (1975). He has
suggested that each entity mentioned in a text evokes a Frame.
This Frame then serves as an active representation searching out
more specific information from subsequent sentences in the text.
The problem of identifying any particular reference to this entity
is resolved in terms of the markers on each Frame terminal which
accept or reject information as appropriate for assignment to that
Frame. While the representational problem is solved in terms of
the Frames themselves.

In contrast to more specific models such as Haviland and Clark's
"Given - New" model the Frames account attributes to context an
active role in interpreting current information. The way in
which the contextual information is represented directs the way
in which the current information is interpreted.

There are of course several shortcomings of the Frames approach
which have not been considered in this paper. The way in which
a particular Frame is evoked is not very clear, what determines

the particular terminals on the Frames is not clear and so on. However, as a way of approaching text comprehension it has great utility. It suggests solutions to problems that other more specific accounts do not even entertain.

Footnotes

1 If you ask someone to list exemplars of some prespecified category, there are certain items that will always be listed (High Conjoint Frequency) and other items that will be listed very rarely (Low Conjoint Frequency). For instance, if you give the class "vehicle" many people will put down "bus" but few will put down "tank".

2 The F's both by subjects and by materials for this inter-action were <1.

3 The "Given - New" distinction derives from account such as Halliday (1967).

4 If (8) had read "the gradient was dangerously steep for some vehicles" it would be necessary to appreciate that "some vehicles" was to be related to the "bus" or "tank" Frames.

5 Since the second sentences were different in each case a separate analysis was performed for each condition.

References

Halliday, M.A.K. (1967). Notes on transitivity and theme in English: II. Journal of Linguistics, 3, 199-244.

Haviland, S.E. and Clark, H.H. (1974). What's new? Acquiring new information as a process in comprehension. Journal of Verbal Learning and Verbal Behaviour, 13, 512-521.

Minsky, M. (1975). A framework for representing knowledge. In Winston, P.H. (Ed.) The Psychology of Computer Vision. McGraw Hill.

Rosch, E. (1973). On the internal structure of perceptual and semantic categories. In Moore, I.E. (Ed.) Cognitive Development and Acquisition of Language. New York: Academic Press.

Sandford, A.J. & Garrod, S. (1975). The processing of class membership information directly and indirectly. Paper presented at a meeting of the Experimental Psychology Society, July 1975 at Cambridge, England.

Wilkins, A.J. (1971). Conjoint frequency, category size and categorisation time. J. Verbal Learning, 10, 382-385.

SENTENCE COMPREHENSION PROCESSES IN THE PRE-SCHOOLER[1]

DONALD J. FOSS, RANDOLPH G. BIAS and PRENTICE STARKEY

University of Texas at Austin

Among the goals of developmental psycholinguistics are these three:
to construct a description of the tacit knowledge of language that
children have at various points in their development; to describe
the processes that children use in producing and comprehending sentences; and to describe the principles of change or development
that govern the first two. It is obvious that these are inter-
dependent tasks.

Over the past decade and a half there have been a large number of
studies whose primary aim has been to describe the knowledge structures that are possessed by children as they are acquiring language.
Most of the grammar-writing projects fall under this heading.
However, these same data have also been used to inspire and support
models of sentence processing. Likewise, much of the experimental
work on acquisition has been aimed at characterising children's
knowledge, although some of the same data have also been used to
support hypotheses concerning comprehension processes.

Theories of linguistic competence are, of course, tremendously
under-determined by the data. Even so, the range of data available to the linguist who is attempting to describe the linguistic
knowledge of adults is much greater than that typically available
to the developmentalist. Linguists find that judgments of synonymy
and grammaticality are indispensible data to use in constructing
grammars. But it is just these sorts of data that are notoriously
hard to get from, say, the two- or three-year-old. Collections of
children's utterances do not put heavy enough constraints upon
models of linguistic knowledge. With respect to theories of language use, the problems are even more serious. About the only
data that we have, aside from the child's utterances, are those

that are gathered in comprehension tests given after the sentences.
A variety of such tests has been used, ranging from verbal responses such as imitation and paraphrase to such non-verbal tasks
as picture pointing and toy moving. As we will have occasion to
point out in somewhat more detail later, the variety of comprehension tests can be something of a problem. When they lead to con-
verging results, all is well; we can go to sleep at night with
some confidence that we are on to something. However, this state
of affairs is not common in our field. There are differences of
opinion in the literature concerning such substantive issues as
the acquisition sequence of various grammatical constructions.
Some of these differences may be due to variation in the techniques
used to assess comprehension.

Developmental psycholinguists have not, for the most part, faced
the following issue: in order to make inferences from response
measures to underlying structures or processes, it is necessary
to have a theory of the task as well as a theory of the structure
or process of direct interest.   In fact, what one really requires
is a theory of the process x task interaction. In a stimulating
paper entitled, "You can't play 20 questions with nature and win",
Allen Newell (1973) made a similar point with what he called the
First Injunction of Psychological Experimentation. It is: "Know
the method your subject is using to perform the experimental task".
It hardly needs saying that we do not have a theory of, say,
paraphrase or of imitation.   If we don't know what is involved in
imitation, how certain can we be about our conclusions concerning
comprehension when they are based upon imitative responses?

Given the above remarks, it may be somewhat surprising to learn
that one of the aims of the work reported in the present paper is
to expand the types of comprehension measures that are available
to the developmental psycholinguist.   This is not as inconsistent
as it may first appear.   We hope to show that the comprehension
measure here to be introduced has certain advantages over others
that have been used. If this is so, then working out a theory
of this task will be as good a bet as working on some others.
(Understanding other measures of comprehension is, of course, also
of great importance).   Thus, the present paper has two foci. One
is methodological in nature.   The project involves an attempt to
expand the types of comprehension measures that are available to
constrain our theories of comprehension and its development.   The
second focus has to do with some substantive issues in the comprehension of complex sentences.

An "On-line" Measure of Comprehension in Children

As stated above, all of the measures of comprehension that have
been used with children involve observations that are taken after
the sentence (or larger unit of discourse) has occurred. In con-

trast, we have tried to develop a technique that can measure the
momentary difficulty that children are having while they are
comprehending sentences. If we could specify the points within
a sentence where a child was having difficulty in processing it,
then we could more readily test, and even construct, models of
the comprehension process.   Furthermore, we could evaluate whether
and how the points of processing difficulty change during the
child's development. Such data would constrain theorizing about
comprehension processes and their development in a way that the
types of data that are presently available can not do.   This
assumes, of course, that we also have some idea about how the
task itself operates. At present, we do not have answers to all
of the questions that one could ask about the task, but some
progress has been made.

The measure of comprehension difficulty that we have devised uses
reaction-time (RT) as the dependent variable. We present children
with recorded sentences and ask them to do two things. First, we
ask them to comprehend the sentences. As a check on whether they
do in fact comprehend them, a question about the content of each
sentence is asked after its presentation.   Second, we ask the
children to listen within each sentence for a particular word and
to push a button as soon as they hear this target word.   We
measure the time that it takes them to respond to the target and
these times are our major dependent variable.   The assumption
behind the word-monitoring task is that the time to respond to the
target word is directly related to the difficulty of processing
the sentence at the point where the target word occurs. Thus,
when the sentence is relatively easy to understand at the point
where the target occurs, the time to respond to the target will be
relatively short since the child will be able to devote more of
his attention to the word-monitoring task.   When the sentence is
giving the processing mechanism difficulties, the time to respond
to the target word will be relatively long.

The retionale and general technique bear a close resemblance to
those that we have used with adults in a technique called phoneme-
monitoring (see Foss, 1969, 1970; Foss and Swinney, 1973).   We
settled on the word-monitoring technique after pilot work with a
few others.   Phoneme-monitoring is obviously too difficult for
the pre-schooler. The word-monitoring task can be grasped by a
late three-year-old.   There is some controversy in the literature
about the mechanisms involved in phoneme-monitoring (Foss, 1969;
Morton and Long, 1976), although space does not permit exploring
them here. For the present we will simply note that such theorizing exists.

Experiment 1: Demonstration Experiment

One of the first things that we had to do when developing this new

technique was to see if it would work in what we might call a
"clear case" situation. If we could not find RT differences in
such a situation, then there would be no hope for using the task
to detect more subtle differences in sentence processing difficulty.
The first experiment used sentences like those shown in Table 1.
In half of the sentences the target word was immediately preceded
by a common word known to the children in the study. In the other
half of the sentences the target word was immediately preceded by
a non-word, an item that could not be known by any of the children.
In addition, we included a second variable, the syntactic function
of the target word in the sentence.   Actually this latter variable
was a complex and purposefully confounded one.   In half of the
sentences the target was a noun which occurred either in the main
clause of the sentence, in a prepositional phrase, or in a for/to
complement clause.   In the other half of the sentences the target
was a verb in a that-complement clause.

We reasoned that children ought to be slower in responding to
targets that occurred after non-words since their processing
mechanisms would be busy trying to decode the unknown item and
hence could not devote as much attention to the word-monitoring
task. At the time when the target word occurred, the child would
be busy unsuccessfully searching his mental lexicon for the nonsense word and/or be busy trying to construct an interpretation
for this item from the information available in the item's context.
In addition, the child may have had a harder time assigning word
boundaries when one of the input items was not a word. For a
number of reasons we also suspected that children would be slower
in responding to targets in that-complements than to those in the
other category. Sentences with that-complements have two surface
agents before the target verb, a fact that may confuse the child;
they have verbs of propositional attitude (e.g., believe, know,
think) which may be psychologically more complex (see Hakes, 1971;
Harris, 1975); and such sentences simply appear to be harder to
process. Given the way that we had stacked the deck in our favour,
we were prepared to toss in the word-monitoring technique if neither
of the variables had any effect.   It would be very unlikely that
any lexical or syntactic variable would lead to a difference in RT
if these two did not.

The subjects in Experiment 1 were 16 children from the University
Nursery School in Austin, Texas.   The eight boys in the study had
an average age of 4.11 and the eight girls averaged 4.10.   Two
material sets were used so that sentences carrying real words just
prior to the target item were counter-balanced with those carrying
nonwords in that position.   Five sentences of each of the four
types were presented to the children, who were tested one at a
time in a small room at the school. In addition to the experimental
sentences we also presented 13 filler sentences, six of which did
not contain the specified target word.   It turned out that the

Table 1. Sentences used in the Demonstration Experiment (Expt. 1)
Target words are underlined.

Item Preceeding the Target	Target	
	Noun in Main Clause, Prep Phrase, or For/to Complement Clause.	Verb in that-Complement Clause
word	Ernie likes to throw a red ball while Bert runs to catch it.	The nice lady believed that the girl watched the baby during nap time.
nonword	...zed ball...	...rul watched...

children liked to have a turn at saying what the target would be,
so these fillers also helped to keep up the child's interest in the
task.   The sentences were tape recorded and presented free field.
A millisecond timer was automatically started at the beginning of
the target word and was stopped by the child's button pushing response.   After each sentence the child was given a question to see
whether he or she understood the sentence, e.g., "Who ran to catch
the ball?"

The results of the study are shown in Table 2.   The effect of the
word vs. nonword variable was large (153 msec.) and highly significant: the F_1 value was 20.07, $p < .001$, while F_2 was 20.37,
$p < .001$.   Min F' (1,29) = 10.11, $p < .005$.   The effect due to the
syntactic function of the target was also large (162 msec.) and
reliable, F_1 (1,12) = 9.63, $p < .01$; F_2 (1,9) = 17.13, $p < .003$;
and min F' (1,21) = 6.16, $p < .03$.   The two variables did not
interact.   In addition neither of the effects due to sex of subject
nor materials was significant;   nor did these two interact with
any of the other variables.

Table 2.   Reaction times (msec) in Demonstration Experiment.
                 (Expt. 1)

Item Preceding	Target		
The Target	Noun	Verb	X̄
word	568	792	680
nonword	783	884	833
X̄	676	838	757

Naturally we found this to be a very rewarding study.   The variables
that should have shown effects on RT did so;   those that shouldn't,
didn't.   The absolute values of RT were not too high, the differences were substantial, and the N needed to get reliable results
was quite modest.   We had had the impression that doing RT work
with children was going against any underground lore but we were
in the happy position of being able to doubt this lore.   A bit of
patience and an enjoyable task (which the children seemed to find
word-monitoring) appeared to be all that was required to get
beautiful data.   Unfortunately, things have not always worked out
quite so nicely;   but let us not get ahead of our story.   In sum,
we felt at this point that the methodological battle had been
carried, and we moved to a more substantive issue.

Processing Relative Clause Sentences

There has been a fair amount of work reported in the literature on
the acquisition and processing of relative clauses. There is also
a controversy about what types of relative clause containing sentences are most difficult to process. It was upon this issue that
we tried to bring to bear data gathered with the word-monitoring
technique. Most of the data in the literature is consistent
with what Slobin (1971) called "Operating Principle D: Avoid
interruption or rearrangement of linguistic units." Slobin
raised this principle to a language processing universal (his
Universal D4), "The greater the separation between related parts
of a sentence, the greater the tendency that the sentence will not
be adequately processed (in imitation, comprehension, or production)"(Slobin, 1971, p.354). We will call this the Interruption
hypothesis. Data in support of this hypothesized universal include the wellknown difficulty of handling multiply self-embedded
sentences such as (1).

(1) The coffee that the mug that Esther purchased holds is
    delicious.

Chomsky (1965, p.14) proposed one reason why this difficulty might
arise:

> To account for the greater unacceptibility of self-embedding
> (assuming this to be a fact), we must add other conditions on
> the perceptual device beyond mere limitation of memory. We
> might assume, for example, that the perceptual device has a
> stock of analytic procedures available to it, one correspond-
> ing to each kind of phrase, and that it is organised in such
> a way that it is unable (or finds it difficult) to utilize a
> procedure ϕ while it is in the course of executing ϕ.

In support of this notion, Baird (1973) found that self-embedding
contributed more than simply nesting to the difficulty of comprehension in subjects ranging in age from four years to college.
He used a verbatim recall task in his study. Also Brown (1971)
found that embedding led to poorer performance in a picture pointing task when he tested three-, four- and five-year olds. And
Smith (1974) reported poorer performance in an elicited imitation task for embedded as opposed to right-branching sentences
when he tested children between 29 and 36 months of age. This
pattern of results seems to fit the criterion for a good night's
sleep mentioned earlier. A variety of tasks have converged upon
similar results. However, not all of the results in the literature are as somniferous.

Sheldon (1974) used a toy moving task to study the comprehension
of various types of relative clauses in children from 3.8 to 5.5.
It is worth looking at her sentence types in detail; they are

shown, along with their labels, in (2 - 5).

(2) SS   The dog that jumps over the pig bumps into the
          lion. (1.58)

(3) SO   The dog that the pig jumps over bumps into the
          lion. (.52)

(4) OS   The dog jumps over the pig that bumps into the
          lion. (.88)

(5) OO   The dog jumps over the pig that the lion bumps
          into. (1.52)

The labels, SS etc. are to be read as follows. The first letter
tells us whether the relative clause modifies the subject (S) or the
object (O) of the matrix sentence; the second letter tells us the
role of the relative pronoun in the relative clause, it is either
the subject (S) or the object (O).   The first two sentence types,
SS and SO, embed the relative clause inside the matrix sentence.
According to the Interruption hypothesis, these sentences should
be harder to process than the OS and OO sentences. Other factors
may be at work, of course. Thus, if sentences which do not conform
to Bever's (1970) primary functional labeling strategy (i.e., NVN
sequences correspond to actor-action-object) are more difficult
to process, then SO sentences should be harder to comprehend than
SS sentences.

The numbers in parentheses after the example sentences give the
mean number of correct sentences (out of three possible) that were
obtained by Sheldon in her toy moving task. These results do not
fit the Interruption hypothesis since the score for the SS sentence
type was high.   Sheldon proposed a new hypothesis to account for
her results, the "Parallel Function" hypothesis. It says that
sentences will be difficult to process when coreferential noun
phrases have different functions (subject vs. object) in their two
occurences.   More explicitly: In a complex sentence, if coreferential noun phrases have the same grammatical function in their
respective clauses (SS and OO), then that sentence will be easier
to process than one in which the coreferential noun phrases have
different grammatical functions (SO and OS).   In her paper Sheldon
gave some reasons for thinking that the Parallel Function Hypothesis
has applicability beyond the relative clause case.   Also, if one
goes back and looks at the data of Brown (1971), his results are
found to be consistent with the Parallel Function Hypothesis. They
fit the latter hypothesis better than the Interruption hypothesis,
in fact.   The same cannot be said of Smith's (1974) data, however.
His results do not fit the Parallel Function hypothesis at all.
On the other hand, they do not give very strong support to the
Interruption hypothesis either. Smith found that the SO sentence
type was the worst imitated.   His data are most consonant with
a combination of the NVN hypothesis and Chomsky's Minimum Distance

Principle.   We can get one more bit of relevant data by inspecting Gaer's (1969) data obtained from three to six year old children
in a picture pointing task.   Among the sentences that Gaer used
were those of the SO and OS type.   According to the Interruption
hypothesis the former should be more difficult to process.   The
differences were very slight and almost certainly non-significant.
This is, of course, what the Parallel Function hypothesis would
predict.

It is possible to get bogged down in all of the experimental minut-
iae and such is not, of course, our intent.   It is important to
note the discrepancies however, so that the problem is set out
clearly.   At the moment we do not know which, if any, of the hypotheses are correct.   You may have a favourite and that favourite
may be conditioned not only by your view of how children comprehend
sentences, but also by the technique used in the experiments.   If
your favourite and ours differ, then we will probably get into a
discussion regarding the merits of the data in support of the
respective positions.   This will lead you to want to dissect the
technique which yielded results supporting our view, and vice
versa.   Such discussions show the importance of the process by
task interaction that we alluded to earlier.

Before we go on, let us do just a bit more dissecting on our own.
According to Sheldon's hypothesis the two sentences in which the
coreferential NPs do not have parallel function should lead to
poorer performance and they did.   However, they did so in somewhat different ways.   If we go back to the data reported in her
dissertation (1972) we can obtain performance measures on a clause
by clause basis rather than just the overall figures given in her
(1974) article.   In (6) and (7) the SO and OS sentences are repeated, this time with the average percent correct toy moving
responses given after each clause.

(6) SO The dog that the pig jumps over (57%) bumps into the
         lion (39%).

(7) OS The dog jumps over the pig (85%) that bumps into the
         lion (34%).

It is apparent that there is a substantial difference in performance on the first clause.   This is the result expected by both
the NVN and the Interruption hypotheses.   In (8) and (9), similar
data for the SS and OO sentences are presented.

(8) SS The dog that jumps over the pig (92%) bumps into the
         lion (54%).

(9) OO The dog jumps over the pig (91%) that the lion bumps
         into (50%).

It is noteworthy that the first clause scores in (8) and (9) are

so similar.   This suggests that the presence of the interruption
has no effect, exactly Sheldon's conclusion.   It is odd, however,
to think that the presence of the relative pronoun has no impact
on the child's comprehension mechanism.   The children are not
simply treating the relative clause sentences like conjoined sentences, Sheldon had control conditions which showed that.   Perhaps
part of the reason why Sheldon obtained the results she did has
to do with the toy moving task itself.   That task is one in which
the response occurs considerably later than does the sentence itself.   In order to see whether there was a local effect due to
the presence of the relative pronoun, we conducted a word-monitoring experiment.

Experiment II: Local Effect of Relative Pronouns

The hypothesis with which we undertook this study was that relative
pronouns provide a signal to the sentence processing mechanism to
enter into a special "sub-routine" for constructing the appropriate
syntactic analysis of the input string.   In entertaining this
hypothesis we were assuming, of course, that we were dealing with
sentence processing mechanisms that had acquired such a sub-routine
and the appropriate knowledge concerning the valid cues to operate
it.   We will return to this matter below.

Pre-school subjects were presented with SS and OO sentences like
(10) and (11).

(10)   SS   The woman that _kissed_ the man delighted the clown.

(11)   OO   The woman _kissed_ the man that the clown delighted.

The target word was the first verb in each sentence, those underlined in (10) and (11).   If there is a momentary increase in
processing difficulty due to the presence of a relative clause,
then word-monitoring RTs should be longer in SS sentences than in
OO sentences.

In the experiment we presented eight examples of each sentence
type, along with a number of fillers, to almost 50 preschool
children from two Austin, Texas, nursery schools.   Again the
content items in the sentences were counterbalanced across sentence types (although any subject saw only one version of any
particular sentence);   the sentences were presented over tape
free field;   and we asked a question after each sentence(e.g.
Who kissed who? or Who delighted who?

This time things did not go quite so well.   In the first place,
fewer subjects gave us enough data to analyse.   The data from
only about 35 of the 50 children were usable, primarily because
a substantial number of children gave very long RTs or else failed
to respond on a large fraction of the trials.   Furthermore, we

did not see a significant difference, overall, of the sort that we
had been expecting.   There was a difference, all right, but the
variability was too great to make it significant.   At this point
we turned to the comprehension data (data which will not be reported in detail here).   We noted the obvious, that is that some
children did much better on the comprehension test than others.   We
also noted that the absolute value of the RT data for the good
comprehenders was much lower than that for the poor comprehenders.
(A good comprehender is a child who gets at least one of the nouns
correct in his answer at least 50% of the time).   We had 26 good
comprehenders and 10 poor ones.   Furthermore, the difference in
RT to the two sentence types was greater (see Table 3) and significant (by subjects) in the case of good comprehenders; t = 2.38,
p < .05.   (Because of the post hoc nature of this analysis, we did
not compute F_2).

To summarize these data, we found some evidence that subjects
who were paying attention to the sentences and who were able to do
relatively well on the comprehension task were affected by the
presence of the relative pronoun in the sentence.   It was as
though the word 'that' signaled to them that they had to do some
extra processing.   To again use the computational metaphor, the
relative pronoun apparently signaled the children to activate a
sub-routine.   This activation takes processing capacity, a conclusion we reach by inspecting the RT data.   Thus, we found some
evidence, albeit slim, that there is an increase in local processing complexity for children when the relative pronoun "that" occurs.

Table 3.   Reaction Times (msec.) in Expt. II

Subject	Sentence Type	
Type	SS	OO
Good	753	680
Poor	929	887

Even the "good" comprehenders were far from perfect however.   They
made many more mistakes than college students or even, say, eight-
year olds would.   It appears, to continue the metaphor, that the
sub-routines are not de-bugged in the case of these preschoolers.
This observation suggested that the children might be error prone
when listening to sentences that bear a superficial resemblance
to the relative clause sentences. In other words, pre-school
children might not yet have learned which surface cues were valid
indicators of a relative clause and which were not.   Experiment III

.s designed to replicate Experiment II and to extend it along these lines.

Experiment III: Local Effects of "That"

Experiment III made use of the sentence types like those shown in (12 - 15).

(12) SS The woman that <u>kissed</u> the man delighted the clown.

(13) OS The woman <u>kissed</u> the man that delighted the clown.

(14) IT That the woman <u>kissed</u> the man delighted the clown.

(15) DOS That woman <u>kissed</u> the man that delighted the clown.

The SS and OS sentences permit a replication of the phenomenon for which we obtained some evidence in Experiment II, namely that sentence processing is momentarily more difficult after the presence of a relative pronoun than when there is no such pronoun.

The other two sentence types were included to test the hypothesis that preschool children may be simply using the word "that" as a cue to enter the relative clause processing sub-routine, i.e., that they would over-generalize by interpreting "that" as an indicator of a relative clause even when it is inappropriate to do so. This appeared to be a reasonable hypothesis since, from the error data, we had good evidence that children around five years of age did not have good control over the relative clause. The children at this age are obviously still acquiring the relative clause construction (especially subject relatives). What this means from the point of view on an information-processing analysis is that they do not yet have the appropriate cues, and the operations attendant upon the occurrence of these cues, well integrated into their sentence processing mechanism.

Consider, for example, the OS sentence which begins with the demonstrative "that". This has been labeled a DOS sentence (=15). We know from investigations of Limber (1973) that the demonstrative is one of the early constructions that enters into the child's productive vocabulary. It seems unlikely that a sentence beginning with a demonstrative would be harder than one beginning with a definite article - unless the child is confused about the cue value of the word "that", sometimes mistaking it for a relative pronoun. Such mistakes seem likely <u>a priori</u> when the child is in the process of acquiring the relative clause construction (i.e., when the child is adding the capability to deal with this construction to his sentence processor). If such confusion exists, then it seems plausible to predict that RTs will be elevated after demonstrative pronouns. Reaction time will be longer in DOS sentences like (15) than in OS sentences like (13).

The IT sentence (14) presents another case. In sentences of this type the word "that" also signals the presence of an embedded sentence, although the construction type is different of course. Numerous possibilities exist for the youthful processing mechanisms and the measurement technique, the IT sentences were included to see which might receive support. One possibility, for example, is that the IT sentences will behave exactly like the SS sentences on the monitoring task. This might suggest, depending upon the results for the other sentence types, that the effect of the word "that" on the processing mechanism was very short-lived. (Recently Cairns and Kamerman (1975) have found that the effects of ambiguous lexical items on phoneme-monitoring in adults are quite transitory.) Other possibilities exist, but we will not explore them here.

In the present experiment we made one major change in methodology; each subject was run in two experimental sessions on two separate days. This was done to get rid of some of the practice efforts and thereby to obtain more stable estimates of RT. Each day the children heard sentences exemplifying all four major types, plus fillers. The subjects were 20 pre-schoolers drawn from a population very similar to that of the earlier studies. The average age of the children was 5.3.

The results of the study showed that there was a very large (177 msec.) and significant difference in RTs across the two days. We will concentrate on the data from Day 2 since they are the more reliable. Once again we divided the subjects into two groups on the basis of their performance on the comprehension test, again focusing upon the good comprehenders since they are the children for whom we have evidence of attentiveness to the task. Most of the children fell into the category of good comprehenders. In order to maximize the validity of the various comparisons that we wanted to make, we selected pairs of sentence types and then chose as subjects just those children who did well on both types. (Recall that by doing "well" we mean that on the majority of sentences the subjects correctly gave one of the two nouns in answer to a question such as, "Who kissed who?". We adopted this criterion since it seemed to include the subjects who were clearly paying attention and trying.) Most of the children were either included in all pairwise comparisons (about 15 - 18 children) or excluded from all of them; but there was some change in the subject composition across the various comparisons of interest.

First question: Did we replicate the earlier finding that "that" leads to a momentary increase in processing difficulty in relative clause sentences? This amounts to asking whether SS sentences led to longer RTs than did OS sentences. The average RTs were 511 msec. and 444 msec., respectively. The SS sentences did lead to reliably longer RTs, F_1 (1,14) = 6.13, p <.03; F_2 (1,17) = 11.06,

$<.01$; and min $\underline{F}'$ (1,27) = 3.94, $\underline{p}<.07$. Although the min $\underline{F}'$ value just fails to reach significance, we are prepared to accept it as convincing given the fact that this replicates the earlier finding (using different subjects and materials).

Next question: Is there any evidence that the demonstrative leads to longer RTs, that it may be leading to confusion? Here the answer is far less certain. The difference in RT between DOS (552 msec.) and OS (418 msec.) sentences is over 100 msec. in the direction predicted by the confusion hypothesis. However, this difference did not reach acceptable levels of statistical significance, $\underline{F}_1$ (1,13) = 3.58, $\underline{p}<.09$; $\underline{F}_2$ (1,17) = 5.78, $\underline{p}<.03$. The other comparison that might have provided evidence in favour of the confusion hypothesis IT vs. OS, had $\underline{F}_1<1$. As noted above, there are other reasons for predicting that this latter comparison would not be significant. None of the other pairwise comparisons reached acceptable levels of significance either.

To summarize the results of Experiment III, we found evidence to support the hypothesis that the presence of a relative pronoun increases the local processing difficulty of the sentence containing it - at least in the preschool child who is in the process of acquiring the relative clause construction. Thus, even though there are no comprehension differences between SS and OO sentences overall, when assessed by the toy moving task, the word-monitoring results lend support to the view that the interruption of a main clause by a relative clause increases processing difficulty. Whether or not one takes this finding to support the Interruption Hypothesis depends to a certain extent upon how one interprets this hypothesis (vide infra).

Continuing our summary of the results, we found some evidence which suggests that children who are acquiring the subject relative construction tend to respond to the word "that" as though it is a cue for activating the relative clause sub-routine - even when "that" is not actually functioning as a relative pronoun (Over-generalisation hypothesis). The evidence was equivocal, however. This issue was, we felt, worth following up. It appeared that we might have found a developmental performance "dip". That is, it might be the case that younger children would do just fine on the demonstrative sentences (DOS) - until they begin to acquire the relative. Then they may go through a period of uncertainty about the significance of the word "that". This uncertainty would lead to poor performance on both the SS and DOS sentences. (Incidentally, these two types did not differ in word-monitoring RTs in Experiment III). Experiment IV was designed to be a follow-up study.

Experiment IV: Testing the Overgeneralisation Hypothesis

Experiment IV tested the Overgeneralisation hypothesis in two ways;

it was, in effect, two experiments combined into one set of materials and presented to the same subjects.   Earlier we remarked
that it seemed unlikely that a sentence beginning with a demonstrative would be harder to process than one beginning with a definite
article unless the child was confused about the cue value of the
word "that".   Perhaps that remark was too hasty. Beginning a
sentence with a demonstrative does seem somewhat unnatural unless
the following noun has occurred earlier in the sentence or unless
the referent of the noun is physically present at the time the
utterance occurs.   Neither of these two conditions obtained in
Experiment III.   The first of the two sub-experiments remedied
this potential deficiency of Experiment III.   Sentences like those
in (16) and (17) were presented to the children. Again the target
word is underlined.

(16)   The dog surprised the cat and the cat <u>scratched</u> the child.

(17)   The dog surprised the cat and that cat <u>scratched</u> the ·child.

In (17) the target verb is once again preceded by the word "that"
functioning as a demonstrative.   However, this sentence meets the
naturalness condition stated above.   If the word "that" leads to
increases in processing difficulty even when it is not functioning
as a relative pronoun, then RT should be longer in sentences like
(17) than in those like (16).

The second sub-experiment contrasted sentences like those in (18)
with sentences such as (19).   Targets are underlined.

(18)   The boy that the lady <u>spanked</u> bothered the girl.

(19)   The boy the lady <u>spanked</u> bothered the girl.

As is obvious, (18) and (19) are identical except for the deletion
of the relative pronoun in (19).   In studies using the phoneme-
monitoring technique with adult subjects it has been found that RT
is longer when the word "that" is deleted than when it is present
(Hakes and Foss, 1970). Note, however, that if the word "that" is
signaling to the subjects that they should enter a (for them) complicated sub-routine, then sentences like (18) might actually be
<u>harder</u> for the preschool child to process than sentences such as
(19), exactly the opposite pattern of results that we obtain from
adults in similar studies.   If this were to be the case, if we
found an interaction of sentence type x age, it would provide us
with some fertile ground for our experimental ploughs.

The subjects in Experiment IV were, again, nursery school children
drawn from the same population as that in Experiment III.   The
children were presented with 40 test sentences and a number of
fillers spread across two testing days.   The children averaged
5.8, a bit older than those in the preceding study.

.e again in reporting the results we will concentrate on the
.ore reliable Day 2 data and the data from those children who gave
evidence of attending to the task.   With respect to the first
sub-experiment, the one that tested the difference between definite
articles and demonstratives in the "natural" condition, we have no
use for statistics.   The mean RTs for these two conditions were
463.6 msec. and 464.2 msec., respectively. To say the least, we
cannot reject the null hypothesis given these results.   Things were
no better for the Overgeneralization hypothesis when the data from
the second sub-experiment were inspected.   Average RT was 504 msec.
in the sentences with "that" present (supposedly the more troublesome sentences for the children) and 553 msec. for the sentences
with "that" deleted.   Hence, these data were in accord with the
results obtained with adults. No age x sentence type interaction
is available here.

Taken together, these two sub-experiments lend no support to the
Overgeneralisation hypothesis. This aruges that the suggestive
results in the DOS condition of Experiment III were due to the un-
naturalness of the sentence-initial demonstrative (or to error).[2]
It further suggests that the children who are acquiring the relative clause construction are finely tuned to the other cues
available that permit discrimination between the various functions
of "that".   These include such cues as intonation and the surrounding lexical items.

Overview

We have now looked at the results from our experiments using the
word-monitoring technique. What conclusions can we draw from them?
One reasonable conclusion is that the word-monitoring technique
works; it can provide information of fine enough grain to help
adjudicate theoretical issues. It must be said however that the
technique is not put forward here as the experimentalist's panacea.
It has a number of built-in limitations. For one thing, the age
range with which it can be used is limited; it almost certainly
will not be suitable for investigating language acquisition before
four years of age. Also, it is a very time-consuming method to
use, hence expensive. In addition, it is important that some
other independent measure of comprehension be gathered along with
the RTs. We need a way of excluding those subjects who are not
closely attending to the task. Further, since the variability in
the RT data is high, one needs to have either a very powerful variable or else large numbers of subjects and sentences. Because of
the attention span of young children, it is difficult to present
large numbers of sentences.

With respect to the issues surrounding the processing of relative
clauses, we have seen that the RT data support the following points
First, the concept of "interruption" needs to be clarified.

One can have short-term processing difficulty, as indexed by the
RT data, without necessarily finding lowered comprehension of the
difficult clause when the comprehension measure is taken after
the sentence.   Second, the data from Experiments II and III
support the idea that children find the presence of a subject relative clause to be momentarily taxing.   This finding would fit
very nicely into the framework suggested by Chomsky which we
quoted earlier.   Recursive calls to the same subroutine hurt the
child's comprehension.   This is a happy state of affairs since
it permits a uniform mechanism to account for this aspect of comprehension in both children and adults.   There are a number of
more precise models of comprehension into which such a finding
would be neatly integrated.   One promising one is that of the Augmented Transition Network (Woods 1973) which has inspired some
psychological experimentation on the processing of relative clauses
by adults (see Wanner and Maratsos, 1974).

Third, we have also obtained evidence that five-year old children
are sensitive to a range of cues which permits them to discriminate
the various functions of "that" very quickly during the processing
of a sentence.   This suggests that the children are sensitive to
such cues as intonation and that they are able to use sentential
context to quickly decide what the function of "that" is in any
particular sentence.   This finding also has developmental implications.   It suggests that children do not attempt a large scale re-
organisation of their processing mechanisms when they are acquiring a troublesome new construction.   It is a conservative system.

It is clear that we have not solved very many problems by utilising
data gathered with the word-monitoring technique.   In fact, it
seems apparent that data of the sort we have here reported are
going to complicate our task of accounting for the acquisition of
comprehension.   But, we would argue, they are also going to make
the stories that we tell about the comprehension process and its
development closer to the truth.

Footnotes

1 The work reported here was supported in part by a grant from
   the National Institute of Education, Department of Health,
   Education and Welfare. However, the opinions expressed herein
   do not necessarily reflect the policy or position of the N.I.E.,
   and no official endorsement by the N.I.E. should be inferred.

   The authors would like to thank the staff and students at the
   All Saints Episcopal Nursery School, Austin, Texas, for their
   excellent cooperation in helping us carry out most of the research reported in this paper. In addition, thanks go to David
   Fay who helpfully discussed various aspects of this research

with us.

2   Another possibility is that the target words in Expt. IV occurred
    too far from the word "that" to reflect the momentary difficulty
    that "that" engendered. By appealing to this explanation we
    could save the Overgeneralisation hypothesis, but it would
    clearly be post hoc to do so.

References

Baird, R. (1973). Structural characteristics of clause-containing
    sentences and imitation by children and adults. J. Psycholing.
    Rsh., 2, 115-127.

Brown, H.D. (1971). Children's comprehension of relativized English
    sentences. Child Development, 42, 1923-1936.

Cairns, H.S., and Kamerman, J. (1975). Lexical information process-
    ing during sentence comprehension. J. Verb. Learn. Verb. Behav.,
    14, 170-179.

Chomsky, N. (1965). Aspects of the theory of syntax. Cambridge,
    Mass: MIT Press.

Foss, D.J. (1969). Decision processes during sentence comprehension:
    Effects of lexical item difficulty and position upon decision
    time. J. Verb. Learn. Verb. Behav., 8, 457-462.

Foss, D.J. (1970). Some effects of ambiguity upon sentence com-
    prehension. J. Verb. Learn. Verb. Behav., 9, 699-706.

Foss, D.J. and Swinney, D.A. (1973). On the psychological reality
    of the phoneme: Perception, identification and consciousness.
    J. Verb. Learn. Verb. Behav., 12, 246-257.

Gaer, E. (1969). Children's understanding and production of sen-
    tences. J. Verb. Learn. Verb. Behav., 8, 289-294.

Hakes, D.T. (1971). Does verb structure affect sentence comprehen-
    sion? Percept. & Psychophys., 10, 229-232.

Hakes, D.T. & Foss, D.J. (1970). Decision processes during sen-
    tence comprehension: Effects of surface structure revisited.
    Percept. & Psychophys., 8, 413-416.

Harris, R.J. (1975). Children's comprehension of complex sentences.
    J. Exp. Chil. Psychol., 19, 420-433.

Limber, J. (1973). The genesis of complex sentences. In T. Moore
    (ed.) Cognition and language development. N.Y.: Academic Press.

Morton, J. and Long, J. (1976). Effect of word transitional pro-
    bability on phoneme identification. J. Verb. Learn. Verb. Behav.,
    15, 43-51.

Newell, A. (1973). You can't play 20 questions with nature and win: Projective comments on the papers of this symposium. In W.G. Chase (ed.) Visual Information processing. N.Y.: Academic Press.

Sheldon, A. (1972). The acquisition of relative clauses in English. Unpublished Ph.D. dissertation, Austin, The University of Texas.

Sheldon, A. (1974). The role of parallel function in the acquisition of relative clauses in English. J. Verb. Learn. Verb. Behav., 13, 272-281.

Slobin, D.I. (1971). Development psycholinguistics. In W.O. Dingwall (ed.) A survey of linguistic science. College Park, Md: Linguistics program/University of Maryland.

Smith, M.D. (1974). Relative clause formation between 29 and 36 months: A Preliminary Report. In E.V. Clark (ed.) Papers and reports on child language development, No. 8 Stanford: Committee on Linguistics.

Wanner, E. and Maratsos, M. (1974). An augmented transition network model of relative clause comprehension. Unpublished paper, Harvard University.

Woods, W.A. (1973). An experimental parsing system for transition network grammars. In R. Rustin (ed.) Natural Language Processing. N.Y.: Algorithmics Press.

REFERENCE AS A SPEECH ART: AN ARGUMENT FOR STUDYING THE LISTENER

S.R. ROCHESTER

University of Toronto

In this paper I am concerned with the speaker as an "artist" who creates coherence by guiding listeners to select precise referents for noun phrases. This guidance is not the referring of symbols to things which characterises all language, but a particular referring which marks out some noun phrases as requiring more information in order to be understood, and other noun phrases as requiring no further information. I wish to argue that it is extremely risky to assess the speaker's artistry solely from an analysis of the speaker's performance. The art of referring, I will argue, lies in the guidance it gives to the listener. Therefore, to judge the art we must examine not only the speaker's production, but also the listener's experience of that production.

This would seem to be an obvious prescription: to evaluate the guidance, determine whether those guided get lost or find their way. That it is not obvious, or at least not obvious enough, will be shown from a consideration of two levels of analysis of noun phrase reference. The first level of analysis occurs with so-called "egocentric reference" in which the speaker fails to take the point of view of the listener into account. On both intuitive and empirical grounds, egocentric reference seems to point to the speaker's lack of skill. Intuitively, it seems that any one who fails to consider the listener must be an inept guide. And there is apparent support for this intuition in the fact that younger children and thought-disordered adults use egocentric

reference more frequently than older children and undis-
turbed adults. However, there are some data and an
argument which suggest that these conclusions must be
validated by studies of listeners.

The second level of analysis concerns the location of
referents in the utterance context. There is putative
support for the view that locating referents in the
situational context of an utterance is a less skillful
procedure than locating referents in the verbal context.
Again, the support derives from opinions about the adequacy of the speakers: situational reference is used
by younger children, by disturbed adults, and by working
class as opposed to middle class children and adults.
At this level of analysis, there are again data which
challenge the easy inference and suggest that the speaker's art can only be assessed adequately by a study of
the listener.

Egocentric Reference

If we try to imagine what sort of guidance might be
helpful to the listener, and what sort might be misleading, one sort of guidance seems clearly misleading.
This is the case where the speaker promises information
which is never actually given. For example, if the
speaker mentions that old hat, the listener expects to
find more information about that hat elsewhere in the
utterance context. The information may be in the immediate environmental situation, or in the listener's recent
memory of the verbal or situational context, or may be
about to be presented. If the referent is in fact recov-
erable, the speaker's guidance to the listener would seem
to be helpful. If the referent is nowhere to be found,
the guidance would seem to be misleading.

Children. If this conception is correct and helpful
guidance requires that promises be fulfilled, then such
guidance is not provided by all native speakers. Young
children in particular appear to break their promises
to listeners, as Piaget has vividly described (1955).
For example, when Piaget asked some children to tell
fairy tales to other children and to explain mechanical
devices to them, he found that his subjects understood
the stories and devices presented to them but gave
inadequate accounts of these stories and devices to
their peers:

> The explainer always gave us the impression
> of talking to himself without bothering about
> the other child. Very rarely did he succeed
> in placing himself at the listener's point
> of view. (Piaget, 1955, p.115).

To demonstrate the children's frequent failures to take
their listener's point of view, Piaget presents the
account of Gio, an 8 year old who tells the story of
Niobe:

> Once upon a time there was a lady who has
> twelve boys and twelve girls, and then a
> fairy, a boy and a girl. And then Niobe
> wanted to have some more sons. Then she
> was angry. She fastened her to a stone.
> (Piaget, 1955, p.116)

In addition to the elliptical predicates and a lack of
continuity between the sentences, use of noun phrase
reference is also misleading. For instance, the pronouns she and her in the last two sentences invite the
listener to select referents, but the precise referents
are ambiguous from the verbal context.

Later work has confirmed Piaget's descriptions, while
suggesting that children do not ignore their listener
needs as thoroughly as he suggests (cf. Maratsos'
excellent review of this literature, 1976). It is clear
from these later studies that young children often fail
to provide referents for noun phrases which require them.
What is not clear, and what has not been studied is the
extent to which this failure affects the listener. We
simply do not know whether our intuitions about broken
promises by the speaker are correct. Or, more precisely,
we do not know how to specify the function relating
speaker's behaviour and listener's experience in this
regard. For example, it certainly seems likely that the
extent of the effect of egocentric reference depends in
part on the characteristics of the listener (child or
adult? friend or stranger?) and in part on the situational context (narrative or conversation?). But we do
not know the parameters of these variables, nor how they
depend on the distribution of egocentric reference in
the speaker's production.

Schizophrenic Speakers. Young children are not the only
speakers who regularly fail their listeners. Egocentric
use of noun phrases has also been found with schizophrenic speakers who are diagnosed as thought-disordered.

In a series of naturalistic studies (Martin and Rochester
1975; Rochester and Martin, in press; Rochester, Thurston
and Rupp in press), we compared 20 thought-disordered
speakers to an equal number of schizophrenic speakers
whom psychiatrists judged to be free of thought-disorder,
and to 20 speakers with no history of psychiatric disturbance.

Thought-disordered speakers are interesting because their
language is impaired only briefly and (in young patients)
typically returns to normal with the remission of the
psychotic episode. Moreover, the impairment itself
is interesting because it is also restricted: the speakers'
clauses are generally well-formed and their lexicon is
adequate but the coherence of their discourse is somehow
impaired. Consider the following sample from a thought-
disordered speaker:

> but what's to say there's nothing up in that
> ice age/the ice age that is yet to come supp-
> osedly this summer and this winter coming up/
> you could see quite a recession of them/ and
> then they come on pretty strong/

Each clause is rather well-formed in itself, but the
text is not fully coherent. What seems to be problema-
tic here is that the final two clauses contain noun
phrases (them and they) which presuppose for their interpretation something other than themselves. But, as
with Gio's story cited above, no clearly appropriate
referents are apparent.

To assess the extent of this use of unclear ("egocentric")
reference, we examined samples from interviews, from
cartoon descriptions and interpretations, and from narratives of the thought-disordered speakers, and from the
two control groups as well (cf. Rochester and Martin,
in press,for a detailed account of these analyses). For
each speaker, we examined about 225 noun phrases. About
70% of these were definite noun phrases which required
referents from elsewhere in the verbal or the situational
context. For each speaker, we determined where the definite noun phrase referents were located by having two
judges categorise referents as being explicit in the
utterance context, implicit, or unclear. We found that
unclear reference was virtually never used by normal
speakers: they never use it in cartoons, and use it no
more than 2% in narratives and 3% in interviews. In
contrast, thought-disordered speakers use unclear reference rather often: in 3% of cases in cartoons, 9% in

interviews and 20% in narratives. Proportions for
schizophrenic speakers who are not thought-disordered
tend to be small, and differ reliably from those for
thought-disordered subjects in narratives and interviews.

It must appear at this point that the use of "egocentric"
reference is clearly a hindrance to the listener. After
all, this sort of reference is used significantly more
often by younger children than by older children, and
by thought-disordered speakers as opposed to those who
appear fully coherent to psychiatrists. Maratsos' findings with normal adults, however, suggest that egocentric reference is not peculiar to poor communicators.

Normal Adults. Maratsos (1976) tested a group of 13
parents in an ingenious experiment. He placed an
opaque screen between himself and the subject and gave
the subject a toy plastic car and several plastic ducks
and wooden rabbits. He asked the subject to put one of
the animals into the car and inquired (somewhat awkwardly)
"Who got into the car?". To his surprise, he discovered
that half the adults tested replied "the duck" or "the
rabbit", rather than using the indefinite article.
That is, the adults failed to recognise that the experimenter's view was blocked by the screen so that he could
not know the particular toy specified. In fact, the
adults performed no better than their young children
(3 and 4 year olds) tested on a similar task: 10 out of
17 children also used direct rather than indirect reference.

It appears from these findings that referential competence does not reach a perfect state. Rather, just
as young children and thought-disordered adults use
unclear reference, so at times do normal adult speakers.
But if this is true, if all speakers at times use 'egocentric' reference, can we be sure that such reference
is misleading to the listener? How much unclear noun
phrase reference does it take to make discourse incoherent for the listener? Or, to say this another way,
how often may the speaker use egocentric reference and
still be a helpful guide for the listener? As Elena
Lieven (1976) observes, since ordinary conversational
exchanges are rarely ideal, we must ask how far from
the ideal a speaker may stray and still be understood
by the listener.

With egocentric reference, differences between speakers
do not necessarily reflect differences in how listeners

experience those speakers. Some egocentric reference
may be acceptable to listeners and may not disrupt the
overall coherence of the discourse. Or perhaps as little
as 2% of definite noun phrases with unclear referents
disorient listeners and interfere with their comprehension. Until we explore this question empirically, we
simply do not know how to map the speakers' behaviour
into a communication function for listeners.

Explicit Reference

In the past 15 years there have been several studies
done to show that speakers differ in their placement
of explicit referents. Much of this work has emerged
from the formulations of Basil Bernstein and his students
Exempletive of this work is Peter Hawkins' (1973) observation that working class children use twice as many
referents in the situational context as middle-class
children. He concludes that working-class children
rely on different elements of the communicative
situation than middle-class children: the former, he
suggests, rely on the listener's awareness of the
situation while the latter rely on differentiations
through language.

In our work with schizophrenic speakers cited above, we
find that thought-disordered speakers use more situational referents than other speakers. The exact results
depend on the utterance context, since the proportion
of situational reference used by all speakers varies
with context. In cartoon accounts, differences are
large and reliable: 33% of thought-disordered speakers'
definite noun phrases contain referents in the situational
context as compared to about 17% for other speakers.

It is instructive to look at some examples in order to
understand the differences. Situational referents, in
our analyses, include all personal pronouns, so references to me, myself, my plans and so on are situational
referents and could account for the differences between
thought-disordered speakers and other subjects. However,
examining the data in detail we find this is not the
case. Instead, the differences are due to the thought-
disordered speakers' extensive references to the actual
situation. For example consider the following descriptions of a single cartoon:

    (1)   here we have a chap in the pillory or in
          the stocks with both his hands and his
          feet being held by the device/and we have

a woman who is seated by his side doing
her knitting/and she has ravelled the yarn
all about his hands and his feet/ and is
now getting ready for knitting.

(2)   she's kni/well he's got yarn in his hands
      and on his feet/and she's winding/imagine
      winding a ball of wool off of a man who's
      in the stocks.

Example (1) is from a normal speaker and (2) is from a
thought-disordered speaker. Definite noun phrases with
situational referents are underlined. These examples
characterise the data nicely. Normal speakers tend to
describe the cartoon in front of them as if the experimenter were unable to see the cartoon. Non-thought-
disordered speakers behave in about the same way. But
thought-disordered subjects describe the cartoon in
recognition of the fact that the experimenter is capable
of seeing the events and persons being described.

Given that working-class children, thought-disordered
patients, and probably younger middle-class children
(Martin, 1976) use an abundance of situational reference
as compared to middle-class children, non-thought-
disordered patients, and older middle-class children,
respectively, what may we conclude? Surely, we may not
conclude that the use of situational referents has any
particular effect on the listener's comprehension of
discourse. The fact that there are statistically significant differences between speakers tells us nothing
about the effects of those differences on listeners. To
repeat the argument made with regard to egocentric reference, until we assess the listener's comprehension empirically, we are ignorant of the communicative significance
of the differences in speakers' usage.

Judging the Speaker

There are many possible ways to assess the effects of
the speaker's performance. One may construct sentences
which contain referents in different locations, and note
how quickly listeners can respond to those questions
(as Haviland and Clark, 1974, have done)or how readily
subjects can recall those sentences (as Clark,1974
suggested). Or, to take a task closer to ordinary discourse, one may present texts which differ in referent
locations and ask listeners to paraphrase the text, or
to perform some action (e.g. path-finding) based on
instructions in that text (as Catherine Garvey and her

colleagues have done, Garvey and Baldwin, 1970, 1971;
Garvey and Dickstein, 1970).

We chose to have judges evaluate the coherence of texts,
in an effort to assess in detail the results of the
interviews conducted with our schizophrenic and control
subects. We asked 10 lay people to serve as 'editors'
for interview transcripts. The editors were given a set
of 3-4 sample transcripts, each divided into about 15
clausal segments, for each of 60 speakers. Using a
double-blind procedure, we gave the judges the 60 sets
to evaluate (cf. Rochester, Martin and Thurston, in
press, for details of this study) and asked that they
read through each sheet quickly, putting a check next
to any segment which seemed to disrupt the flow of the
discourse.

We decided that a segment was "disruptive" where at
least 7/10 judges agreed that it interfered with the
discourse flow. Disruptive segments turned out to reflect
psychiatrists' evaluations very neatly, separating
speakers who were diagnosed as thought-disordered (hit
rate = 75%) from those who were schizophrenic but not
thought-disordered (false alarm rate = 5%) and from
those who were not schizophrenic (false alarm rate
= 0%). Disruptive segments therefore seemed a promising
criterion variable with which to evaluate speaker's
use of noun phrase reference.

We performed a multiple regression analysis to estimate
the effects of "egocentric" noun phrase referents and
situational referents on our judges. The predictor
variables were the proportions of noun phrases with
referents at each location (e.g. in the verbal context,
in the situation, or where the location was unclear),
and the criterion variable was the proportion of disruptive clauses to total clauses. We found that only
"egocentric" referents were a useful predictor of
judges' evaluations of coherence: noun phrases with
unclear referents accounted for 34% of the total variation in the criterion variable (F = 26.2; df = 1,58;
P <.001). Noun phrases with situational references
predicted only 2% of the total variation, and yielded
an insignificant F ratio.

Thus, in our studies, egocentric reference had a decided
effect on evaluations while situational reference was
essentially irrelevant. As Labov and Labov (1976)
might observe, not all differences made a difference.

These findings lend empirical support to the argument
that we must assess the implications of speaker differences for listeners. That is, to tell is a difference makes a difference, test it.

Conclusion

As studies of language move from syntactic analyses
of isolated sentences to more comprehensive semantic
and pragmatic analyses of connected discourse, it is
important that we also shift our experimental approach
from a view of speakers as solitary actors to a view
of speakers as communicators. This shift is necessary
not merely to prevent us from making naive assumptions
about listener effects which do not exist, but to open
our vision to new questions which we ought to be asking.
In general, we ought to be asking about the functions
which map speaker differences into listener (and reader)
experiences. In particular, three questions seem
interesting at this time.

First, are there uniformly successful strategies for
speakers? "Successful" in this case implies a successful communication of the speaker's message to the listener, and may be defined in terms of evaluations of
coherence, listener's reaction time, readers' paraphrase,
listeners' recall and so forth. For example, a strategy
which may be successful across various contexts and
listeners might be: Never use definite noun phrases with
unclear referents. A corollary to this would specify
the strategic range and circumstances. For example:
Use inexplicit referents only with close friends, and
then only once per discourse frame.

Secondly, are there strategies which differ across
speakers and are irrelevant to listeners? Situational
versus verbal context locations for noun phrase referents
would seem to be an instance of a dimension which is
irrelevant for listeners. However, this may depend
strongly on the interpersonal roles of the speaker and
listener, on the listeners' characteristics, and on the
purpose of the discourse. A corollary question to this
would be to explore whether such speaker strategies are
differentially successful for speakers. For example,
the speaker's rate of speech or hesitation patterns or
ability to process incoming information might vary
depending on his or her predominant strategy.

Finally, it seems important to ask whether there are

coincidences between strategies which are successful
for listeners and for speakers. For example, the use
of unclear referents might not only impede the listener's
comprehension but might also make the speaker's formulations of new clauses more difficult. Or, long hesitations
between clauses might aid the listener by giving extra
processing time for comprehension and aid the speaker
in the same way.

In summary, to assess the "art" of the speaker, whether
in referring or in other speech acts, we cannot simply
gaze on the artist's production and compare it to the
productions of other artists. On the contrary, we must
use a procedure which is anathema to most aestheticians
and poll the audience.

Acknowledgements

I gratefully acknowledge the continued support of the
Benevolent Foundation of Scottish Rite Freemasonry,
Northern Jurisdiction, U.S.A., and the Clarke Institute
Associates' Research Fund, and the invaluable contributions of my colleague Jim Martin. I am also indebted
to Dr. Mary Seeman and Dr. Alexander Bonkalo for their
generous help in conducting these studies.

References

Clark, H.H. (1974). Semantics and comprehension. In
    T.A. Sebeok (ed.) Current trends in linguistics, Vol.
    12: Linguistics and adjacent arts and sciences. The
    Hague: Mouton.

Garvey, C., and Baldwin, T. (1970). Studies in convergent
    communication: I. Analysis of verbal report. Maryland:
    The Johns Hopkins University Press.

Garvey, C., and Baldwin, T. (1971). Studies in convergent
    communication: III. Comparisons of child and adult
    performance. Maryland: The Johns Hopkins University
    Press.

Garvey, C., and Dickstein, E. (1970). Levels of Analysis
    and social class differences in language. Maryland:
    The Johns Hopkins University Press.

Haviland, S.E., and Clark, H.H. (1974). Acquiring new
    information as a process in comprehension. Journal of
    Verbal Learning and Verbal Behavior, 13, 512-521.

Hawkins, P.R. (1973). The influence of sex, social class,
    and pause location in the hesitation phenomena of
    seven-year-old children. In B. Bernstein (ed.)

Primary socialization, language and education: Class, codes and control (Vol. 2). Boston, Mass.: Routledge & Kegan Paul.

Labov, W., and Labov, T. (1976). Learning the syntax of questions. In these Proceedings, (Vol. 2).

Lieven, E.V.M. (1976). Verbal interaction between adults and young children: the implication of individual differences. In these Proceedings, (Vol. 1).

Maratsos, M.P. (1976). The use of definite and indefinite reference in young children. London: Cambridge University Press.

Martin, J.R. (1976). Learning how to tell: Semantic systems and structures in children's narrative. Manuscript. University of Essex.

Martin, J.R. and Rochester, S.R. (1975). Cohesion and reference in schizophrenic speech. In A. Makkai and V.B. Makkai (eds.) The First LACUS FORUM 1974. Columbia, S.C.: Hornbeam Press.

Piaget, J. (1955). The language and thought of the child. Cleveland: Meridian Books. The World Publishing Co.

Rochester, S.R., and Martin, J.R. (in press). The art of referring: The speaker's use of noun phrases to instruct the listener. In R. Freedle (ed.) Discourse comprehension and production. New York: Erlbaum Associates.

Rochester, S.R., and Martin, J.R., and Thurston, S. (in press). Thought disorder in schizophrenia: The listener's task. Brain and Language.

Rochester, S.R., Thurston, S., and Rupp, J. (in press). Hesitations as clues to failures in coherence: Studies of the thought-disordered speaker. In S. Rosenberg (ed.) Sentence Production: Developments in research and theory. New York: Erlbaum Associates.

GESTURE AND SILENCE AS INDICATORS OF PLANNING

IN SPEECH

BRIAN BUTTERWORTH and GEOFFREY BEATTIE

University of Cambridge

A characteristic of human talk is that it is typically accompanied by bodily movements, most noticeably of the arms and hands. It is a matter of common observation that a subclass of these hand and arm movements appear intimately linked with the process of speech production: they are rhythmically timed with the speech, and often seem to reflect the meaning which the speech expresses. We call these movements Speech Focussed Movements (SFMs). These can be distinguished from other movements which accompany speech, scratches, twitches and the like, since the latter are not timed with speech, and do not in any apparent way reflect the meaning of what is said.

Previous studies of movements accompanying speech generally have not distinguished SFMs from other kinds of bodily movement and have concentrated on two main lines of thought. First, studies in the psychoanalytic tradition have looked upon movements as revealing the speaker's emotional or affective state (Freud, 1905, p. 77; Deutsch, 1947, 1952; Feldman, 1959; etc.) Secondly, many authors have supposed that these movements constitute an alternative channel of communication either augmenting the verbal message or substituting for the verbal message (Mahl, Danet and Norton, 1959; Baxter, Winters and Hammer, 1969), though a demonstrable benefit to the listener of these 'signals' has not been found except for communication of shape information (Graham and Argyle, 1975). And indeed, there is evidence that, for at least the kinds of conversation studied, communication and conversation are no way impaired, and in some ways improved if the speakers cannot see each other, e.g. on the telephone (Butterworth, Brady and Hine, in press). However, speakers do seem to adjust the verbal message somewhat in such situations (Moscovici, 1967) and where gestures are eliminated, especially where the communication of spatial information is required (Graham and Heywood, 1975).

There have been attempts to link the occurrence of movements to
the structure of the verbal message. Dittmann and Llewellyn (1972)
found there was a significant relationship between the occurrence
of movements and the second position in phonemic clauses and argued
that they were linked, encoding these clauses. However, they counted
all movements of the head, hands and feet as measured by an acceler-
ometer, and the relationship accounted for little of the data.
Kendon (1972), with a very small sample of data ($1\frac{1}{2}$ min. from 1
speaker), associated movements with a more detailed and hierarchical
linguistic analysis; he concluded that "speech-accompanying movement is produced along with the speech, as if the speech production
process is manifested in two forms of activity simultaneously".

We decided to look in detail at the relationship between SFMs and
speech; in particular, to see if there was any connection between
gestures and speech planning. There were two reasons for trying
to look at SFMs in this way. First, a number of previous studies
have shown that the temporal patterning of speech output - where
delays in it occur - are indicators of underlying planning processes: so if gestures have anything to do with planning they should
distribute in some interesting way with respect to these delays.
Second, one of us had, inadvertently, got hold of some data that
already obliquely pointed in this direction. For purely educational
purposes, he had videotaped one of his lectures, and playing it back
had noticed that some gestures could not only unequivocally be
linked to a word in the speech - like, raising one hand during the
utterance of "when certain problems can be raised", but more interestingly the gesture occurred strikingly before the word "raise" -
in this case, on the word "certain". So it looked very much as if
the speaker knew what the word would be, or at least had a pretty
good idea, well before he uttered it. And the third reason is
that, if gestures do reflect planning, could they tell us more
about the planning processes than we had already surmised from the
hesitation data?

The claims for particular planning processes can be summarized
briefly. Goldman-Eisler (1958) found that some pauses are associated with the lexical selection process, in that there was a
significant relation between words which were unpredictable in
context with words which were preceded by a pause. She argued
that the delay is caused by the extra time required to select from
a larger ensemble of potential continuations. Henderson, Gold-
man-Eister and Skarbek (1966) noticed that longish stretches of
uninterrupted spontaneous speech are characterised by alternating
phases of hesitant and fluent output, and speculated that the temporal rhythm reflected an underlying cognitive rhythm, where the
hesitant phase is used to plan the speech in the immediately
succeeding fluent phase: thus there would be a rhythmic alternation

of Planning phases and Execution phases.

Our first two experiments are designed to explore this last speculation. These experiments are described in detail elsewhere
(Butterworth, 1975; Beattie and Butterworth, forthcoming), so
we shall sketch in just those finding relevant to experiment III
on SFMs.

Hesitation and Planning

Experiment I

We wondered if each temporal cycle of a hesitant and a fluent
phase coincided with some well-understood linguistic unit like a
sentence or a clause, since it seemed a priori unlikely that speakers would plan ahead in some other units. We recorded two-
person conversational arguments between speakers and an experimenter. Each speaker had to pick one proposition which he strongly agreed with from a list of sixteen. We made the proposition
fairly complex and such that the speaker couldn't rely to too
great an extent on previous practice with it, or prior knowledge,
since we wanted the task to be cognitively fairly demanding so
that the speech would be thoughtful rather than recitative.
(Examples: "Socialism, at least in the English version, is designed
to elevate the lazy and incompetent to the level of the industrious
and able. "The moral example of the Americans in Vietnam is not
as iniquitous as that of the British, almost anywhere." "From each
according to his ability, to each according to his need." "Marriage
is an immoral institution: thank God it's almost extinct.")

A total of $3\frac{1}{2}$ hours speech was recorded from the conversation. The
recordings were fed through a signal detector, and its output recorded on a pen-oscillograph. Periods of phonation time and
silent time were plotted as a cumulative step-function, with pause
time on the ordinate. The points at which changes in the accelerat-
ion of the pause/phonation ratio occurred were estimated by inspection and straight lines fitted using the Method of Averages. Of the
speech which showed the cyclic pattern, 12.5 mins, from 3 speakers
was subjected to a detailed analysis.

A transcript of the speech was matched to the **pen-oscillograph** record so that the exact temporal location of the words could be
determined. Points of transition between the cycles were marked
on the transcript. Cycles lasted on average about 18.00 secs.,
(S.D. = 5.29). Out of a total of 43 cycle transitions, 32 corresponded to the beginning of a new clause; however there were 183
clauses in all (56 sentences) so each cycle would compromise, on
average, about 4.5 clauses. We thought that a more semantic
notion than 'clause' might give a better match.

We asked 8 judges (who were not the original speakers) to divide
up a transcript - in normal orthography and punctuation, but with
no additional temporal information marked - into "ideas". Judges
had to decide for themselves what these were, and would presumably
draw upon intuitions employed for precising texts, or taking notes
at lectures. Where more than half the judges agreed on a location
as a transition betwen one "idea" and the next, we took that as
an Idea division. Out of 32 Ideas, 17 corresponded to cycle transitions (1,736 words in all), and cycle and Idea transitions could
theoretically occur betwen any two words: $G = 42.23$, $p < 0.001$. (G
statistic in Sokal and Rohlf, 1973).

It might be argued that, since all Idea boundaries and all but 11
cycle starts are coincident with clause starts, the proper candidates for these locations is not between words but between
clauses. However, when the relation between Ideas and cycles for
clause start only is examined, the association still holds ($G =
103.52$, $p < 0.001$).

These data support the speculation of Henderson et al, that there
is a cognitive rhythm consisting of alternating phases of planning
and execution, and further, that what is planned is an 'idea' -
a fair-sized semantic unit - which will be realized linguistically
as several (surface) clauses. And it is probably more useful,
therefore, to think of syntax as an output constraint rather than
part of the higher planning processes.

Hesitation and Planning

Experiment II

This experiment was an attempt to establish the same kind of conclusion using a rather different method. Instead of conversations,
we used monologues. More importantly perhaps, instead of defining
Ideas retrospectively we tried to define Ideas in advance by instruction. The task for 12 speakers was to describe five things
only loosely connected (the Low Cohesion condition), without talking about any connections which might hold among them: five rooms
in his or her parent's home; five spectators at a football match.
As a comparison, the speaker had to describe the relations among
the parts of an object or a sequence of interrelated actions, without describing the parts, or the participants, themselves in any
detail (the High Cohesion condition): the front quad of a Cambridge
college (all speakers were Cambridge undergraduates), or the interactions of a single male at a discotheque. Each speaker had to do
the four description tasks, and each description was to last for
about three minutes.

The proportion of pausing in successive 15 sec. samples were calculated, and the following results obtained. In the Low Cohesion

condition there was a wide variation in proportion of pausing in
the 15 sec. samples; in the High Cohesion condition the variation
was much less marked.   Sudden increases in the proportion of pausing in the Low Cohesion task coincided with the beginning of a new
'idea', in this case when the speaker moved on to a description of
the next room, or the next spectator.   A comparison between the
first 15 secs. and the second 15 secs. of each of the five descriptions showed a significant decrease in the proportion of pausing
(Wilcoxon Matched-Pairs Signed Ranks Test, two tailed; $P < 0.001$).

A second striking feature of these data, is that the mean duration
of each description is 45.9 secs. for rooms, and 53.4 secs. for
actions, where they were long enough to permit this analysis.   (Some
were very short: there seemed no way to get speakers to stick to a
standard length).   Typically, though not invariably, successive
15 sec. samples became more and more fluent, indicating that there
is higher level of organisation than the 18 sec. cycle mentioned
above, that can be revealed by temporal analysis.   We hope to be
able to look more closely at this temporal and ideational hierarchy.

Thus we can hypothesis, with varying degrees of confidence, two,
or possibly, three planning processes:

1.   lexical planning;
2.   Idea planning
(3.   higher Idea planning).

We know, however, little about the relations between them.   In
particular, we do not know whether Goldman-Eisler's lexical pauses
are caused by the time to search the lexicon when the speaker has
the semantic specification of the item he needs, or whether it is
caused by the need to create the specification when the speaker
has only a general idea of what he will say.   Secondly, we do not
know what the temporal relationship is between having a general idea
of what to say, and having the semantic specification of the words
one needs to say it.

Speech Focussed Movements and Planning

Experiment III

Procedure

We studied samples of monologues and dialogues. The dialogues consisted of videotaped naturally-occurring academic interactions
(3 supervisions, 1 seminar); the monologue was a video record of
one subject in the previous experiment.   Approximately 4 hours of
dialogue data were available and half an hour of monologue.   All
subjects were either undergraduates, postgraducate or staff members
of the University of Cambridge.   The total duration of speech
analysed was 849.8 secs. chosen in a fairly random fashion.

The only constraints on this selection were that the speaker's turn
in the conversation had to be at least 40 secs., so that temporal
cycles could be identified and secondly that some SFMs had to occur.
Data were available on 7 speakers. Forty-four switching pauses
(pauses bounded by vocalisations of different speakers) totalling
83 secs. were also analysed.

The temporal analysis was carried out as follows: temporal cycles
were identified in the manner described in Experiment I, and thereby
independently of the location of the gestures since only sound could
be detected by the signal detector.   All speakers showed temporal
cycles of alternating Planning and Execution phases, except one,
who had a mean pause rate around 10%.   A timer - to one hundredth
of a second - was mixed on to the videorecordings, so that precise
timing of words and gestures could be achieved, and the result
matched to the temporal cycle data.

The present analysis concentrates on three classes of hand and arm
movement:

(1)   Speech-focussed movements - all movements of the arm or hand
      except self-touching (e.g. finger-rubbing, scratching). This
      class includes gestures, "batonic" movements and other simple
      movements.

(2)   Gestures - more complex movements which appeared to bear some
      semantic relation to the verbal component of the message.

(3)   Changes in the basic equilibrium position of the arms and
      hands, that is, changes in the position where the hands return
      to after making a SFM.

The classification of each movement into one of the categories was
performed by each experimenter independently, and disagreements
resolved by rechecking the videotape together and arguing about it.
The exact time of the initiation of each SFM and equilibrium change
could be obtained by utilising the slow motion facility of VTR.
The points were located on the pause/phonation plots, and on transcripts of the verbal output.   In the case of gestures the exact time
between the initiation of the gesture and the first phone of the
word with which it was associated was noted.   The time of each
equilibrium change was also noted.

Results

The number of SFMs, gestures, and SFMs-gestures occurring per unit
time during pauses or periods of phonation in planning and execution phases were analysed. (See Table 1).

Table 1. The rate of production (per 1,000 seconds) of SFMs,
          Gestures and SFMs-Gestures during pauses (H) and
          phonations (S) in Planning (P) and Execution (Ex)
          phases.

		H	S	Mean Rate
SFMs	P	118.4	191.9	153.0
	Ex	341.0	199.7	226.2
	Mean Rate	196.5	197.5	
Gestures	P	59.2	44.3	52.2
	Ex	280.1	106.9	139.4
	Mean Rate	136.7	89.6	
SFMs-Gestures	P	59.2	147.6	100.9
	Ex	60.9	92.8	86.8
	Mean Rate	59.8	107.9	

The analysis revealed that SFMs occurred most frequently per unit
time during execution phases. The highest incidence of this class
of behaviours was in pauses in the execution phases. SFMs were
approximately 3 times more frequent per unit time during such
pauses than during pauses in the planning phases.

Gestures yielded an essentially similar distribution but in the
case of gestures the trends were much more pronounced. Gestures
were approximately 5 times as frequent per unit time during pauses
in the execution phases than during pauses in the planning phases.
Gestures were almost 3 times as frequent during pauses in the
execution phase as during periods of phonation in the execution
phase.

The residual class of SFMs-gestures displayed a very different
distribution. This time there was no overall difference in the
number occurring per unit time during pauses in the planning and
execution phases and these behaviours were most common during
periods of phonation, particularly in planning phases.

Table 2. Number of Gestures and SFMs-Gestures occurring during
         each half of the planning and execution phase.

	P		Ex	
	1	2	3	4
Gestures	6	8	31	31
SFMs-Gestures	19	11	20	17

The distribution of gestures and SFMs-gestures across sections of
the rhythm was also analysed (see Table 2).   Some difference in
the distribution of these 2 classes of phenomena was apparent.
In the planning phase non-gestural SFMs tended to occur more frequently in the first half than in the second half, whereas gestures
displayed the opposite tendency, being more common in the second
half than in the first half. Similarly non-gestural SFMs tended
to occur more frequently in the first half of the execution phase
than in the second half whereas gestures showed no particular tendency towards bunching in either half.

One point should be made about the trends described. Although the
differences in distribution across parts of the cognitive rhythm
are very marked, when the data is decomposed, the differences are
not consistent across all 6 subjects. Two subjects who contributed
least data failed to show the trends described.

From our Analysis of Variance we find that speech-focussed movements
are significantly more frequent per unit time in the execution
phases of the rhythm than in the planning phases ($F(1,12) = 8.65$,
$p < 0.05$).   Further there is a significant interaction effect: in
the planning phase SFMs are more frequent during periods of phonation whereas in the execution phase SFMs are significantly more
frequent during periods of hesitation ($F(1,12) = 7.60$, $p < 0.05$).

When SFMs are decomposed into gesture and non-gestural SFMs, an
Analysis of Variance reveals a significant interaction between
movement type (SFM-G/$\bar{\text{G}}$) and specific location in the rhythm. Non-
gestural SFMs are most frequent per unit time in periods of phonation in the planning phase whereas gestures are most frequent per
unit time during periods of hesitation in the execution phases.
Gestures are least frequent during periods of phonation in the
planning phase.

This distributional difference between Gestures and other SFMs
suggests a functional difference. Other SFMs consist mainly of
simple batonic movements and their close relation to periods of

actual phonation in both Planning and Execution phases indicates
that a common-sense interpretation of them as emphasis markers is
well-founded. The asymmetry in the distribution of Gestures
suggests, on the other hand, that these are not mere emphatic markers, but are functionally related to planning. Since they are relatively infrequent in the Planning phase itself, they are not
connected with the ideational planning process but with the lexical
planning process. This hypothesis is supported by their close
association with pauses in the Execution phase.

Further evidence of the functional distinction between Gestures
and SFMs is to be found by comparing their distributions in respect to the form-class of the words they are associated with (see
Table 3). Gestures are heavily concentrated on Nouns (41.3%),
Verbs (23.8%) and adjectives (15.9%) - classes which contain most
of the unpredictable lexical items. Other SFMs however, are much
more evenly spaced over form-classes.

The initiation of Gestures usually precedes, and never follows,
the words they are associated with. The mean delay being around
.80 secs., with a range of .10 secs. to 2.50 secs., (see Table 4).
The length of delay seems unaffected by the position in that
clause: again arguing for the connection of Gestures with lexical
selection, independent of higher level plans that determine the
syntactic shape of the output.

The relationship between changes in the basic equilibrium position
of the arm and hands, phasal transition points and clause junctures
was analysed (see Table 5). A significant tendency for changes in
the basic equilibrium position to correspond to both the terminal
points of planning phases ($G = 22.118$, $p < 0.001$), and to the
terminal points of execution phases ($G = 39.336$, $p < 0.001$) was observed.

Changes in the equilibrium position were also found to coincide with
junctures between clauses ($G = 61.448$, $p < 0.001$). These results
provide further evidence that the planning and execution phases,
identified from changes in the gross temporal patterning of the
speech,have some underlying psychological significance.

One observation made during the course of the study was that SFMs
and gestures were particularly infrequent in switching pauses (those
pauses bounded by the vocalisations of different speakers) even when
such pauses preceded fluent speech and we can thus presume were
given over to ideational planning, in similar fashion to pauses in
planning phases. Table 6 lends support to this observation, moreover there were significantly fewer SFMs in switching pauses than
in pauses in either the planning or execution phases (Sign test
$p < 0.05$). One could not test the reliability of this difference
in the case of gestures because some subjects did not gesture at all

Table 3. Proportion of Gestures and SFMs-Gestures associated
with syntactic classes.

	Unadjusted	
	Percent Gestures	Percent SFMs-Gestures
Noun	41.3	28.6
Verb	23.8	21.4
Adjective	15.9	7.1
Adverb	1.6	4.8
Pronoun	6.3	4.8
Preposition	6.3	4.8
Conjunction	3.2	7.1
Dem. Adj.	1.6	9.6
Relative Pron.	0	7.1
Quantifiers/ Determiners	0	0
Modal Vb.s	0	0
Interjections	0	2.4
Etcetera	0	2.4
	100.0	100.0

	Adjusted*	
	Percent Gestures	Percent SFMs-Gestures
Noun	26.20	11.05
Verb	16.88	9.13
Adjective	28.61	8.02
Adverb	2.71	4.98
Pronoun	8.73	3.88
Preposition	5.72	2.76
Conjunction	4.82	6.63
Dem.Adj.	6.33	22.93
Relative Pron.	0	30.62
Quantifiers/ Determiners	0	0
Modal Vb.s	0	0
Interjections	**	**
Etcetera	**	**
	100.00	100.00

* The percentage of Gestures, SFMs-Gestures adjusted for the
number of occurrences of items in the associated syntactic
classes.

** The number of occurrences of Interjections and of the word
"Etcetera" were too small for a reliable estimate to be made.

Table 4.   Mean duration of the delay (in seconds) between Gestures
           and the associated word, analysed by syntactic class,
           clause position and clause length.

	Syntactic class.					
	N	V	Adj.	Pron.	Prep.	Dem. Adj.
M_1	.915	.736	.747	1.20	2.33	1.06
M_2	.770	.661	.664	.905	.583	1.06

	Clause position (in words)				
	1-2	3-4	5-6	7-8	8+
M_1	1.29	.534	.881	1.82	.804
M_2	.737	.393	.672	1.22	.746

	Clause length (in words)				
	1-4	5-8	9-12	13-16	17+
M_1	.498	.868	.744	.810	1.26
M_2	.498	.706	.650	.540	.990

M_1 = all Gestures which show some delay.

M_2 = M_1 + Gestures which are initiated with or during the word.

Table 5. Position in speech of changes in the basic equilibrium
          position of the arm and hand.

	Number of changes in equilibrium position corresponding to each category.	Number of occurrences of each phenomenon not accompanied by change in equilibrium position.
End of P Phase	5	20
End of Ex Phase	8	20
Clause Juncture	24	180
Other	14	1,620

during pauses in the small samples studied. It was also noted
that no SFMs or gestures occurred in switching pauses preceding
planning phases in the case of any of the 5 subjects.   This observation is consistent with the hypothesis that little verbal
planning occurs during such pauses, and is delayed until the
speaker's turn begins.   These results suggest that SFMs and
gestures are not a direct product of verbal planning. It is hypothesised that some affective activity or process needs to interact
with verbal planning to produce SFMs and gestures and this activity
commences when a speaker claims the floor by uttering some word
or sound.

Conclusions

A number of conclusions may be advanced tentatively.

1.   There are two fundamentally distinct kinds of SFM: Gestures
     and other SFMs.

2.   The distinction between Planning and Execution phase put for-
     ward by Henderson et al. (op. cit.) is well founded, on the basis
     of the distribution of equilibrium changes and of Gestures.

3.   Gestures are products of lexical preplanning processes, and

4.   seem to indicate that the speaker knows in advance the semantic
     specification of the words he will utter, and in some cases
     has to delay if he has to search for a relatively unavailable
     item.

5.   The selection of lexical items does not appear to be part of
     the ideational planning process.

6.   Lexical planning is a necessary, though not sufficient con-
     dition for the occurrence of Gestures.

We should perhaps offer our guesses as to why gestures should
precede lexical items, rather than occur simultaneously.   Lexical
items have to be drawn from a large ensemble of potential continuations since the mental lexicon probably consists of 20-30,000
entries.   Not all, of course, will be candidates, at each choice
point, but for unpredictable lexical items, which typically require
a hesitation to access, the available ensemble will be pretty large.
However, there can only be a much smaller set of gestures and,
therefore, a relatively very small subset available at each choice
point.   And by Hick's Law the word selection process will take
longer than the Gesture selection process.

The source in ontogenesis of the intimate relationship between
speech and gesture has been mentioned by von Raffler Engel (1975)

and explored in some detail by McNeill in his paper, "Semiotic
Extension" (1975). According to McNeill, gestures arise from the
conceptual elaboration and adaptation (semiotic extension) of
sensory-motor action schemas for use in adult speech production.
The gestures so generated represent basic Objects, Actions and
Locations, which constitute the categories in the action schemas.
This view is consistent with the concentration of Gestures on Nouns
and Verbs (and Adjectives?), but from his analysis of data from
speakers describing their own actions, he argues that gestures
should be synchronous with the associated verbalisations and not
preceding them. He offers no grounds for distinguishing Gestures
from other SFMs, nor SFMs from changes in equilibrium position.
Nevertheless, like McNeill, we feel that the study of gestures is
an important addition to the presently small repertory of techniques
available for probing the mechanisms underlying the production of
speech.

References

Baxter, J.C., Winters, E.P. and Hammer, R.E. (1968). Gestural
Behaviour during a brief interview as a function of cognitive
variables. J. Pers. Soc. Psychol., 8, 303-7.

Beattie, G.W. and Butterworth, B. (forthcoming). Manipulation
by instruction of the content and temporal structure of spontaneous speech.

Boomer, D.S. (1964). Speech disturbance and body movement in
interviews. J. Nerv. Ment. Dis., 136, 263-266.

Boomer, D.S. (1965). Hesitation and grammatical encoding. Language and Speech, 8, 145-158.

Boomer, D.S. and Laver, J.D.M. (1968). Slips of the tongue.
Br. J. Disord. Commun., 3, 2-12.

Butterworth, B. (1975). Hesitation and semantic planning in
speech. Jnl. Psycholing. Res., 4, 75-97.

Butterworth, B., Hine, R.R. and Brady, K.D. (in press). Speech
and interaction in sound-only communication channels. Semiotica.

Deutsch, F. (1947). Analysis of Postural Behaviour. Psychoanalytic Quarterly, 16, 195-213.

Deutsch, F. (1952). Analytic posturology. Psychoanalytic Quarterly, 21, 196-214.

Deutsch, F. and Murphy, W.F. (1955). The clinical interview.
New York: International Universities Press.

Dittman, A.T. (1972). The body-movement-speech rhythm relationship
as a cue to speech encoding. In Siegman, A.W. and Pope, B.
Studies in Dyadic Communication. New York: Pergamon Press.

Feldman, S.S. (1954). Mannerisms of speech and gestures. New
York: International Universities Press.

Freud, S. (1905). Fragments of an analysis of a case of hysteria.
In the Standard edition of the Complete Works of Sigmund Freud,
Vol. 7, London: Hogarth, 1953.

Goldman-Eisler, F. (1958). Speech production and predictability of
words in context and the length of pauses in speech. Language
and Speech, 1, 96.

Graham, J.A. and Argyle, M. (1975). The communication of extra-
verbal meaning by gestures. Int. J. Psychol. (in press).

Graham, J.A. and Heywood, S. (1975). The effects of elimination
of hand gestures and of verbal codability on speech performance.
Eur. J. Soc. Psychol., 5, 189-195.

Henderson, A., Goldman-Eisler, F., and Skarbek, A. (1966). Sequen-
tial temporal patterns in spontaneous speech. Language and
Speech, 9, 207-216.

Kendon, A. (1972). Some relations between body motion and speech.
In Siegman, A.W. and Pope, B. Studies in Dyadic Communication.
New York: Pergamon Press.

McNeill, D. (1975). Semiotic Extension. In Solso (Ed.) Information Processing and Cognition. Hillsdale, N.J.: Lawrence
Erlbaum.

Mahl, G.F., Danet, B., and Norton, N. (1959). Reflection of
major personality characteristics in gestures and body movement.
Paper presented at Annual Meeting American Psychological Assoc.,
Cincinnati, Ohio, Sept. 1959.

Moscovici, S. (1967). Communication process and language. In
Berkovitz, L. (Ed.) Advances in Experimental Social Psychology,
Vol. 3. New York: Academic Press.

Sokal, R.R. and Rohlf, F.J. (1973). Introduction to Biostatistics.
San Francisco: W.H. Freeman.

Von Raffler Engel, W. (1975). The correlation of gestures and
verbalisations in first language acquisition. In Kendon, A.,
Harris, R.M. and Key, M.R. (Eds.) Organisation of Behaviour
in Face-to-Face Interaction (World Anthropology, Vol. II). The
Hague: Mouton.

SOUND PATTERNS AND SPELLING PATTERNS IN ENGLISH

ROBERT G. BAKER and PHILIP T. SMITH

University of Stirling

Introduction

The relationship between spoken and written English has for a long
time been a subject of discussion and dispute. If orthography is
to be seen merely as a transcription for speech sounds, a one-to-
one letter-to-phoneme cipher devised to help children and foreign-
ers pronounce the language, then the English system presents many
shortcomings. Few English sounds are uniquely represented by a
single letter, few English letters have a unique pronounciation,
and there appear to be many redundant letters, such as unpronounced
final -e. There has been a continuous history of attempts to
change English spelling since the thirteenth century. One of the
most recent attempts at reform, and almost certainly the most
widely tested one, the "initial teaching alphabet" for beginning
readers, will be discussed below. Throughout this history of dis-
content, however, there have been a number of distinguished voices
raised in favour of the traditional system, stressing in particular
the desirability for orthographic differentiation of homophones
("right" vs. "rite" vs. "write" vs. "wright") and the maintenance
of orthographic identity for lexical roots and their derived forms
in spite of phonemic variation ("divine" vs. "divinity"; "serene"
vs. "serenity"; "profane" vs. "profanity"). In this case alphabetic orthography was not seen primarily as a phonemic transcription.
Rather, although initially based on a phonological level of language,
it is designed for people who already know the language and do not
generally require help in pronouncing it. It would be contended
for example that predictable morphophonemic variations, such as the
phonemically differing plural morphemes in "cats" and "dogs" need
not be signaled in the orthography, and indeed it may be misleading
if they are. Just as language is analysed by linguists as a hier-

archical structure or "system of systems" (Firth, 1957), by means
of which information is conveyed on different linguistic levels,
(phonetic, phonological, morphemic, syntactic, semantic etc.), so
written language itself arbitrarily selects one or more of these
different levels for conveying information to the eye, resulting
in the various types of writing systems (alphabetic, syllabic
logographic, pictographic, etc.) that are in evidence in the
literate world.

It is unclear which linguistic levels, or which particular units
on these levels, are most "naturally" represented in written
language (in terms of speakers' linguistic awarenesses), which
levels of representation are most efficiently utilised by fluent
readers, and which levels might in turn be more accessible to
children learning to read. It appears that words are intuitively
more obvious units than morphemes, morphemes more obvious than
syllables, and syllables more obvious than phonemes (Savin, 1972).
Research quoted in Leontiev (1975) suggests that under different
conditions e.g. different noise levels, people employ quite
different linguistic levels in their speech perception strategies,
and we should expect such differences of strategy to be reflected
also in reading performance. The differing emphasis on different
levels is to be noted in the assumptions underlying different
methods of teaching children to read, viz. whole-word vs. phonic
methods.   There may furthermore be individual subject differences
in the optimal levels of representation. Whatever the results of
continuing inquiry in these areas, the question still remains as
to whether, having once chosen a particular level of representation
for orthographic reform, one should then adhere to that level with
rigid consistency (as the reformers advocate) or whether information
from other levels, should, for whatever reasons, be allowed to in-
trude into the written medium.

A principled defence of traditional English spelling as a mixed-
level orthography is presented by Chomsky and Halle (1968; referred to henceforth as CH).   "The fundamental principle of orthography", they write, "is that phonetic variation is not indicated where it is predictable by general rule" (p.49).   The
weak form of this principle is that minor allophonic variations
are not shown in the spelling - the stronger form, intended by CH,
is that higher level rules (e.g. morphemic and syntactic rules) are
operative in reducing the need to specify certain phonemic differentiations.   Thus, for example, the phonetically (and phonemically)
differing second vowels of -

a) divine vs. divinity

b) photograph vs. photography

may be accounted for by general vowel alternation rules dependent
on morphemic affixation processes; in the first case (a)) in

terms of derivational suffixation, and in the second case (b))
in terms of suffixation and concomitant predicatable stress
placement change.   Thus, although the words in each pair contain
phonetically (and phonemically) different vowels, there will only
be one abstract "underlying phonological representation" (UPR).
CH imply furthermore that the internal representation of a word
(in our heads) will most likely, in accordance with the principle
of economy (= saving space in our heads) take a form whereby only
those pronunciation features are stored which are not predictable
by general rules, and that these general rules will also be somehow
incorporated into our heads.   In their analysis of the English
sound system CH find that such minimal abstract representations,
which may or may not be isomorphic with internal representations
in lexical storage, bear a striking resemblance to the words'
traditional spellings (as in the examples above) - hence the ex--
alted psychological status of English orthography. Many of the
general rules operate on far from obvious UPRs in far from obvious ways, and CH admit to the existence of many exceptions to
their rules.   Nevertheless English orthography is claimed to be
near-optimal in terms of both economy and psychological reality.

In response to this claim two extreme standpoints can be identified
in the literature.   Firstly there is the obvious objection raised
by Lotz (1972) that the modern English orthography, with its "underlying morphophonemic system", does indeed have something "underly
ing" it - namely an earlier "phonetic" stage in the history of the
orthography.   The rules that relate underlying forms to today's
pronunciation are diachronic rules. (For Lotz the principal
advantage of the standard spelling is that it remains relatively
constant and universal in the face of historical change and social
and regional variation.)   On the other hand, Lotz's argument may
be turned back to front; for example, as O'Neil (1969) states,
"the resistance of standard spelling to change reflects ... an important fact of historical linguistics: the underlying phonological
spellings of languages that are represented by alphabetic writing
systems are themselves quite resistant to change." If we leave
aside for the time being issues concerning the precise neuropsycho-
logical nature of internal representations of human language, and
concentrate primarily on what aspects of linguistic structure people
are able to make use of in translating between the written and
spoken norms of language, then the foregoing remarks can be seen
as a good starting point for a series of psycholinguistic experiments.

The specific question to which our studies are addressed is derived
directly from CH's treatment of the interaction between underlying
phonological forms and the placement of lexical stress in English
words, and as such constitutes an evaluation of some particular
claims about English orthography.   The studies also represent a
more open-ended examination of people's use of different levels

of linguistic information in their performance in specific tasks.
If our results can be explained in terms of the rules outlined in
the "Sound Pattern of English", then we have at least gained
support for the counter-reformist argument concerning the rich
information content of traditional English orthography (if not for
the reality of UPRs), and we may indeed postulate that rules of
the CHian type form part of a native reader's linguistic competence. If not we may still sensibly ask whether any consistent
effect of spelling conventions can be detected in our subjects'
performance.

There are four main studies to be reported. In the first three
studies subjects are presented with a nonsense word embedded in a
real sentence context and asked to read the whole sentence aloud.
Syntactic function of the nonsense words is varied by means of
context and spelling of the nonsense words is manipulated along
various dimensions suggested by "The Sound Pattern of English".
Detailed descriptions and results of the first two experiments
are available in Baker and Smith (1976), and of the third experiment in Smith and Baker (1976). In the fourth study two groups
of children, one group taught to read by means of the initial
teaching alphabet, the other by means of traditional spelling,
take part in a simplified version of the reading-aloud task. The
basic independent variables for all the studies are provided by
CH's analysis of the factors involved in the location of primary
lexical stress in English nouns and verbs. These factors are
summarized in Table 1.

In Table 1 stress assignment is seen to depend on four sources of
information:

(1)   the distinction between lax and tense vowels; in the face of
      imprecise phonetic specification and considerable dialectal
      variation we define as lax (L) those vowels which occur in
      "bid, bed, bad, bod, bud", and as Tense (T) the vowels in
      "bide, bead, bade, bode,Bude";

(2)   weak and strong syllables: a syllable containing a lax vowel
      followed by not more than one consonant is said to be weak
      (W), otherwise it is said to be strong (S);

(3)   part of speech: noun or verb;

(4)   word length: 2 or 3 syllables.

These four sources of information interact to give the standard
pronounciation in Table 1.

Table 1. Rules for the assignment of primary stress in two-and three-syllable nouns and verbs without internal morpheme boundaries.

	WORD LENGTH IN SYLLABLES	QUALITY OF FINAL SYLLABLE (S)	SYLLABLE RECEIVING PRIMARY STRESS	EXAMPLES
Nouns	2	-L	First	ínfant
	2	-T	Second	políce
	3	-WL	First	jávelin
	3	-SL	Second	agénda aréna
	3	-T	First	ántelope
Verbs	2	-W	First	cáncel
	2	-S	Second	eléct eráse
	3	-W	Second	consíder
	3	-S	First	gállivant híbernate

Two further points must be made.   Firstly there is CH's use of
morphemic boundaries to limit the application of rules in Table
1.   For example verbs such as "defer" and "comprehend" should,
according to the rules, receive 1st syllable stress, like "edit"
and "implement" respectively.   However, by marking a morpheme
boundary between de and fer and com and prehend, the rule is
blocked.   The set of prefixes (de, com, re, per, etc.) and roots
(pel, mit, fer, etc.) to which this condition applies would be
marked in the lexicon.

Secondly there are a number of apparent exceptions to the rules
(e.g. "eclipse" and "umbrella") which CH account for by positing
underlying phonological forms which differ significantly from
their surface phonetic forms but which resemble their standard
orthographic forms in a non-fortuitous fashion.   Thus the final
-e of "eclipse" will be present in the UPR (to be deleted by a
later "-e deletion" rule) and will function as a third syllable.
Similarly, the double -ll- in "umbrella" will have an underlying
representation as "two consonants", while phonetically only one.
The application of the rules of Table 1 to these UPRs will then
produce the correct stress pattern.   Words of this type of
structure present a strong test case for CH's near-optimality
claim.

Experiment 1

Method

The subjects, all psychology undergraduates and native speakers
of English, were instructed to read aloud a series of sentences
in which one noun or one verb had been replaced by a nonsense
word.   The nonsense words were all based on two, three and four
syllable English words used by CH to illustrate the operation
of their stress placement rules, but transformed by the addition,
removal or substitution of one letter, in such a way as to produce a word in which the critical orthographic factors for stress
placement in the model word were either preserved or lost.
Examples are shown below:

English	Nonsense	
	Structure preserved	structure altered
maintain	mainkain	maintin
cinema	conema	cinempa
edit	egit	edite
veranda	zeranda	verada

Example sentences:

The $\begin{smallmatrix}\text{egit}\\\text{edite}\end{smallmatrix}$ above his head glowed.

The inspector began to $\begin{smallmatrix}\text{egit}\\\text{edite}\end{smallmatrix}$ that Albert was involved.

Subjects' readings were tape-recorded, and their pronunciations of
the nonsense words transcribed and classified according to phonemic
structure and stress pattern.

Results

The results of Table 1 would predict that, depending on the interaction between the orthographic form and the perceived syntactic
function of the word, primary stress would either remain the same
as in the real English model or be shifted.   In the latter cases
it is possible to **compute** an index of the degree to which rules
are followed, i.e.

$$\frac{\text{Number of rule pronunciations}}{\text{Number of rule pronunciations + number of model pronunciations}}$$

We refer to this as the "CH-score". A score of 1.0 means that the
rules are being followed in every instance, a score of 0.0 means
that the rules are never being followed. Scores for individual
words cover the whole range from 1.0 (e.g. "agapt" from "adapt",
when presented as a noun) to 0.0 (e.g. "javerin" from "javelin"
when presented as a verb) with an overall mean score of 0.40.
Shown below are CH-scores for words subcategorized according to
number of syllables and which syllable should receive primary
stress. Two syllable words achieve significantly higher CH-scores
than three and four syllable words. A hypothesis that a pure guess-
ing strategy is being used, which will naturally work more efficiently with only two syllables to choose between, may be discounted,
since there are more individual words in all categories with very
high or very low CH-scores than would be predicted from such a
strategy.

NUMBER OF SYLLABLES	STRESSED SYLLABLE ACCORDING TO RULES	EXAMPLE	MEAN CH-SCORE
2	First	maintin	0.502
2	Second	edite	0.617
3	First	astonize	0.308
3	Second	cinempa	0.205
4	Second	balalaka	0.125
4	Third	metropolsis	0.321

There is also a significant effect for whether the word is presented
as a noun or as a verb, and a significant interaction between syntactic category and predicted direction of stress placement change.
Thus nonsense nouns for which a forward shift of stress is predicted
(e.g. maintin, collap) more often follow the rules than similar nonsense verbs, and vice versa when a backward shift is predicted
(e.g. edite, cancele).

Although there are several systematic features of these results,
the overall conclusion is that the circumstances under which subjects use general rules (high CH-scores) depend somewhat haphazard-
ly on a variety of non-linguistic factors. It would be dangerous
to make inferences from studies using nonsense words without the
most careful control of the similarity of the nonsense words to
real English, and this additional control was achieved in our
second experiment.

Experiment 2

Method

The same procedure was repeated with a new set of two and three
syllable nonsense words. This time we set out to construct
words as dissimilar as possible to existent real words, while
still conforming to English spelling conventions and pronounce-
ability restrictions. Twenty subjects were used to check on
similarity to real words. If any of the words we had constructed
suggested the same real word to more than two of the twenty judges,
it was dropped from the list. Examples of the words are shown
below.

Accepted	Discarded
palomp	nuzelt (suggested "nuzzled")
tupaivend	nushreamope (suggested "mushroom")
ollanteam	basimug (suggested "basement")

Pronunciations were classified according to phonemic structure,
and scored according to whether location of primary stress followed
the rules or not.

Results

The main factors influencing performance are shown in Table 2, in
this case for two syllable words only. The degree to which subjects follow CH's rules clearly varies a great deal. The main
significant effects here are

(i)   which syllable should correctly receive stress, and

(ii) an interaction effect between which syllable should receive
      stress and whether the word is presented in a noun or a verb
      frame.

Table 2. Percentage first syllable stress in two-syllable words
          in Experiment 2 as a function of syntactic category,
          strength of first syllable, and which syllable correctly
          receives stress according to rules.

	NOUN		VERB	
	STRONG FIRST SYLLABLE	WEAK FIRST SYLLABLE	STRONG FIRST SYLLABLE	WEAK FIRST SYLLABLE
FIRST SYLLABLE PREDICTED	98.0	83.0	97.3	68.0
SECOND SYLLABLE PREDICTED	86.0	46.0	70.8	30.5

Verbs tend to receive first syllable stress less often than nouns
irrespective of which syllable should receive stress. Furthermore,
the strength of the first syllable exerts a significant influence
on performance.   If the first syllable is strong it is more likely
to be stressed irrespective of which syllable should receive stress.
The results for three syllable words present a similar picture except that in this case a strong second syllable significantly re-
duces the tendency towards first syllable stress.   These effects
are in conflict with CH's rules, according to which the nature of
the penultimate syllable is predictive only for three-syllable
nouns with a lax vowel in the third syllable.

The general conclusion of this experiment is that subjects take
account of all of CH's linguistic distinctions, and that all the
rules in Table 1 are influencing performance, even though there
are some over-generalizations to linguistically irrelevant penultimate syllables.   Since our original choice of spellings for
the nonsense words in this experiment had not been undertaken in
any methodical way, but rather according to our own somewhat impressionistic predictions as to how subjects would pronounce the
words, and since the effects of spelling do not emerge unambiguously from the data, we decided to carry out another version of
the experiment in which spelling is varied independently and
systematically within individual words.

Experiment 3

Method

A third set of nonsense words was constructed. These were all
two syllable words built on two different patterns, i.e. CVCVC
and $CVCVC_1C_2$, where C represents a consonant letter selected ran-
domly from a set of twelve (p,t,k,b,d,g,m,n,f,v,s,z), V represents
a vowel letter from a set of five (a,e,i.o.u), and C_1C_2 a final
consonant cluster from a set of seven (-sp, -st, -sk, -mp, -nt,
-nd, -ns).   Examples are:

> bivib dinump vapag nekog gident

The spelling of each basic word was then modified in three different ways

(i)   doubling of medial consonant,
(ii)   addition of final -e, and
(iii) substitution of second vowel by an orthographic diphthong
       (i.e. ee, ai, oa, oo) in the hope that subjects would thus
       be constrained to pronounce a tense vowel in the second
       syllable.

Examples of the transformations are:

Basic word	-CC	-e-	$-\bar{V}-$
vapag	vappag	vapage	vapaig
gident	giddent	gidente	gideent
bivib	bivvib	bivibe	***
nekog	nekkog	nekoge	nekoag
dinump	dinnump	dinumpe	dinoomp

*** There is no regular and unambiguous means of representing
    interconsonantal /aI/ in English spelling.

These transformations were carried out in order to manipulate in
a more regular and consistent fashion some spelling conventions
which are of particular predictive value in CH's rules (viz. -CC-
in "umbrella", -e in "eclipse").

The experimental procedure was the same as in the first two experiments.

In a subsidiary task a different group of subjects were instructed
to carry out a morphemic analysis on the same nonsense words. The
procedure was illustrated to them by means of the following examples, in which the number of oblique lines between morphemes
may be said to represent, in an impressionistic fashion, the
strength of the morpheme boundaries.

E.g.   towel   sofa   yellow (no morpheme boundaries)

         Black/guard   black//bird   grey///bird

         light/house//keeper   (= a beacon warden)
         light///house/keeper   (= a slim domestic)

Subjects were asked to use the same techniques to indicate whether
they considered that any of the nonsense words contained more than
one morpheme, and if so to mark the location and strength of the
morpheme boundary or boundaries.   The aim of this subsidiary
task was to gauge the homogeneity of our material and to ascertain
whether perceived morpheme structure exerted any influence on the
pronunciation of the nonsense words.

Results

There is a great variety of possible pronunciations of the nonsense words.   However, since by far the most common pronunciation
involved both a lax first vowel and a lax second vowel (this
accounts for 49.9% of all pronunciations), we will concentrate
on this subset of the data.

The frequency of first syllable stress in LL pronounced words in
relation to word structure, spelling convention and syntactic
category, (but ignoring $\bar{V}$ forms due to their small numbers) is
presented in the upper part of Fig. 1.   The complete figure represents a comparison of the results of this experiment with those
of Experiment 4 (for which see below).

All the main effects are significant; number of final consonants,
syntactic category, and spelling convention.   The spelling convention effect is produced by the generally lower frequency of
first syllable stress with -e spelling.   There is also a significant interaction between spelling convention and syntactic
category, this being produced by the much smaller difference between nouns and verbs with -e spelling than with the other two
spelling conventions.

Thus, in general, the tendency to stress the first syllable is
significantly reduced by presenting the word as a verb, by strengthening the second syllable (i.e. introducing a final consonant cluster), and by adding a final -e.

In the morphemic analysis task the number of boundaries in each
position in each word was counted, and the means and the variances
of the distribution of boundaries in different categories of words
were calculated.

One result is of particular interest.   There was a significant
negative correlation between the variance in the location of

boundaries within words and the size of noun v. verb stress differences discovered in the main experiment (see Fig 1) - that is, low
variance in the location of morpheme boundaries is correlated with
large differences in the frequency of first syllable stress between nouns and verbs, and vice versa.   Furthermore, far greater
variance (i.e. far more indecision) in morpheme boundary location
was associated with the final -e spelling convention than with any
of the other conventions.   As was seen in Fig. 1 words with final
-e spelling also show the smallest differences between nouns and
verbs.

This result is plausibly interpreted as a morphemic effect.   Where
a morpheme boundary exists in the middle of a real English word
(e.g. permit) CH's rules give rise to a large difference in stress
between the noun and the verb form (pérmit v. permít), but this
effect does not appear when there is no morpheme boundary (e.g.
worship): we are obtaining very similar results with our nonsense
material.

Experiment 4

Method

In order to compare the performance of beginning readers with the
adult data already collected, Anne Groat (at the time a final year
psychology undergraduate) carried out a version of experiment 3
with two groups of six and seven year old children.   One group
had been taught to read with the initial teaching alphabet (ita)
but had since (albeit recently) transferred to traditional spelling; an otherwise comparable group had never been exposed to ita.

One interesting facet of ita is that it is not claimed to be a
perfectly uniformly phonemic orthography.   On the contrary, many
concessions to traditional spelling are made.   This is done in
order to ease the transitional stage during which traditional
spelling is introduced.   According to Haas (1970) many of the
particular concessions made are arbitrary and without sound empirical motivation.   With respect to the theoretical background
of this paper it is to be observed that the double consonants of
traditional spelling are preserved in ita, whereas unpronounced
final -e is not.

Results

The results are shown in Fig.1.

Both groups of children showed the usual overall reduction in
first syllable stress for verbs and for words with a strong final
syllable.   However both groups of children differed from the adult
group in two significant ways.   Whereas in adults there was a

general reduction of first syllable stress for final -e spellings
as against Normal and -CC- spellings, there was no such effect in
the children's performance. Again for final -e spelling alone
adults showed no effective stress differences between nouns and
verbs, but large differences for the other spelling conventions.
This effect is not observed in the two groups of children. Both
these results suggest that some subtle effects of final -e have
not yet developed in young children.

One further interesting, if surprising, trend in the data seems
to provide an example of children following CH's rules more consistently than adults. In the children's data there is a reduction in first syllable stress for verbs with Normal spelling and
two final consonants which is relatively larger (in relation to
all other factors) than the corresponding reduction for nouns
with the same spelling. This is the first time that we have found
in our data support for the CHian distinction between two syllable
nouns ending in two consonants but with a lax second vowel and
consequently with first syllable stress (e.g. "témpest"), and two
syllable verbs with the same phonemic structure but with second
syllable stress (e.g. "tormént"). The distinction is not apparent in the adult data.

Only one significant effect distinguishes between the two different groups of children. For both the adult group and the traditional orthography children words with -CC- spelling and two
final consonants exhibit much larger stress differences between
nouns and verbs than any other spelling does. This effect,
which on the basis of the results of Experiment 3, we argue is a
morphemic effect, is significantly diminished in the ita group:
apparently ita does not aid the perception of morphemic structure
as much as traditional orthography.

Discussion

Most of the linguistic factors in CH's rules appear to have an
influence on our subjects' pronunciations of nonsense words.
Apart from one minor trend in the children's data we did not observe noun/verb differences relating to CH's phonological distinction of Table 1 (i.e. where presence of two final consonants
following a lax vowel does not attract stress in two syllable
nouns, but does in two syllable verbs). In our results stress
in nouns and verbs alike is attracted to strong syllables, and
furthermore this effect is not restricted to the final syllable.
There is a tendency for a strong first syllable to attract stress,
irrespective of what follows.

We are unable to draw any firm conclusion about under-lying forms.
The use of final -e in Experiment 3 does indeed reduce the frequency of first syllable stress (in LL words); but unfortunately

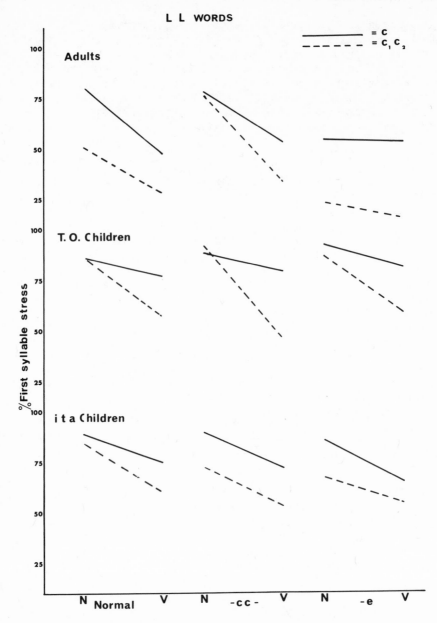

<u>Fig. 1</u> Percent first syllable stress for LL words (2-syllable
   words containing 2 lax vowels) as a function of spelling (Normal,
   medial double consonant, or additional final <u>e</u>) and number of
   final consonants (1 final consonant = C, 2 final consonants =
   C_1C_2).   The results for Adults refer to Experiment 3, and for
   children to Experiment 4.

not only for verbs ending in one or two consonants (such as the
real words "allege" and "elapse") and nouns ending in two consonants (real word "eclipse"), but also for nouns ending in one consonant (real word "endive") where the rules still predict first
syllable stress.   Indeed the differences involving final -e can
best be explained as morphemic effects, not UPR effects. However
this failure to detect UPR effects should not be judged harshly.
UPRs, if they exist as such, will necessarily be arrived at by
the language learner during a process of abstraction and generalisation from surface phonetic forms, and indeed after considerable
exposure to and familiarity with the latter.   Our subjects are
not given time to become familiar with the nonsense words we
present, and different techniques are required to explore this
point further.   Nevertheless we have gone some way towards
illuminating the rich potentialities of English spelling on a
number of different linguistic levels. It is clear that we should
seek a fuller understanding of the ways in which these levels
of representation interact before committing ourselves in favour
of spelling reform. In any case it appears that any reform deliberately restricting orthography to the phonemic level may well
be misguided.

Acknowledgements

This research was supported by a grant from the Social Science Research Council.

References

Baker, R.G. and Smith, P.T. (1976). A psycholinguistic investigation of English stress assignment rules. Language and Speech, 19, 9-27.

Chomsky, N. and Halle, M. (1968). The Sound Pattern of English. New York: Harper & Row.

Firth, J.R. (1957). Papers in Linguistics: 1934-1951. London: O.U.P.

Haas, W. (1970). Phonographic Translation. Manchester: M.U.P.

Leontiev, A.A. (1975). The heuristic principle in the perception, emergence and assimilation of speech. In Lenneberg, E.H. and Lenneberg, E. Foundations of Language Development. New York: Academic Press.

Lotz, J. (1972). How language is conveyed by script. In Kavanagh, J.R. and Mattingly, I.G. Language by Ear and Eye. Cambridge, Mass.: M.I.T. Press.

O'Neil, W. (1969). The spelling and pronunciation of English. In Morris, W. The American Heritage Dictionary of the English Language. Boston: Houghton Mifflin.

Savin, H.B. (1972). What the child knows about speech when he starts to learn to read. In Kavanagh and Mattingly (op cit).

Smith, P.T. and Baker, R.G. (1976). The influence of English spelling patterns on pronunciation. Journal of Verbal Learning and Verbal Behavior, 15, 267-285.

CHILDREN'S KNOWLEDGE OF SELECTED ASPECTS OF
SOUND PATTERN OF ENGLISH

ROSEMARIE FARKAS MYERSON

Harvard University

Introduction

This study questioned the inner knowledge that children
between the ages of eight and seventeen have about the
structure underlying the spoken form of certain complex
derived words of their language. Complex derived words
are those suffixed words involving systematic changes
in the sounds of the base word in the derived word.
For example, in the pair of words distort-distortion,
the addition of the suffix -ion has changed the final
[t] sound of the base verb distort to an [š] sound in
the derived noun distortion. Such changes between base
and derived word are predictable; whenever -ion is suffixed onto a base verb, the final consonant of the base
becomes palatalised in the derived noun. This study
investigated when children learn that distort bears the
same systematic relations to distortion that relate
bears to relation. By testing word learning and recall
behaviour and comparing this to production and intuition
performance, the study shows how and when children learn
that complex derived words are not unanalysable wholes
but rather can be derived from their underlying morphemic constituents.

N. Chomsky and M. Halle in Sound Patterns of English
(SPE) (1968) showed that one can describe the process by
which base words are transformed into complex derived
words by ordered sets of phonological rules. SPE
hypothesised that knowledge of the pattern of sound
changes between base and derived words is part of the

377

adult English speaker's phonological competence. The
phonological rules in SPE have as input both syntactic
and phonological information about the underlying morphemic structure of the derived word; the rules require
that the speaker have internalised a representation of
the morphemes that is abstract in relation to the
phonetic (spoken form) of the word.

Novel derived words like <u>cran-orange juice</u> or <u>chicken-
burger</u> show that adults know how not only to pronounce
but also to assign meaning to newly encountered derived
words formed with neutral (nonsound changing) word
derivational suffixes. The adult's language processing
mechanism enables him to recognise the underlying morphemic structure of such words and to use etymological
cues for deriving meaning. Aronoff's theory of word
derivational rules (1974) described the processes by
which base words of a specific syntactic category are
transformed into derived words of another syntactic
category. He found that word derivational processes
can only be described in terms of the morphological
structure of subsets of base words. English word formulation processes are such that the syntactic label,
the pattern of sound changes, and a semantic reading
for the derived word are predictable from the morphological constituents. Since word analysis is like
backwards word formation processing, Aronoff theorised
that the speaker who possessed inner knowledge of complex word formation rules has the ability both to create
new derived words and to analyse newly encountered
words. Aronoff did not believe that every time a
speaker used a derived word that the word formation
rules had to be used to generate the word; rather, he
saw word formation rules as "once only" rules, for
once a word is learned, it takes on a psychological
identity of its own.

Bogoyavlenskiy (1973) studied the ability of five and
six year old Russian children to understand certain
Russian suffixes that create a new word from an existing word, to use these suffixes creatively, and to
explain their meaning. He found the following developmental sequence to children's acquisition of knowledge
of Russian suffixes:

1.   Child learns suffixed word as a whole.
2.   Child develops inner appreciation of the semantic
     meaning of the suffix.
3.   Child applies word formation process in creating
     novel forms.

4. Child has metalinguistic awareness of the process;
    i.e. the ability to describe and/or discuss inner
    knowledge of the word formation process.

Prior Research on Knowledge of English Complex Word Formation Processes

The research on the inner knowledge that either children
or adults have about English complex word formation
processes is not consistent. The results that a researcher drew from his data were dependent upon the
nature of the task that was required of the subjects.
Subjects have alternative strategies for solving different tasks. If the task required that the subject
produce by creating and saying aloud new words, then
the data failed to show that even college students
consistently applied the required pattern of sound
changes (Robinson, 1967; Steinberg, 1973). However
studies of word segmentation ability that presented
different kinds of tasks to the subjects have shown
certain knowledge of such sound pattern changes. Using
an associative learning task for investigating the
knowledge that children had about English tense-lax
vowel shift relations, Moskowitz (1973) found that
children between the ages of seven and twelve had inner
knowledge of $[\varepsilon y] \sim [æ]$ and $[iy] \sim [\varepsilon]$ vowel shift sound
changes before the suffix -ity. Since the age of the
children who showed inner knowledge of vowel shift
alternations was such that all had also been exposed to
the written language, and since Read's (1971) work with
younger children's writing had shown no knowledge of
tense-lax vowel shift pairings, Moskowitz concluded
that the source of her children's vowel shift phonological knowledge was the English spelling system.

Ladefoged and Fromkin (1967) showed that adults attending a lecture on linguistics could correctly mark the
pronunciation of nonce suffixed words which had been
printed in sentences. From this evidence, Ladefoged
and Fromkin concluded that adults know the systematic
sound pattern relationships between base and complex
derived word pairs. Moskowitz's and Ladefoged and
Fromkin's studies show that psycholinguistic tests that
do not require that a subject create and say novel
suffixed words, can show certain knowledge of complex
word formation sound patterns.

Design of the study

Selection of subjects

The English orthographic code requires considerable
phonological sophistication of the reader for the written code tends to preserve morphemic identity rather
than represent the surface sounds of complex derived
words. As shown in SPE, sound changes that are predictable from the general rules of phonology are not
given visual representation. Thus, one can hypothesise
that the level of a child's inner knowledge of word
formation rules would relate to his reading.

72 subjects, 18 from each of grades three, six, nine
and twelve were selected. The mean ages of the four
groups were:

8 years 2 months for the third graders.
11 years 1½ months for the sixth graders.
14 years 3½ months for the ninth graders.
17 years 3½ months for the twelfth graders.

All subjects were monolingual speakers of English attending local suburban schools; all had normal or better
intelligence. The youngest children were approximately
eight years old since the pilot testing had shown that
children much younger than this had great difficulty
performing word suffixing operations. Within each grade
the children were selected on the basis of their scores
on standardised tests of silent reading paragraph comprehension. One third of the children in each age
group were poor readers, one third average, and one
third were good readers.

Selection of the Sound Patterns

Moskowitz (1973) found that changes in children's
abstractions about English tense-lax vowel pairs occurred at the same age as when the children learn to read
and to write. In order to question Moskowitz's hypothesis that the spelling system was the source of the
children's knowledge of English sound pattern relations,
this study investigated children's acquisition of inner
knowledge of sound patterns that are and are not part
of the orthographic code. Of the five sound patterns
selected for this study, three were patterns that are
abstractly encoded in the English writing system and
two were sound changes that are not given any visual
representation. Below are listed five sound patterns
used in the study and a discussion of their orthographic

representation in relation to the sound pattern. The three suffixes used were -ion, -ity, and -ical.

1. Palatalisation before -ion as in <u>relate-relation</u> or <u>distort-distortion</u>. Whenever the suffix -ion is added on to a base verb, the final consonant of the base word is changed to a palatalised sound. The palatalisation of the final dental is abstractly encoded orthographically, for one consonant letter, "t", is used to represent the [t] sound of <u>distort</u> and the [š] sound of <u>distortion</u>.

2. Vowel shift before -ity as in <u>sane-sanity</u> or in <u>grave-gravity</u>. The orthographic representation of the [εy] ~ [æ] tense-lax vowel sounds in each word pair is abstract in relation to the phonetic sounds, for one vowel letter, "a", represents the [εy] sound of <u>sane</u> and the [æ] sound of <u>sanity</u>.

3. Vowel shift before -ical as in <u>meter-metrical</u> or in <u>gamete-gametical</u>. As in vowel shift before -ity, the visual representation of the tense-lax vowel sounds is abstract in relation to the spoken sounds. One vowel letter, "e", represents the [iy] sound of <u>meter</u> and the [ε] sound of <u>metrical</u>, two sounds that acoustically (and articulatorily) are qualitatively different.

4. Stress shift before -ity as in <u>moral-morality</u> or in <u>human-humanity</u>.

5. Stress shift before -ical as in <u>history-historical</u> or in <u>method-methodical</u>. In contrast to the abstract representation of palatalisation sound changes and of tense-lax vowel shift sound changes, stress placement is never marked in the writing system. The written language does not formally give the child any visual encoding of the English pattern of stress shift rules.

Therefore, while at the beginning stages of learning to read and to write, children are taught that the letter "a" represents [εy] and [æ] sounds and that the letter "e" represents [iy] and [ε] sounds, they are not formally taught English stress shift sound changes as part of their reading and writing education. Thus, the spelling system can not be the source of children's knowledge of English stress shift rules.

<u>Three Tests of Oral Language Competence</u>

The children's knowledge of the five sound patterns was investigated by three types of tasks:

1. A production test which required that the child
    create and say novel derived forms.

2. A conscious judgment test in which the child had
    to decide which of two derived words sounded better.
    One choice was the phonologically correct sound pat-
    tern and the second choice was incorrect because
    the required sound changes had not been made.
    The incorrectly formed derived word sounded phoneti-
    cally similar to its base word; the phonetics of
    each incorrect choice are like those of the T type
    words of the word recall tests (see Table 1).

3. A word recall test which measured the child's
    ability to recall derived words over a period of
    weeks.

All three tests used nonce words whose sound patterns
and meanings were modeled on real words. All nonce
words were verbally defined and accompanied by pictures;
all were used in verbal contexts such that the syntactic
category of the derived word was clear. In the production test, in the conscious judgment test, and in the
teaching session for the derived words of the word
recall test, the subjects were shown a suffix card on
which was printed the suffix to be affixed to that base
word.

The production test measured children's ability to
create and say novel derived words. This test was
similar to that used by Steinberg (1973) for questioning
vowel shift knowledge of college students. The production test had ten items, two for each of the five
sound patterns.

The conscious judgment test measured children's intuitions about which of two derived words sounded better.
There were ten items, two for each of the five sound
patterns. A zero correct score for a sound pattern
meant that the child twice chose the incorrect (i.e.
the phonetically similar) form and not the phonologically correct derived word. A score of one phonologically correct choice showed random behaviour with that
sound pattern. Only a score of two phonologically
correct choices showed that the child's intuitions
about that sound pattern were consistently correct.

The word recall test involved teaching each child ten
suffixed words, two for each of the five sound patterns.
Recall ability was tested one day (Recall I), one
week (Recall II), and six weeks (Recall III) after the

teaching session. Words were to be recalled as taught.
Piaget (1967) showed in a study on the development of
visual memory in children, that the child's memory
code depended on inner operations. In the word recall
study, it was hypothesised that the ability to learn
and recall novel derived words could reveal what the
child knew about word formation processes. It was expec-

Table 1. Nonce based and derived words for Word Recall
          Test.

Version A

	Base	L-type form	T-type form
Palatalisation before -ion:	prezate delort	prezašən	delorDīən
Vowel Shift before -ity:	verane trave	verænity	trɛvity
Vowel shift before -ical:	dəreter məgeet	dər ɛ trical	magītical
Stress shift before -ity:	túrəl rómməl	turálity	rómməlìty
Stress shift before -ical:	gáthod néttory	gathódical	néttorìcal

Version B

	Base	L-type form	T-type form
Palatalisation before -ion:	prezate delort	deloršən	prezaDīən
Vowel shift before -ity:	verane trave	trævity	verɛnity
Vowel shift before -ical:	dəreter məyeet	m əy ɛ tical	dərītrical
Stress shift before -ity:	túrəl romməl	rommálity	túrəlìty
Stress shift before -ical:	gathod nettory	nettórical	gáthodìcal

ted that a child's derived word recall performance would
be affected not only by the pattern of the sounds of
the derived words in relation to those of its base word
but also by the knowledge that the child had internalised
with respect to that sound pattern and by the child's
metalinguistic awareness of the sound pattern. Pilot
testing (1973-1974) had shown that the task of learning
and later recalling ten new derived words formed according to ten different rules was too difficult to allow
any subject to memorise the sounds of these ten words
in a mechanical fashion.

Two types of words were taught for each sound pattern.
One was the L type word, the phonologically correct
word. Its sounds differed systematically from those
of its base word since the required phonological rules
had been applied. The second word taught for each
pattern was called the T type; it sounded phonetically
similar to its base word. (see Table 1).

The T word was not correct because the sound changes
required by English grammar had not been made. The T
palatalisation words are possible English sound sequences; for example, delorDiən rhymes with accordian. The
T vowel shift words are possible English sound sequences
trɛvity rhymes with brevity, and məgitical with hermiti-
cal. These T words are incorrect derived words because
given the sound structure of the base words and the
suffixes added, English phonology requires specific
sound changes. The stress pattern of rómməlìty and
néttorìcal is found in words like alligator, but there
are no words rhyming with T stress shift words because
whenever -ity or -ical are suffixed onto multisyllabic
base words, the stress is always on the syllable preceding the suffix.

Two versions of the word recall test were created with
half of the subjects in each grade being taught each
version. Version A differed from Version B in that a
suffixed word that was taught as a T sound pattern in
Version A was taught as an L word in Version B. This
controlled any possibility that one word of a sound
pattern was easier than another. A randomised order
was used for teaching the ten derived words.

Below is an example of the ritual used for teaching the
ten derived words of the recall test. The story was
accompanied by a picture of a fishing knife and by a
suffix card with -ity printed on it.

> The next word is trave. Trave means sharp.
> My fishing knife is trave; it is sharp. Now
> add 'i' 't' 'y' to trave to make a new word
> meaning sharpness. My knife is trave. I
> like its travity. What do I like about the
> knife? I like its ..... My knife is sharp;
> I like its .....

The child was to say the derived word as taught in each
pause. The story ritual for each word was reviewed
two times during the teaching procedure.

At each recall session, the child heard the same story,
saw the same pictures, but was not shown the suffix
cards. Thus the ritual provided him with the base word
and its meaning. The recall task differed from the
production task not only because it was a word recall
test, but also because it forced the child in recreating
the derived words to provide the suffix. Recall performance showed the relative ease with which a child
recalled the pair of words for each sound pattern. The
procedure of teaching each child one L and one T word
for each sound pattern allowed each child to act as his
own control. If recall performance was not affected
by a child's phonological knowledge, then each T word
should be easier to recall than the L word of that
sound pattern since recalling the T word, a word that
sounded similar to its base word, would require fewer
operations than recalling the L type word. However,
is recall performance depended upon inner lexical
knowledge, then performance with each L and T word pair
of words would reflect the level of the child's knowledge of English phonology.

There were four possible categories of word recall responses for each pair of L and T words: (-L, -T), (-L,+T),
(+L, -T), (+L, +T) (- means not recalled as taught,
and + means recalled as taught).

(-L, -T) provided no evidence of any knowledge of the
word suffixing process (-L, +T) showed knowledge of the
suffix but not of the sound changes. While all T words
sounded similar to their base words, the nature of the
T words and of English phonological constraints is such
that (-L, +T) response is most probable for stress shift
words. -ity and -ical are suffixes used on monosyllabic
base words. Therefore, a child could discover these
suffixes without knowing that on multisyllabic base
words, stress must be shifted onto the syllable preceding the suffix. A child who knew the suffix and not

the stress shift rule could recall the T word correctly
and also err on the L word by changing it to a T type
word. (-L, +T) was unlikely for palatalisation because
English has no words suffixed with -ion that do not
involve palatalisation. Therefore, a child is unlikely
to know the -ion suffix without also knowing the palatalisation sound change. (-L, +T) is unlikely for vowel
shift pairs because each T vowel shift word differed
from the other T words in that it had a lax vowel that
sounded phonetically similar to that of the tense
vowel of the base word. Thus L and T vowel shift words
questioned whether it was easier to learn $[\varepsilon y] \sim [\varepsilon]$
and $[iy] \sim [i]$ alternations or $[\varepsilon y] \sim [æ]$ and $[iy] \sim [\varepsilon]$
English vowel shift pairings. Moskowitz (1973) showed
that removing the glide from the tense vowel of the
base was a difficult task for English speaking children
under the age of twelve. If a child knew -ity and
-ical and not the sound pattern, he would be more likely
to make no change in the base vowel and so give a
(-L, -T) response.

Evidence of inner knowledge of a word formation process
could be provided when the subject could correctly
recall the L word of the pattern: (+L, -T) or (+L, +T)
categories of responses. It was expected that the
child would first be correct only in his recall of the
L words and would err on the T word because in recalling
that word, due to his inner schema for that sound pattern, he would apply this to the T taught word as well
as to the L taught word; he would err in his recall of
the T word because he has to change it to its phonetically correct sound pattern. (+L, +T) was expected to
be a more mature level of word formation knowledge.
It was expected that only when the child could recognise that the T word's sounds deviated from the expected
pattern could he then place a special memory tag on
the word and so be able to correctly recall both the
L and the T word of that sound pattern.

The subjects heard the sounds of the ten derived words
only at the teaching session. However, there were two
types of subject recall errors that elicited special
teaching at Recall I and at Recall II. Such special
teaching never changed the subject's recall score for
that session, but it did give him specific information
that he might be able to use at later recall sessions
to help in a more successful reconstruction of the correct derived word. The first type of subject error
that elicited special examiner help was due to a subject's shortening a multisyllabic base word to a mono-

syllable in the derived word. This apocopation error
was found with the stress shift words. By shortening
the base to one syllable, the subject was able to
retain in the derived word only that syllable, which
in the base word, had been stressed. He avoided the
problem of shifting stress. For example, given
gathod as the base word, the subject who gave gáthical
as the derived word provided the examiner with no
evidence as to whether or not he knew the stress shift
rule. His response was correct in that the stress was
on the syllable before the suffix, but he had not
derived the word correctly from its base word, gáthod.
Whenever a child made a gáthical type recall error,
he was told that he had used the correct ending, but
was asked if he could keep more of the base word in
making the new word. Not all subjects were able to
become aware that they were shortening the base word
to a monosyllable in making their derived word.
Giving this extra help at an earlier recall session
allowed the later recall responses of those subjects
who knew the suffix but not the correct stress shift
rule to be distinguishable from those subjects who knew
the suffix and the correct stress shift rule. Those
who knew only the suffix would give, at a later recall
session, either a gáthical or a T type response while
those who knew the word formation rules were able to
give the L sound pattern.

The second type of error that elicited special teaching
at Recall I and Recall II was a subject's failure to
use the suffix that had been used in the teaching
session. For example, if a subject used -ty instead
of -ity in forming the derived word, one can not know
whether or not he knows vowel shift before -ity.
Therefore, after testing recall for all ten words, the
examiner showed the child a suffix card for any word
which had not been formed with the required suffix.
The child made a new derived word using the correct
suffix, and this was scored separately. This special
teaching was expected to help some subjects to use the
required suffix at the next recall session.

Results

Table 2 compares performance on all 3 tasks. The Recall
test scores are based only on recall of the L words because, as was noted above, for different phonological
reasons the T words for palatalization, vowel shift,
and stress shift differ in their probability of being
recalled as taught if inner knowledge of English phono-

logy affects word recall behaviour. Table 2 shows dramatic differences between the tasks. The table demonstrates that conclusions about a child's linguistic knowledge should not be drawn from a single test. There
were comparatively few phonologically correct responses
in the production task, and although the conscious judgment tasks produced a higher level of performance, the
developmental trends are somewhat uneven. The most interesting analyses were obtained from the Recall data,
and the remainder of the paper will concentrate mainly
on the Recall task.

Table 2. A Comparison of Correct Responses on Produc-
          tion, Conscious Judgment, and Recall III by
          Grade and Sound Pattern.

The table compares for each sound pattern the percent of
subjects in each grade getting the L word correct at
Recall III to the percent of subjects giving two phonologically correct production responses to the percent of
subjects twice selecting the phonologically correct
derived word.

Palatalisation -ion	3rd	6th	9th	12th grade
Recall III	89%	89%	100%	100%
Production	22%	57%	50%	55%
Conscious Judgements	49%	94%	78%	67%

Vowell shift -ity				
Recall III	39%	72%	83%	78%
Production	0%	0%	0%	0%
Conscious Judgments	39%	56%	39%	44%

Vowel shift -ical				
Recall III	22%	39%	56%	67%
Production	0%	0%	0%	0%
Conscious Judgments	33%	44%	44%	44%

Stress shift -ity				
Recall III	17%	44%	78%	94%
Production	0%	6%	28%	33%
Conscious Judgments	67%	72%	89%	50%

Stress shift -ical				
Recall III	17%	39%	87%	89%
Production	0%	0%	22%	33%
Conscious Judgments	56%	94%	83%	94%

Word Recall Test

Word Recall Test - Grade and Word Type Effects

The word recall test was a difficult one; only one child, a ninth grade good reader, was able to correctly recall all ten derived words as taught. The mean number of words correctly recalled rose from third to ninth grades; there was also an increase in recall accuracy from Recall I to Recall III.

At Recall III, the mean for third grade was 2.89, for sixth grade 4.11, for ninth grade 6.11, and for twelfth grade 5.89. A repeated measures two way analysis of variance on the total number of derived words recalled as taught by subjects in each of the four grades showed significant grade effects $F(3,68) = 15.84$, $P < .001$, significant recall effects $F(2, 136) = 16.30$, $P < .001$ and nonsignificant interaction effects.

An important type of error responses was that in which the subject changed a T taught word to an L pattern or an L type taught word to a T type. The mean number of T words changed to their phonologically correct L form increased with grade. In contrast, the mean number of L words changed to T type in recall decreased with grade. A repeated measures two way analysis of variance on the total number of errors on T words in which the subject changed the T taught word to its phonologically correct L pattern showed significant grade effects, $F(3,68) = 5.07$, $P < .01$, significant recall effects, $F(2, 136) = 9.67$, $P < .001$, and nonsignificant interaction effects. A repeated measures two way analysis of variance on the number of subjects changing L taught words to T type showed nonsignificant main and interaction effects. However, the level of significance for grade effects was .054 so that one can be fairly certain that the differences were real.

Word Recall - Sound Pattern Effects

The significant grade effects shown by the analysis of variance on the total number of derived words recalled correctly may be due to three processes.

1. Children in the higher grades may remember more.

2. Children in the higher grades may be better at discriminating L words from T words.

3. Children in the higher grades may differ in their biases towards L and T responses.

The ten derived words questioned knowledge of five different sound patterns. If children in the higher grades simply remembered more, then there should be no differences in changes in recall over grade in relation to the type of word or the sound pattern or in errors due to recalling L words as T types or vice versa. Tables 3 and 4 clearly show that the children's improvement with grade differed for the five sound patterns.

A measure of discrimination of L and T words (A) and a measure of L bias (B) were computed (this analysis was suggested by P. Smith).

$$A = \frac{\text{number of L and T words correctly recalled}}{\text{total number of L and T responses.}}$$

$$B = \frac{\text{total number of L responses}}{\text{total number of L and T responses}}$$

For example, at Recall III Grade three vowel shift -ity L=L:7, T=T:3, L→T:0, T→L:7.

$$A = \frac{7 + 3}{7 + 3 + 0 + 7} = 0.59$$

$$B = \frac{7 + 7}{7 + 3 + 0 + 7} = 0.82$$

The results of these computations are shown in Table 5. By definition, the A and B measures in Table 5 include the performance only of that proportion of the children who gave L and/or T responses. Therefore, A and B measures exclude data on children who made suffix errors and/or apocopation errors on both the L and T words of a sound pattern or who, on the vowel shift words, failed to change the tense vowel of the base for both the L and T words of a pattern.

Per sound pattern, A showed only a slight increase with grade despite large increases in the total number of L and T responses, but the changes in B are more significant. For palatalisation and for vowel shift, B was substantially greater than 0.5 for all grades. At third grade, this was based on the responses of 89% of the children for palatalisation but on less than 50% of the third graders' responses for vowel shift. Vowel shift L bias did not change significantly over the grades, but from sixth grade on the figure is based on data from larger proportions of the 18 subjects per grade.

The slight trend for B to decrease with increase in grade may reflect the effect of A, i.e. that older children show a slight increase in their ability to discriminate between L and T words.

The B stress shift measures differed from those for palatalisation and vowel shift in that there was a shift in bias from $B < \frac{1}{2}$ to $B > \frac{1}{2}$; i.e. the younger children, as already noted, preferred the T words and the older children the phonologically correct L sound pattern. The shift in stress shift -ity occurred between third and sixth grades and for stress shift -ical it was between sixth and ninth grades. This stress shift bias switch is consistent with the acquisition of a new rule.

Word Recall - Consistency per Subject per Sound Pattern

(-L,-T), (-L, +T), (+L, -T), (+L, +T) were the four possible categories of responses per subject to each pair of L and T words. Chi square tests showed significant grade and category of response differences for all sound patterns except palatalisation -ion. Palatalisation showed nonsignificant grade differences since almost all the subjects knew this sound pattern. The significant Chi Square for vowel shift reflects the rise in the total number of children giving (+L, -T) and/or (+L, +T) responses. These results are shown in Table 6.

The mean grade level characteristic of each category of response for each sound pattern was calculated from the data. For example, for palatalisation, 2 grade three subjects, and 2 grade six subjects were the only ones to give the (-L,-T) pattern of response. Thus the mean grade level associated with this category is

$$\frac{2 \times 3 + 2 \times 6}{2 + 2} = 4.5$$

This calculation gives a cumulative density mean grade level characteristic of a particular category of response (Table 7). The error term, calculated by assuming a Poisson distribution on random data, was

$$\frac{3\sqrt{5}}{2\sqrt{(\text{number of students in the category})}}$$

Table 3; Recall Test: Number of subjects in each grade getting each L and T word correct at Recall I, II, III. (18 subjects per grade)

Grade 3	Recall I L=L	Recall I T=T	Recall II L=L	Recall II T=T	Recall III L=L	Recall III T=T
Palatalisation						
+ion	7	3	12	3	16	0
Vowel +ity	9	4	9	5	7	3
Vowel +ical	6	2	6	2	4	2
Stress +ity	1	2	2	7	3	8
Stress +ical	1	4	2	8	3	6
	24	15	31	25	33	19
Grade 6						
Palatalisation						
+ion	15	2	15	2	16	2
Vowel +ity	12	3	12	5	13	3
Vowel +ical	6	2	6	1	7	2
Stress +ity	8	2	8	5	8	5
Stress +ical	7	4	7	9	7	11
	48	13	48	22	51	23
Grade 9						
Palatalisation						
+ion	13	6	17	6	18	4
Vowel +ity	14	7	15	7	15	7
Vowel +ical	9	4	9	7	10	9
Stress +ity	12	3	13	5	14	7
Stress +ical	15	5	13	10	16	10
	63	25	67	35	73	37
Grade 12						
Palatalisation						
+ion	16	6	18	5	18	4
Vowel +ity	14	8	14	10	14	9
Vowel +ical	10	6	11	4	12	4
Stress +ity	14	2	16	3	17	3
Stress +ical	13	5	16	7	16	9
	67	27	75	29	77	30

Table 4.   Recall Test: Number of subjects in each grade changing L→T and T→L for each sound pattern.

Grade 3	Recall I L→T	Recall I T→L	Recall II L→T	Recall II T→L	Recall III L→T	Recall III T→L
Palatalisation +ion	1	5	1	8	0	13
Vowel +ity	0	7	1	6	0	7
Vowel +ical	0	4	0	5	1	4
Stress +ity	2	1	5	1	4	2
Stress +ical	3	1	5	3	5	3
	7	18	12	23	10	29
Grade 6 Palatalisation +ion	1	13	1	12	0	11
Vowel +ity	0	5	0	7	0	7
Vowel +ical	3	2	0	5	1	5
Stress +ity	1	2	3	2	2	4
Stress +ical	2	0	6	1	5	0
	7	22	7	27	8	27
Grade 9 Palatalisation +ion	1	9	1	11	0	11
Vowel +ity	1	7	1	7	2	8
Vowel +ical	0	3	2	2	1	4
Stress +ity	3	8	4	10	3	8
Stress +ical	0	8	1	5	0	7
	5	35	9	35	6	38
Grade 12 Palatalisation +ion	0	9	0	13	0	14
Vowel +ity	1	5	1	5	1	6
Vowel +ical	2	5	0	7	0	7
Stress +ity	0	10	1	11	1	14
Stress +ical	1	7	0	8	0	8
	4	36	2	44	2	42

Table 5. Recall Test: Ratio of total L and T correctly
         recalled to the total number of L and T
         responses (A) and ratio of all L responses
         to the total number of L and T responses (B)
         for each sound pattern and for each grade at
         Recall III.

Palatalisation	A	B
Grade 3	.55	1.00
6	.62	.93
9	.67	.88
12	.61	.89

Vowel +ity	A	B
Grade 3	.59	.82
6	.70	.87
9	.69	.72
12	.77	.67

Vowel +ical	A	B
Grade 3	.55	.73
6	.60	.80
9	.79	.58
12	.70	.83

Stress +ity	A	B
Grade 3	.65	.29
6	.68	.63
9	.60	.69
12	.57	.89

Stress +ical	A	B
Grade 3	.53	.35
6	.78	.30
9	.79	.70
12	.76	.73

Table 6. Recall III: Cross Tabulations of Responses of 18 Subjects in Each of Four Grades according to 4 Stages for Each Pair of L and T Words of Each Sound Pattern.

Palatalization +ion

	Grade	-L -T	-L +T	+L -T	+L +T
	3	2		16	
	6	2		14	2
Grade	9			14	4
	12			14	4

$(\chi^2 = 8.61$, not sign.)

Vowel Shift +ity

	Grade	-L -T	-L +T	+L -T	+L +T
	3	10	1	5	2
	6	4	1	11	2
Grade	9	1	2	10	5
	12	3	1	6	8

$(\chi^2 = 19.67$, $p < .05$)

Vowel Shift +ical

	Grade	-L -T	-L +T	+L -T	+L +T
	3	12	2	4	
	6	9	2	7	
Grade	9	5	3	4	6
	12	5	1	9	3

$(\chi^2 = 19.48$, $p < .05$)

Stress Shift +ity

	Grade	-L -T	-L +T	+L -T	+L +T
	3	7	8	3	
	6	7	3	6	2
Grade	9	1	3	10	4
	12	1		14	3

$(\chi^2 = 30.65$, $p < .001$)

Stress Shift +ical

	Grade	-L -T	-L +T	+L -T	+L +T
	3	10	5	2	1
	6	6	5	1	6
Grade	9	1	1	7	9
	12	2		7	9

$(\chi^2 = 32.31$, $p < .001$)

Table 7.   Mean grade for subjects giving (-L, -T),
           (-L, +T), (+L, -T), (+L, +T) categories of
           responses at Recall III to each sound pattern.

	(-L, +T)	(-L, +T)	(+L, -T)	(+L, +T)
Palatalis-ation	4.5 ±1.68	-	7.3 ±.44	9.6±1.06
Vowel +ity	5.5 ±.79	7.8±1.5	7.6 ±.59	9.4 ±.81
Vowel +ical	6.3 ±.60	7.1±1.18	8.3± .68	10.0±1.12
Stress +ity	5.3 ±.83	5.0±.90	9.2 ±.58	9.3±1.12
Stress +ical	5.2 ±.76	4.9±1.01	9.4 ±.81	9.1±.67

The mean grade levels for the four possible categories
of responses to the pair of words for each sound pattern
showed that the nature of the five pairs of words was
such that there was no one pattern of change. For
palatalisation(-L, +T), as expected, was not a possible
category of response; i.e. no child knew the suffix
-ion who did not also know the sound pattern.(-L, +T)
for vowel shift overlapped or was later than (+L, -T),
showing that children did not know the laxing rule any
earlier than they knew the vowel shift rule.

That some children were able to give T vowel shift
responses is important to note. The two T vowel
shift words involved the [ɛy]~[ɛ] sound alternation
on the -ity word and the [iy]~[i] sound alternation
on the -ical word. Moskowitz (1973) found that children
between the ages of seven and twelve did not learn to
say to criterion these sound alternations. In this
study, while the T vowel shift -ity word was difficult
for third and for sixth graders, nevertheless, three
third and three sixth graders correctly recalled the
T word at Recall III. Seven ninth graders and nine
twelfth graders (50%) were able to recall correctly
the T vowel shift -ity word. Despite the fact that
vowel shift -ical was the least well known of the five
sound pattern, two third graders, two sixth graders,
and nine ninth graders and four twelfth graders were
correct at Recall III on the T vowel -ical word. Thus,
while only a few children in the lower two grades were
able to recall a word involving only the laxing of the
base vowel, up to half of the older two grades were
able to learn the T word correctly. Smith (1975)
believed that showing that children could learn to give

only the lax sound for a base with a tense vowel suffixed with -ity would provide "strong support" for SPE
analysis of English phonological rules. Ninth and
twelfth graders' performance with the two T vowel shift
words can be interpreted as showing that tense-lax
alternations and vowel shift changes can be separate
psychological processes. The mean grade for (+L, +T)
category of response for each of the five sound patterns
was about ninth grade. Despite the fact that most
children knew palatalisation by third grade, the mean
grade for (+L, +T) palatalisation is not lower than
that for the other sound patterns. This implies that
there is a maturational component affecting the minimum
age at which a child has sufficient metalinguistic
awareness of a sound pattern to be able to give (+L,+T)
responses. For stress shift, the mean grade for knowledge of each sound pattern (+L, -T) coincided with the
mean grade for metalinguistic awareness of a sound pattern.

Recall Test -Non L→T or T→L Recall Errors

The recall tests required that the subjects create
derived words from nonce base words. Non L→T and
T→L error responses can provide clues to how children
acquire full knowledge of a sound pattern by giving
evidence of incorrect or partially correct hypotheses
about a word formation process. The gáthical type
response was a category of errors found on the stress
shift words. A few subjects in each grade reduced
the multisyllabic base word to a monosyllable. As the
grade level of the subjects increased, fewer subjects
made this error at Recall III, but the number never
reached zero for the stress shift -ical words. Gathical
errors showed that the subject knew that the stress
belonged on the syllable preceding -ity or -ical suffix;
gathical was an incorrect derived word because the base
word was multisyllabic. Therefore, those subjects per-
sisting in gathical type errors over the three recall
sessions despite the special teaching provided at the
two earlier sessions showed incomplete knowledge of the
word formation process with -ity or -ical.

The addition of -ality, or sometimes -idity, -icity,
-acity, or -erity instead of only -ity, to base words
was another error. This error sometimes persisted over
the three recall sessions with the vowel shift words,
despite the special examiner suffix teaching at the
earlier recall sessions. Walker's "Rhyming Dictionary"

(1936) lists five vowel shift -ity words that end in
...avity and ten words ending in ...anity, but approximately 150 words ending in -ality, 50 ending in -ility
(plus 300 ending in -ability or -ibility), 50 in -idity,
100 in -acity or -icity, and 10 in -erity. While there
were few ...avity and ...anity vowel shift words (the
pattern of the two nonce words), three of the five
...avity pairs are in Thorndike-Lorge G list (1944) and
five of the ten ...anity word pairs are also on this
list. "Ality" type error responses may show that the
child's ability to create and say new words not in the
lexicon is affected by the frequency of and the number
of applications of the suffix to specific morphological
subclasses of base words.

Some subjects gave as the derived word for the base word
néttory, not nettórical as taught but norétical. Since
the stress placement was correct, this type of metathesis
of the two syllables of the base was not counted as an
error, but the subject was asked if he could make his
derived word more like that of the base word. Some
subjects never became aware of the fact that they had
shifted the order of the two syllables of the base in a
way that allowed them to keep the stress on the "et"
syllable. Walker lists six pairs of base-derived words
ending in ...orical but 25 ending in ...etical. Again,
the subjects' responses seemed to reflect an influence
of word frequency on linguistic performance data.

These non T→L and L→T recall error responses imply that
word formation knowledge (as hypothesized by Aronoff
1974) is sensitive to the subclasses of base words involved in specific suffixing operations. The tests of
this study questioned the psychological reality of
children's ability to create new words not in the lexicon; the children's performance on the recall test
showed that children acquire systematic knowledge of
complex word formation processes by learning first to
extend a sound pattern to one (or a few) morphological
subclasses of base words. Word frequency and frequency
of application of a sound pattern seemed to be factors
affecting the child's extension of knowledge of a sound
pattern to the full set of subclasses of base words to
which it can be applied.

Word Recall and Conscious Judgment Tests

Relative to the conscious judgment task, the production
task was shown to be a poor method for testing inner
knowledge of complex word formation processes (Table 2).

Also the analyses of the recall data showed that performance reflected phonological development. It is now important to analyze why conscious judgment and word recall tasks did not always show the same response patterns.

With palatalization -ion, third graders did poorer on conscious judgment than on the recall test, but on the stress shift items the reverse was true, and on the two vowel shift items the third graders performed similarly on each test. The explanation of such differences lies in the fact that the children could solve the conscious judgment items using different strategies depending upon the sound pattern. To decide about stress shift words, the child only had to check as to whether the derived word had its stress on the syllable before the suffix; he did not need to compare the sound pattern of the derived word to its base. In contrast, a correct decision on palatalization (or on vowel shift) conscious judgment items required that the child compare each choice with its base word. The two choices for conscious judgment palatalization were such that they required more complex mental operations than did the stress shift choices. Therefore, children having inner knowledge of word formation palatalization rules could pass on the word recall test but still fail on the conscious judgment task because of the complexity of the conscious judgment decision process. Thus, despite the fact that the ability to hear a correct sound pattern would be expected to precede the ability to use the correct word formation rule in learning and recalling derived words, the correct decision in the conscious judgment task required more complex mental comparisons between the derived word and its base word, a process difficult for many third graders. Therefore, the word recall performances is believed to provide more accurate insight into children's knowledge of English rules for forming complex derived words.

Guttman Scale Ordering of Difficulty of the Five Sound Patterns

Each child's performance on the five sound patterns can be ranked with a pass-fail score on the basis of his success or failure on the L words at Recall III. This allows one to compare the five sound patterns for probable order of acquisition based on the data from the 72 subjects. The Guttman scale asks whether one can predict which sound patterns a child will know given that he knows 0, 1, 2, 3, 4, or 5 of them. The individual children's performance with the five sound patterns

showed that knowledge of the five sound patterns for
children between the ages of eight and seventeen is not
completely age dependent. While one third grader, the
poorest reader, showed no knowledge of any of the sound
patterns, another third grader, a good reader, was correct on all five L words at Recall III while he changed
all five T words to their phonologically correct L sound
pattern.

Table 8. Guttman Scale Analysis of the Accuracy at
         Recall III on the L Words (72 Subjects).
         Numbers indicate number of Subjects correct
         for each category.

Sound Patterns

	Palatal-ization -ion	Vowel Shift -ity	Stress Shift -ity	Vowel Shift -ical
24 children got 4 L words correct	24	24	24	24
13 children got 3 L words correct	13	11	11	4
24 children got 2 L words correct	23	13	7	5
9 children got 1 L word correct	8	1	0	0
two children got 0 L words correct	0	0	0	0

The scaleogram analysis showed that stress shift -ity
and stress shift -ical were not discriminable for relative order of acquisition. Therefore the final scaleogram analysis used only four sound patterns (stress
shift -ical was excluded). The coefficient of repro-
ducibility was .88 and the coefficient of scalability
was .62. Since over .6 shows that the scale is uni-
dimensional and cumulative (SPSS Manual 1970), the
ordering of the sound patterns is hierarchical. The

Guttman scale ordering from easiest to most difficult
sound pattern was:

1. Palatalization of dental before -ion
2. Vowel shift before -ity
3. Stress shift before -ity (and implied here is stress
                              shift -ical)
4. Vowel shift before -ical.

The results of this study have shown that some children
as young as eight have acquired knowledge of all five
sound patterns. Given that both Moskowitz and this
researcher found that below a certain minimum age one
could not test for genrealizations about complex word
formation processes and since this study also showed
that the acquisition of knowledge of the five sound
patterns was related to word frequency and to the number
of tokens of a particular type, it seems that children's
oral language competence with the five sound patterns
reflects the effects of both maturational and language
experience. Since the study showed that children
acquire inner knowledge of word formation rules independent of whether or not the sound changes are encoded
in the spelling system, it seems difficult to attribute
children's inner knowledge of vowel shift -ity to the
English spelling system if one can not attribute inner
knowledge of stress shift to the same source. Nevertheless, once a child has inner knowledge of complex
word segmentation processes and is also reading material
that is linguistically more complex than that used in
his oral language experiences, it is logical to expect
that experience with the written language will result in
vocabulary enrichment and that this should facilitate
the acquisition of more complete knowledge of the subclasses of base words to which specific word formation
rules are applicable.

Acknowledgement

This research was funded in part by a grant from the
Milton Fund, Harvard Medical School.

References

Aronoff, Mark H. (1974). Word Structures, M.I.T. Thesis,
     Cambridge, Mass.

Bogoyavlenskiy, D.N. (1973). The Acquisition of Russian
     Inflections. In C. Ferguson and D. Slobin (Eds.)
     Studies of Child Language Development. N.Y.: Holt,

Rinehart, and Winston, Inc.

Chomsky, Noam and Halle, Morris (1968). The Sound
    Pattern of English. New York: Harper Row.

Ladefoged, Peter and Fromkin, Victoria (1967). Experi-
    ments in Competence and Performance, I.E.E.E. Vol.
    AU 16-1.

Moskowitz, Bryene (1973). On the Status of Vowel Shift
    in English. In T. Moore(Ed.) Cognitive Development
    and the Acquisition of Language. New York:
    Academic Press.

Piaget, Jean (1968). On the Development of Memory and
    Identity. Worcester, Mass.: Clark University Press.

Read, Charles (1971). Preschool Children's Knowledge of
    English Phonology, Harvard Educational Review, 41,
    1-34.

Robinson, Joanne (1967). Development of Certain Pronun-
    ciation Skills. Thesis, Harvard Graduate School
    of Education, Cambridge, Mass.

Smith, Carlotta (1975). Review of T. Moore's book in
    J. Child Language, 2, 314-317.

S.P.S.S. Statistical Package for the Social Sciences
    (1970). Norman Nie, Dale Bent, C. Hull, New York:
    McGraw Hill.

Steinberg, Danny (1973). Phonology, Reading and Chomsky
    and Halle's Optimal Orthography. J. Psycholing.
    Research, 2, 239-258.

Thorndike, Edward and Lorge, Irving (1944). The Teacher's
    Word Book of 30,000 Words. New York: Columbia
    Teacher's College Press.

Walker, J. (1936). Walker's Rhyming Dictionary.
    Revised and enlarged by L. Dawson. New York: Dutton.

"ENVIRONMENTAL INVARIANTS" AND THE PROSODIC

ASPECTS OF REVERSED SPEECH

J. KIRAKOWSKI and T.F. MYERS

University of Edinburgh

The study we are going to present is part of a larger program which explores the relationship between prosody and syntax in the perception of continuous speech. Our studies to date have led us to the conclusion that the relationship is an interactive one. In this paper, we will be concerned with the possibility that some prosodic features may nevertheless be perceived independently of syntax.

To start with, we will give a working description of those prosodic entities we are going to be dealing with: the syllable, the foot, and the tone group. We will not try to define the syllable using such terms as "boundary" and "nucleus" but in terms of an operational definition, whereby the listener simply reports how many syllables he can hear in a given stretch of utterance. We will take it, following Abercrombie (1965) that a foot consists of one stressed syllable followed by zero or more unstressed syllables, and that an unstressed syllable at the start of a tone group is to be treated as if it belonged to a foot in which the stress is silent. A tone group consists of one or more feet, roughly corresponding in extent to the grammatical phrase (Halliday, 1967) and it may have certain boundary markers in the acoustic signal.

Although a reader may be able to use his knowledge of the grammar to predict, from a text, where the tone group boundaries would occur, the number of syllables, and the number and nature of the feet in each tone group, we would like to hypothesise that he can perceive these features in speech even if his access to the grammatical levels of processing has been denied. The experimental hypothesis is that the prosodic features of syllable, foot, and tone group boundary may still be perceptible when the speech signal

is transformed in a way which denies a listener access to grammatical
levels of processing.

Our motivation for presenting such an hypothesis comes from some
earlier work we did on intelligibility (Kirakowski and Myers, 1975).
Since that paper was presented at a conference of which the proceedings were not published, and since it bears considerable relevance
to what is to follow, we will briefly summarise it in a couple of
paragraphs.

We set out to test whether the effects of grammar and prosody on
the intelligibility of speech were additive or interactive. In
the first experiment, we made up an ungrammatical version of a
long (approximately 100 words) test passage by ordering the words
in reverse. We then recorded each of the two passages (grammatical
and ungrammatical) in two conditions: in one, each passage was
read on a monotone, and the timing of the words was isochronous
(i.e. the interval between successive word onsets was constant),
in the other, the passages were read with normal intonation and
rhythm. (Producing an ungrammatical sequence in the latter condition created considerable technical and conceptual problems, the
solution of which we hope to publish shortly). The readings were
embedded in white noise, and Ss'task was to shadow one of the recorded passages. The data was scored in terms of the number of
words shadowed correctly. To control for variation in individual
response to the shadowing task, Ss all shadowed the same control
passage (which was meaningful and recorded in normal style of
delivery). The results for each subject were calculated according
to the following formula:
(overall performance) = (performance in control) - (performance in
   test)
Prediction of Ss' behaviour derived from two models. An "additive
effects" model predicts that performance slopes of the grammatical
and ungrammatical passages between each reading will be parallel.
An "interactive effects" model predicts that the performance
slopes will not be parallel. The additive effects model is thus
a sort of null hypothesis.

Analysis of variance disclosed that the interactive model was
strongly supported ($\underline{P} < .01$). Figure 1 shows the performance
means in the four conditions.

In the second experiment, we used five word sentences and pseudo-
sentences (made up of the same words as the sentences but with the
words in an ungrammatical and meaningless order). These stimuli
were recorded in four conditions of prosody: normal intonation,
monotone but foot rhythm, monotone but syllable-timed rhythm,
and separately pre-recorded words spliced together producing an
arhythmic effect. We used 50% time compression to control for
ceiling effects. Responses were scored in terms of words correctly

Figure 1

Performance means from Experiment 1
Calculated by:
(No. of words shadowed correctly in control) -
(No. of words shadowed correctly in one of the four test passages)
 ... for each subject.

	prosodic	unprosodic	significance (F test)
Grammatical	-3.35	15.13	P < .01
Ungrammatical	23.33	20.24	NS
Significance	P < .01	P < .01	

Model IE 2 supported (grammar only gives improvement when prosody
is present).

Key:

prosodic   : read with normal "reading" intonation.
unprosodic : read without pitch variation, pausing, each word
               occurring at equal intervals of time.
Source   : Kirakowski and Myers (1975).

Figure 2

Performance means from Experiment 2.
Calculated by number of words reported correctly per subject.

	normal	flat	syllable
Grammatical	70.59	65.46	56.67
Ungrammatical	24.05	27.62	25.52
Significance	p < .01	p < .01	p < .01

	spliced	significance (F test)
Grammatical	43.00	p < .01
Ungrammatical	24.17	NS
Significance	p < .01	

Model IE 3 supported (prosody only produces an improvement when
the stimulus is grammatical).

Key:

Normal   : natural reading of stimuli
Flat   : reading of stimuli without pitch variation
Syllable: reading of stimuli without pitch variation and with
             isochronous syllable rhythm.
Spliced :   stimuli made up of pre-recorded words spliced together
             (unnatural intonation - if any - and arhythmic effect).
Source   : Kirakowski and Myers (1975).

reported.   Ss' task was to write down each sentence after one
presentation. Once again, the statistical analysis confirmed the
interaction hypothesis. Recall means for words in the eight conditions are summarised in Figure 2.

That is, there was an intelligibility loss associated with the
reduction of each of the following prosodic features: tone (see
Halliday - op. cit.), foot regularity and syllable regularity
(see Abercrombie - op. cit.), but only when the stimulus was a
meaningful sentence.

Now, to return to our present experiment. In the first part, we
were concerned with the segmentation of a continuous stream of
speech into tone groups. A female speaker read a paragraph from
Evelyn Waugh's "Scoop" at a normal reading rate, into a Revox two-
channel tape recorder, using a moving coil monocardioid microphone.
Two years later, the speaker was asked to decompose the written
script into syllables according to the way she thought it should
go.   This data was used as the basis for a syllable count in all
subsequent analysis of the written version. We employed one transformation in this part: that of playing the speech backwards. We
accomplished this by running the tape back to front, and playing
back from the track opposite to the one we had recorded on. We
checked spectographically that this procedure simply reversed the
signal without loss of information.

We compared performance in the transformed (backwards) version with
that in the untransformed (forwards) version.

Rees (1975) discusses certain phonetic cues that a tone group provides as to the location of its boundaries. There are four separate
but potentially combinatory categories which he mentions: pause
is his first. He notes that in his corpus of Welsh informal conversation, the pause marker normally co-occurs with the boundaries
of large syntactic units, such as the sentence or the clause, and
that this is the easiest to perceive.

In our experiment, we instructed three Ss to tap whenever they
heard a pause. They went through the recording six times in both
experimental conditions. In scoring, we could readily determine
which interword positions the Ss had marked with a tap, and a summary
of the number of taps at each interword was made. (Even with a
delay ascribable to reaction time, most taps occurred within a
pause.)   Interword positions with no taps in either condition
were ignored in the subsequent analysis. Since the average
length of each pause boundaried group was near enough to that of
an average tone group (Laver,1970, cites "seven or eight syllables")
it was decided not to search for other boundary markers. The
average length of a group in our sample was 7.54 syllables, with
a standard deviation of 2.33 syllables. The correlation between

forwards and backwards conditions was high: Pearson's r = .986
(P <.01).   The stretches of utterance delimited by pauses we
shall refer to as segments. Sixteen segments over which maximum
agreement was reached were separated by splicing and they were
used for the second part of the experiment.

In the second part, using six subjects, we were concerned with the
Ss' ability to pick out the component syllables of each segment,
to differentiate between stressed and unstressed syllables, and to
recognise the pauses occurring between syllables, if any there were.
In effect, we had Ss operating a two-level system of stress
(stressed vs. unstressed). Lieberman (1965) found that linguists
transcribing utterances which had been so transformed as to render
the words unintelligible but still retaining some pitch and intensity information could reliably agree about only two levels of
stress.

We used two experimental transformations. One was the reversing
technique which we had used in part one.   The other was a band-
pass filtering below 250 Hz.   For this we used a Barr and Stroud
variable filter type EF 2, with an attenuation slope of 72 dB per
octave.   This was the highest pass which left listeners unable
to report any of the words or syllables of the original recording.
This level was established beforehand.

Each segment was put on a 95 cm. tape loop running at 4.75 cps.
for the reverse transformation and at 19 cps. for the filtered
and untransformed versions (Ss found the reverse transformation
very hard to concentrate on).   Ss' task was to listen to the
tape loop as many times as they wished, and to write down a "l"
sign for a stressed syllable, "⌣" for an unstressed syllable, and
"/" for a pause between syllables.   We ran the untransformed
version last, so at no time before that were any of the Ss aware
of the meaning of any of the segments.

In the analysis, we were concerned with two things. Firstly, was
the average number of syllables reported for each segment significantly the same between conditions and when compared with the
number of syllables reported by the speaker (according to the
procedure described earlier)?   Figure 3 gives the inter-correlation
matrix for these four situations.

As can be seen, all correlations are high and significant beyond
the .01 level. A similar matrix was also computed for each subject,
and although there was more variability in the coefficients obtained, the pattern was essentially the same, and all correlations
were still statistically significant.

Secondly, how closely are Ss in agreement about the pattern of

Figure 3. Intercorrelation matrix for syllables.

	transcript	untransf.	filtered
reversed	.834	.860	.907
filtered	.957	.969	
untransf.	.995		

Key:

"Transcript": number of syllables indicated by informant on the transcript of passage she recorded.

"reversed" and "filtered": number of syllables reported for each of these two experimental conditions.

"untransf".   : number of syllables reported for the untransformed condition.

stressed and unstressed syllables for each segment of the three conditions?

To answer this question, a measure of similarity between Ss' reports for each segment had to be obtained. In order to compare the reports of six Ss, the method considered them as (5+4+3+2+1=15) pairs.   It worked out the minimum number of points of disagreement between each pair of reports in the segment.

The method of analysing the pairs of reports was as follows. Both reports were considered together, a syllable at a time.   Every time there was a disagreement between reports as to the identity of one syllable, three hypotheses were entertained:

i.   one S had reported as unstressed the syllable the other S had reported as stressed (substitution error).

ii.   one S had omitted to report a syllable (omission error),

iii.   one S reported one syllable too many (commission error).

Implementing each of these hypotheses has different consequences for the subsequent analysis. The hypothesis whose implementation involved the fewest subsequent disagreements was taken as the basis for the remaining decisions, as there was no other criterion whereby the rightness or wrongness of either report could be defined.

Applying the above procedure left us with a minimum disagreement
score for each of the fifteen pairs of reports in each segment.
These could be combined to give a total disagreement score per
segment, and segment means for the three conditions are shown in
Figure 4.

Figure 4.   Mean disagreement scores per segment in the three
             conditions.

condition:	Reversed	Filtered	Control (Untransformed)
score:	59.12	55.37	42.06

If we consider six fictitious reports of one syllable, the summed
disagreement score for the fifteen pairs ranges from zero (no
disagreements) to nine (three stresses and three syllables unstressed).   The average disagreement score is 5.   If Ss are
responding on a random basis, the average disagreement score per
syllable should be 5/15 = 0.33, and if the average number of
syllables in each pair is greater than one, the expected disagreement score per pair of reports under the null hypothesis of random stress assignment is found by multiplying the average syllable
length by the constant .33.

On this basis, the mean expected disagreement score per pair
under the null hypothesis was found for all the pairs in each
segment.   The differences between these and observed scores were
tested and compared against the "t" distribution.   It was found
that the null hypothesis could be rejected more often in the
control condition than in either of the experimental conditions
(see Figure 5).

Figure 5.   Results of tests for differences between observed
             disagreement scores and those expected under the null
             hypothesis.

condition:	Reversed	Filtered	Control
No. cases H_0 rejected:	1	2	12
No. cases of failure to reject :	15	14	4

$\underline{P}$ <.05, d.f. = 14, student's "t" distribution, 1 - tailed test.

Since the patterns for the segments in the experimental conditions
do not have internal consistency, there is no point in carrying the
investigation any further to examine cross-condition similarities.

This is a strange finding, since several Ss remarked that in
listening to the transformed stimuli, they were aided by using
a strategy whereby they picked out the stressed syllables first,
and added the unstressed syllables on subsequent repetitions. It
is also surprising, in that Lehiste (1970, Ch.4) summarised the
outcome of many experiments devoted to isolating the phonetic
correlates of perceived stress by saying it appeared that in all
studies, fundamental frequency provided relatively stronger cues
for the presence of stress than did intensity. Duration also
appeared to play a larger role than did intensity.

If indeed this information is sufficient for perceiving stress in
continuous speech, Ss should have experienced no difficulty in the
stress-assignment task, since this information was preserved in
the two experimental conditions.

A number of interpretations are compatible with the above data on
foot-structure patterning. One is that stress assignment is only
partly (if at all) given by a simple acoustic cue, being adequately
specified by the grammar (Fodor, Bever and Garrett, 1974). If
that were so, we would have predicted that the S would report
either no stresses or very few in the experimental conditions. Yet
correlations in terms of average number of stressed syllables reported per segment between all 3 conditions were high and significant (Figure 6).

Figure 6.   Intercorrelation matrix for stresses.

	untransf.	filtered
reversed	.854	.838
filtered	.950	

Another interpretation, which on close inspection may reduce to
the former, still concedes a major role to grammar, namely that
grammar enables the listener in the control condition to form a
higher-level (syntactic and semantic) representation of a segment, thereby facilitating performance (in ways which need yet
to be spelt out) in a task which is unnatural and difficult,
even after many repetitions of the stimulus. A third interpretation considers the possibility that there is a multiplicity
of acoustic cues underlying stress, of which few (if any) are in
a one-to-one relation with perceived stress. Further, as a
consequence of the listeners having diverse assumptions about the
task there might well have been little unanimity between Ss in
their decisions as to what acoustic cues constituted a stress,
in the experimental conditions. In the control condition, on the
other hand, one role of grammar may have been to bring about a
greater agreement in selection of cues, (in ways we have yet to

formulate).   All three of these interpretations assign some role
to grammar in the perception of stress.   A fourth interpretation,
which may also merit further investigation, considers that stress
may be encoded in the signal at the onset of the nuclear vowel
and more particularly, as a formant transition which makes its
appearance above 250 Hz, and that such a cue also requires presentation in the forward temporal direction.   In other words,
our experimental conditions may have eliminated normal cues for
the perception of stress in continuous speech.

Returning to our quest for invariants in the speech signal, the
experiment suggests that there may be acoustic correlates of
syllable segmentation and tone group boundary, which are invariant
over the signal transformations studied, and which enable perception of these features independently of grammar.   There are,
however, two considerations against inferring to the possible
role of acoustic invariants in the perception of speech from the
data gathered using our experimental paradigm.   Firstly, there
is the danger that the regularities our subjects picked up were
not the regularities they attend to normally in listening to
speech.   Secondly, that the cues which signaled regularities in
our transformed versions were not the cues which normally signal
these regularities.   It is for this reason that we call these
regularities "environmental invariants", following the work on
visual perception by J.J. Gibson (1966).   That is, there are
invariant properties of the stimulus sequences under both experimental transformations, but whether the human perceptual system
is equipped to pick them up (or "resonate" to them) in the normal
course of events, is quite a different matter.   To some extent
this question has, of course, been treated in our earlier work
(op.cit.).

We cannot as yet conclude that such invariants operate at the
level of stress assignment. According to our first interpretation,
this level of phonological organisation interacts in perception
with higher levels of grammar.   This is quite compatible with,
and reinforces, an interactive hypothesis discussed earlier.

References

Abercrombie, D. (1965). Studies in Phonetics and Linguistics.
    London:O.U.P.

Fodor, J.A., Bever, T.G. and Garrett, M.F. (1974). The Psycho-
    logy of Language. New York: McGraw-hill.

Gibson, J.J. (1966). The Senses Considered as Perceptual Systems.
    London: Allen & Unwin.

Halliday, M.A.K. (1967). Intonation and Grammar in British
    English. The Hague: Mouton.

Kirakowski, J. and Myers, T. (1975). The effect of intonation on message intelligibility. Paper presented at the Spring Meeting of the British Acoustical Society, Nottingham.

Laver, J. (1970). The production of speech. In Lyons (ed.) New Horizons in Linguistics. Middlesex: Penguin.

Lehiste, I. (1970). Suprasegmentals. Cambridge, Mass.: MIT Press.

Lieberman, P. (1965). On the acoustic basis of the perception of intonation by linguists. Word, 21, 40-54.

Rees, M. (1975). The domain of isochrony. Edinburgh University Department of Linguistics. Work in Progress, 8, 14-28.

IMPLICATIONS OF STUDYING REDUCED CONSONANT

CLUSTERS IN NORMAL AND ABNORMAL CHILD SPEECH

J. R. KORNFELD

University of Sussex

In observing the behaviour of both normal preschool children who simplify initial consonant clusters, and children with language disorders, we have noted that both groups:

1) Produce "unacceptable" obstruents in cluster-contexts (e.g. $[t^x{}_\wedge k]$ for truck), but "correct" segments in matched # CV-environments ($[t_\wedge k]$ for tuck);

2) Seem to distinguish more phonological contrasts than they reliably produce; and

3) Often do not accept adult imitations of their own speech as genuine imitations. (Kornfeld, 1971; Kornfeld and Goehl, 1974).

In order to account for these phenomena in a principled way, we have proposed that the children are re-analysing adult utterances in terms of their own phonological system. Their lexical feature matrices, which could serve as the common basis for speech production and recognition, would not necessarily be identical to mature underlying forms for cluster-words. These children could then be perceiving, representing, and producing distinctions that might be imperceptible to adult listeners whose abstract forms would differ. Similarly, the subjects themselves might not accept adult imitations as true copies, if the imitations failed to preserve the distinctions found in their own speech.

Empirical support for this explanation has been provided by the results of spectrographic studies, revealing non-adult acoustic regularities in the production of cluster-simplifying children. (Menyuk and Klatt, 1968; Hawkins, 1973). According to our position, illustrated in FIGURE 1, these acoustic patterns (in

"E" FIG.1) are the reflexes of distinctions in the child's abstract representations ("C").   These underlying forms in turn
capture the output of perceptual decisions the child has made
about the identity of segments in initial clusters ("B").   We
feel the child's choice of particular phonological features is
determined largely by the acoustic cues present in adult utterances
("A"), because of his still incomplete knowledge of the morpheme
structure constraints and phonological redundancies in his language ("F").

In arguing for this analysis, we first present a summary of recent
spectrographic and behavioural findings; then consider adult
utterances in terms of acoustic regularities that serve as cues to
the correct perceptual identification of initial consonants.
Finally, we discuss phonetic contexts where important cues are
neutralised or in conflict; and show that common patterns of
cluster simplification can be predicted in just these environments.

I.   Summary of recent evidence: production and perception data

To date, the strongest confirmation of the spectrographic findings reported a few years ago (Kornfeld, 1971) comes from a study
by G. Shafer (1974), involving the acoustic analysis of utterances
from groups of normal and language-delayed, cluster-cimplifying
children.   Her data include comparisons of voiced and voiceless
initial plosives and liquids in clusters, with productions of
these segments in non-cluster contexts;   i.e., $C_1C_2V_1$-C_1V_1 words
like drum-dumb, drawer-door, truck-tuck.   Detailed examination of
spectrograms of such pairs revealed regularly occurring, acoustic
patterns, which indicate her subjects' means of marking the second
segment (C_2) in CCV- words.   Of particular interest are the
effects of C_2 on C_1, the initial plosive.   These include the
following:

1.   Extended duration of C_1.

2.   Marked aspiration of C_1 (e.g., [t^hi:] for tree).

3.   Frequent replacement of the Cl plosive burst by aspiration,
     and/or transformation of C_1 into a fricative (e.g., [fi:]
     for tree).

More direct correlates of C_2 appeared in the formant patterns
between C_1 and V_1.   Notably, for # stop-/l/clusters, F_2 and F_3
formants showed the presence of an /l/-like segment: and for
#stop-/r/ clusters, shifting resonance in the F_1 region indicated
an /r/-like pattern. (Yet typically these acoustic /r/'s and
/l/'s were heard as /w/'s by adult listeners!)

The most general tendency of both subject groups in the Shafer
study was to change the spectral properties of C_1 by creating

turbulence or aspiration, especially in contexts where mature
speakers do not; namely, in #stop-liquid clusters. Moreover,
when the experimenter asked adults to imitate the children's productions, and then examined the acoustic properties of the imitations, she found that the adult renditions failed to preserve all
the acoustic cues present in the children's forms.

The upshot of this work is that cluster-simplifying children are
producing some non-adult distinctions. It is plausible that they
are also using these differences to discriminate among similar
sound sequences in their own speech, as our original model (FIG.1)
would allow. There is, in fact, behavioural evidence to this
effect. In 1974, Goehl and Golden at Temple University tested
language-delayed, cluster-simplifying subjects who produced
/l/'s and /r/'s as (what speech clinicians heard as) /w/'s. They
found that the disordered subjects performed just as well as
normals, on a discrimination test of adult-produced words beginning with /r/'s and /w/'s; and had significantly greater success in
recognising their own productions of /r/-words than did control
groups of normal children and adults.

Given these findings, as well as the production data cited earlier,
there is sufficient reason to pursue the theory that children may
use non-adult means to classify and produce sounds. If this is
the case, one would like to find out what determines the choice of
features that the child selects perceptually, and how he arrives
at a lexical form (that could serve as a basis for both speech
perception and production).

As far as recognition is concerned, it is reasonable to think that
children in the earliest stages of acquisition will be the most
dependent on acoustic detail. Older children and adults need not
rely so heavily on acoustically marked distinctions, since they
can appeal to their knowledge of the morphological, syntactic, and
semantic redundancies in the language to fill in information, which
the acoustic record fails to convey. The performance model proposed here is one of interaction: the content and extent of one's
knowledge of constraints in his language influence - if not direct -
the process of perceptual recognition of phones, both for the adult
and the child ("F" in FIG.1). In the next section, we therefore
turn our attention to the decisions that must be made for correct
identification of initial consonants; and consider the effects
of acoustic parameters in directing these decisions. Our prediction is that phonetic sequences having conflicting or neutralised values for the primary acoustic cues should force the adult to
appeal to higher levels of linguistic knowledge. The linguistic-
ally-naive child, on the other hand, should be led astray in such
contexts. He might well attend to acoustic properties that the
adult does not. He may further represent distinctive feature
correlates of these properties in his abstract forms, and produce

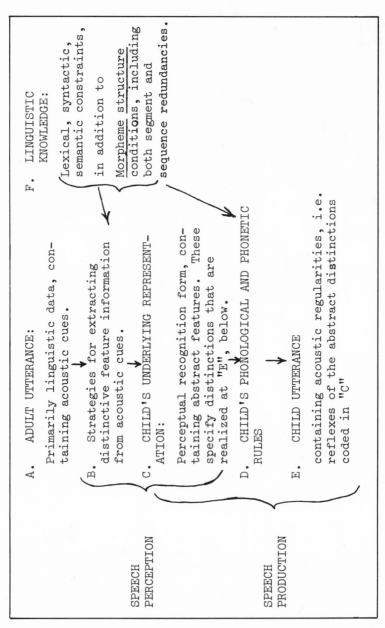

Figure 1.

non-adult distinctions in his phonetic outputs.

II. Acoustic cues in adult utterances

In order to identify initial clusters accurately, one must in principle locate the morpheme boundary relative to the segments; determine the number of segments in the cluster (1-3 in English); and
decide on the feature composition of each particular segment.
This decision making task is simplified by the high predictability
of morpheme structure. In English, if there are three consonants,
the first must be /s/, the second a **voiceless** plosive, and the
third /r/ or /l/.   These regularities can be captured by the
following morpheme structure condition: (After Schane, 1973).

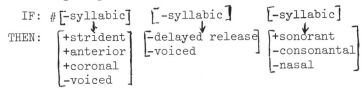

We will assume this is an adequate description of the listener's
knowledge for three-segment contexts, and that such information
is accessible during ongoing perception.

For two-segment clusters not involving the initial /s/, the identity of the plosive is less constrained.   It can be either voiced
or voiceless (drip vs. trip, for instance).   The listener in this
case must make a decision about the values of the voicing and place
features. For purposes of this discussion, the focus will be on
the voicing distinction, mainly because that contrast is more
suspect to disruption in child speech than is the place aspect,
and because voicing contrasts have been more extensively studied
in the acoustic phonetic literature.

II.a Acoustic correlates of voicing

The parameter receiving the most attention recently has been
voiced-onset-time (VOT), defined roughly as the interval between
release of the stop occlusion and the onset of voicing. VOT
differences have been shown to be effective correlates of voicing
contrasts for initial stops, in terms of both spectrographic data
and perceptual tests with synthesized speech, across a number of
unrelated languages.   In brief, the results of a series of studies
by Lisker and Abramson (1965, 1967, 1970) show that for English
listeners, VOT less than 25-40 msec (known as "short lag") correlate
with judgments of a voiced stop; whereas VOT greater than 25-40
msec. ("long lag") cue the presence of a voiceless stop.

Actually, the situation is much more complex.   There is an ongoing
debate about whether it is the absolute value of VOT that is res-

ponsible for the reported categorisations. The issue arises primarily in situations where two otherwise distinctive plosives have
overlapping VOT values; e.g., the bilabial stops in the English
phrases, "this boy" and "the spy"; or the TYPE II and TYPE III
stops in Korean (Kim, 1965; Han and Weitzman, 1970). Attention to
such cases, where canonical VOT differences are neutralized, has
revealed a host of other variables that signal voicing distinctions,
ranging from presence or absence of F_1-transitions following voicing onset, to the direction of the fundamental frequency contour.
The best way to understand the effects of these factors is to
realize that VOT is literally a time interval, resulting from
changes in glottal width, tension, and pressure (both above and
across the glottal opening). And it's the reflex of these
laryngeal factors that gives rise to changes in particular acoustic
parameters.

There are at least six distinguishable acoustic patterns associated
with judgments of voicelessness that can be traced to the glottal
changes mentioned above. These include the following:

1. Greater separation time per se between burst release and the
    first glottal pulse. (Due to the time it takes the stiffened
    vocal cords to relax and allow vibration to occur. (Halle and
    Stevens, 1971.))

2. Absence of rising F_1 transition: no rapid spectral change
    immediately following voicing onset. (Due to the fact that
    movements of supraglottal articulators are completed before
    the vocal cords get in an appropriate position to vibrate.
    (Stevens and Klatt, 1974.))

3. Higher onset frequency of F_1. (Lisker, 1974)

4. Presence of aspiration, manifested by an aperiodic glottal
    waveform and accompanied by the apparent disappearance of F_1.
    (Due to the glottis being spread, and the vocal cords stiff.
    (Halle and Stevens, 1971; Klatt, 1973.))

5. Greater intensity and duration of the burst phase, which would
    be perceived as increased loudness. (Due to more air flow,
    made possible when the glottis is spread; also due to the
    greater differences in pressure across the glottis. (Umeda and
    Coker, 1974.))

6. High pitch on adjacent vowels, and pitch skip immediately after
    voicing onset. (Due perhaps to overall stiffening of the phar-
    ynx as well as the vocal cords. (Halle and Stevens, 1971;
    Hardcastle, 1973.))

Although phoneticians have demonstrated that virtually each of

these acoustic dimensions is capable of cueing a voiceless judgment for an initial stop in #CV- contexts, there still remains some
doubt about the relative primacy of these cues. As a working
definition, we will adopt Klatt's (1973) statement concerning
primacy: "... the most reasonable candidates for primacy cues are
those whose presence would override conflicting values in other
parameters."

In #CV-contexts, both VOT and F_1-transition information would
appear to meet these criteria. But in more complex syllables,
notably two- and three-segment clusters, values of even these parameters may be neutralised. And if listeners, in particular
children, base their decisions on the values of VOT and F_1-trans-
ition duration, one might expect erroneous judgments on voicing and
possibly on other features in cluster contexts. In the following
section, we discuss some striking examples, in which values for
the proposed primary cues deviate from their canonical #CV-norms.

II.b Cases of perceptual difficulty

One such example arises from the influence of the liquid in #stop-
/r/ clusters. Klatt (1973) and Davidsen-Nielsen (1974) have independently reported that VOT for a voiceless initial consonant
before /r/ is increased by 30-50 msec. over its value in a #CV-
syllable. The way that the /r/ affects the VOT value of the
stop is by lengthening the F_1-transition duration, which effectively lowers the F_1-onset frequency; and both of these changes are
cues to a voiced plosive. Greater VOT than normal is apparently
needed to perceive the segment as voiceless. Now, if a child is
using F_1-transition data as a primary cue, he may indeed analyse
voiceless stop-liquid clusters as voiced. Forms such as those
below turn out to be quite frequent among cluster-simplifying
children, and suggest that they do base their perceptions on VOT
and transition information.

E.g., [gak] for truck
       [gisdan] for Kristen
(Ingram, 1971)

       [dau] for trowel
       [gai-iŋ] for crying
(Smith, 1973)

Another possibility in voiceless stop-liquid clusters is that the
delayed onset of voicing may be accompanied by turbulence prior to
the first glottal pulse on the following vowel. This is especially true of the homorganic cluster, #/tr-/, as is argued by
Haggard (1973), Klatt (1973), and Hawkins (1973). All report
tendencies for the stop burst to become affricated in adult
utterances, due to an open glottis which allows greater airflow,

hence greater acoustic intensity, and perceived loudness. If a
child turned to turbulence and loudness as cues when VOT failed,
he might re-analyse an adult #/tr-/ cluster as an affricate or
as a single initial /t/, marked by heavy aspiration. In his productions we would expect things like [tʰi:] for tree, [pʰIti] for
pretty, [kʰi:m] for cream, etc.   Such forms are again borne out
in both the transcription and spectrographic data cited earlier.

Additional evidence that children may construe turbulence as an
important perceptual cue comes from a curious result of Charles
Read's (1975) study, involving the analysis of spontaneously
invented spelling systems by preschool children. He found that
many #/tr-/ words were represented as starting with /č/, the
corresponding affricate. Furthermore, in perceptual testing of
4-6 year old children, he discovered that some subjects felt that
the /t/ in #/tr-/ words was more similar to /č/ than to the
first sound of other words beginning with a "non-cluster" /t/.

Still a different special case is the velar cluster, #/gr-/, where
the VOT value of the stop goes into the nominally voiceless category.   (Klatt, 1973.)   So if a child depended on canonical VOT
values,he might perceive the stop as voiceless; and may in addition, mark it as aspirated, due to the increased interval preceding
voicing onset.   Thus, grape might be rendered as [kʰeip], and
green as [kʰi:n], which again are common types of child outputs.

Finally, consider the case of the nominally voiceless unaspirated
stop in clusters beginning with /s/, such as in spoon, station,
scream, etc.   This environment would appear to be the most
serious example of conflict, because both VOT and F_1-transition
cues for voicing go in the wrong direction; i.e. signal a voiced
segment. (Klatt, 1973; Davidsen-Nielsen, 1974.)   Moreover, the
so-called secondary cues of burst intensity and duration would
likewise suggest a voiced stop. (Umeda and Coker, 1974.)   These
observations are supported by perceptual tests, in which the /s/
segment is removed from natural speech, and the resulting stimuli
presented to listeners for identification.   Davidsen-Nielsen
(1969) reports over 92% of the responses fell into the voiced
category.

In the face of these results, one wonders how adults manage to
correctly perceive words beginning with three segments. Klatt
(1973) maintains that the other cues - aspiration, F_0 shift, etc. -
would take over in this phonetic environment. But a more realistic
answer is that the adult's knowledge of the extremely limiting
morpheme structure constraints comes to the rescue.   It's no
accident that in exactly the phonetic contexts where primary
acoustic cues are at odds, higher level linguistic information
steps in to direct perceptual recognition.

Since young children might not yet have mastered such linguistic
regularities and could not therefore appeal to them to resolve
conflicts in three-segment clusters, we would predict more non-
adult simplification in these environments than in the other cluster contexts, chiefly in terms of mis-perceived voicing of the
plosive; to wit:

[bun] for spoon

[geip] for scrape

[dʰaik] for strike

Indeed, the high frequency of forms like these does seem to be
borne out in the literature. (Cf. Smith, 1973).

One last word of qualification.   Children will be expected to
differ, with respect to patterns in their cluster simplifications,
depending on which acoustic features they take as primary, and
how far along they are in learning the sequence redundancies of
their language. A child may first be observed to say [bun] for
spoon; and only later, [pʰu:n], showing that he has progressed in
his knowledge that aspiration automatically marks initial voiceless consonants.   (These progressions are quite neatly documented
under the #/s/- section of Appendix C, Smith, 1973.)

These last examples bring us back to the original model (FIG.1),
which is intended to reflect ways that the influence of linguistic
knowledge can affect feature-choices in both the perception and
production aspects of children's language behaviour. It is hoped
that further research on children's use of acoustic information
will advance our understanding of these processes.

References

Davidsen-Nielsen, N. (1969). English stops after initial /s/.
  Eng. Studies. 50, 4, 321-339.

Davidsen-Nielsen, N. (1974). Syllabification in English words.
  J. of Phonetics, 2, 15-45.

Goehl, H. and Golden, S. (1974). A psycholinguistic account of
  why children do not detect their own errors. Unpublished MS,
  Dept. of Speech, Temple University, Phil. Penn.

Haggard, M. (1973). Abbreviation of consonants in English pre-
  and post-vocalic clusters. J. of Phonetics, 1, 1, 1-8.

Halle, M. and Stevens, K.N. (1971). A note on laryngeal features.
  M.I.T., R.L.E., Q.P.R., 101, 198-213.

Han, M.S. and Weitzman, R.S. (1970). Acoustic features of Korean
  /P,T,K/, /p,t,k/, and /pʰ,tʰ,kʰ/. Phonetica, 22, 112-128

Hardcastle, W.J. (1973). Some observations on the tense-lax distinction in initial stops in Korean. J. of Phonetics, 1, 263-272.

Hawkins, S. (1973). Temporal coordination of consonants in the speech of children. J. of Phonetics, 1, 181-217.

Ingram, D. (1971). Phonological rules in young children. Papers and Reports on Child Lang. Develop. 3, Committee on Ling., Stanford U., Stanford, Calif., 31-50.

Kim, C.W. (1970). A theory of aspiration. Phonetica, 21, 107-116.

Klatt, D. (1973). Voice onset time, frication, and aspiration in word-initial consonant clusters. M.I.T., R.L.E., Q.P.R., 109, 124-135.

Kornfeld, J.R. (1971). Theoretical issues in child phonology. CLS 7 (Proc. 7th Mtg. Chicago Ling. Soc.) U. of Chicago, 454-468.

Kornfeld, J.R. and Goehl, H. (1974). A new twist to an old observation: kids know more than they say. CLS 10, Proc. of the Parasession on Natural Phonology. U. of Chicago, 210-219.

Lisker, L. (1974). Is it VOT or a first-formant transition detector? Unpub. MS, from 1974 Mtg. of the Am. Assoc. Phon. Sci., St. Louis, Mo.
Lisker, L. and Abramson, A. (1965). Voice onset time in stop consonants: acoustic analysis and synthesis. Proc. 5th Inst'l. Cong. Acoust., Liège, 1-4.

Lisker, L. and Abramson, A. (1967). Some effects of context on voice onset time in English stops. Language and Speech, 10, 1-28.

Lisker, L. and Abramson, A. (1970). The voicing dimension: some experiments in comparative phonetics. Proc. 6th Inst'l. Congr. Phon. Sci., Prague, 1967, 563-567.

Menyuk, P. and Klatt, D. (1968). Child's production of initial consonant clusters. M.I.T., R.L.E., Q.P.R., 91, 205-213.

Read, C. (1975). Children's categorization of speech sounds in English. N.C.T.E. Res. Reports, 171, Urbana, Ill.

Schane, S. (1973). Generative phonology. New Jersey: Prentice-Hall.

Shafer, G. (1974). A spectrographic investigation of acoustic cues in reduced consonant clusters in normal and abnormal child speech. Dept. of Speech, U. of Witwatersrand, Johannesburg.

Smith, N.V. (1973). The acquisition of phonology. Cambridge: C.U.P.

Stevens, K.N. and Klatt, D. (1974). Role of formant transitions
  in the voiced-voiceless distinction for stops. JASA, 55, 3,
  653-659.

Umeda, N. and Coker, C.H. (1974). Allophonic variation in Am.
  Eng. J. of Phonetics, 2, 1-5.

CONTRIBUTORS TO VOLUME 2
(Senior Authors Only)

Baker, Robert G. University of Stirling, Department of Psychology,
    Stirling FK9 4LA, Scotland.

Bartsch, Renate. Universiteit van Amsterdam, Central Interfaculteit,
    Roeterstraat 15, 8e en 9e verdieping, Amsterdam, The Netherlands.

Booth, D. A. University of Birmingham, Department of Psychology,
    Elms Road, P. O. Box 363, Birmingham, B15 2TT, England.

Butterworth, Brian. University of Cambridge, Psychological Labor-
    atory, Downing Street, Cambridge CB2 3EB, England.

Derwing, Bruce L. University of Alberta, Department of Linguistics,
    Edmonton, Alberta T6G 2G1, Canada.

Foss, Donald J. University of Texas at Austin, Department of Psy-
    chology, Mezes Hall 330, Austin, Texas 78712, U. S. A.

Garrod, Simon. University of Glasgow, Department of Psychology,
    Adam Smith Building, Glasgow G12 8RT, Scotland.

Kempen, Gerard. University of Nijmegen, Department of Psychology,
    Erasmuslaan 16, Nijmegen, The Netherlands.

Kirakowski, J. University of Edinburgh, Department of Psychology,
    60 Pleasance, Edinburgh EH8 9TJ, Scotland.

Kronfeld, Judith R. University of Sussex, Laboratory of Experimental
    Psychology, Falmer, Brighton BN1 9QG, England.

Labov, William. University of Pennsylvania, Department of Linguistics,
    3812 Walnut/B2, Philadelphia, Pennsylvania 19174, U. S. A.

Levelt, W. J. M. University of Nijmegen, Department of Psychology,
    Erasmuslaan 16, Nijmegen, The Netherlands.

Longuet-Higgins, H. C. University of Sussex, Laboratory of Experimental Psychology, Centre for Research on Perception and Cognition, Falmer, Brighton BN1 9OG, England.

Markova, Ivana. University of Stirling, Department of Psychology,
   Stirling FK9 4LA, Scotland.

Miller, J. University of Edinburgh, Department of Linguistics,
   15 Buccleuch Place, Edinburgh EH8 9LN, Scotland.

Myerson, Rosemarie F. Harvard University, Graduate School of
   Education, Royal E. Larsen Hall, Appian Way, Cambridge,
   Massachusetts 02138, U. S. A.

Noordman, Leo. Rijksuniversiteit te Groningen, Instituut voor
   Experimentele Psychologie, p a Biologisch Centrum, Vleugel D.,
   Kerklaan 30, Haren, The Netherlands.

Palermo, David S. Pennsylvania State University, Department of
   Psychology, 441 Moore Building, University Park, Pennsylvania
   16802, U. S. A.

Reeker, Larry H. University of Arizona, Department of Computer
   Science, Tucson, Arizona 85721, U. S. A.

Rochester, S. R. Clarke Institute of Psychiatry, 250 College Street,
   Toronto M5T 1R8, Canada.

Schank, Roger. Yale University, Computer Science Department, 10
   Hillhouse Avenue, New Haven, Connecticut 06920, U. S. A.

Steedman, Mark J. University of Warwick, Department of Psychology,
   Coventry CV4 7AL, England.

Wexler, Kenneth. University of California at Irvine, School of
   Social Sciences, Irvine, California 92664, U. S. A.

AUTHOR INDEX

Abelson, R.P., 93
Abercrombie, D., 403
Abramson, A., 417
Anderson, D.B., 181
Anderson, J.M., 257,248
Anderson, J.R., 74
Andrew, C.M., 213
Argyle, M., 347
Aronoff, M.H., 378,398

Baird, R., 321
Baker, R.G., 364
Baker, W.J., 194,198,208,212
Baldwin, J.M., 108
Baldwin, P., 228
Baldwin, T., 342
Ballmer, T., 266,267,269
Bartsch, R., 206,265,268,278,
  280,287
Baxter, J.C., 347
Baylor, G.W., 228
Beattie, G.W., 349
Becker, J., 104
Benedict, H., 227
Benveniste, E., 254
Bever, T.G., 103,119,173,177,
  197,206,207,212,244,246, 247,
  255,260,410
Bickerton, D., 41
Bierwisch, M., 47,164
Bloom, L., 38,212,235
Bloomfield, L., 11,194,195,250,
  257
Blumenthal, A.L., 193,213,247
Bock, J.K., 108,109
Bogoyavlenskiy, D.N., 378
Bolinger, D., 256,258
Boomer, D.S., 105

Booth, D.A., 239
Botha, R.P., 197
Bower, G.H., 74
Brady, K.D., 347
Braine, M.D.S., 59, 245
Bransford, J.D., 45,46,209
Bratley, P., 173
Bresnan, J., 189
Brown, H.D., 321,322
Brown, R.W., 5,18,35,61,211,
  243
Bruner, J.S., 236
Burstall, R.M., 181
Butter, R.R., 41
Butterworth, B., 347,349

Cairns, H.S., 327
Campbell, R.N., 228,235,258
Card, S.K., 298
Catlin, J., 51
Cedergren, H., 15
Chafe, W.L., 289
Charniak, E., 91
Chase, W.G., 156,298
Chomsky, N., 11,55,56,67,87,
  125,195,207,211,212,244,
  251,254,289,321,362,377
Clark, E.V., 47,228
Clark, H.H., 156,178,289,295,
  298,310,341
Clifton, C., 212
Cofer, C.N., 198
Cohen, L.B., 50
Cohen, P., 5,40,41
Coker, C.H., 418,420
Collins, J.S., 181
Collitz, K.H., 138
Culicover, P., 61,62,63
Cullingford, R.E., 96

427

SUBJECT INDEX

Acquisition
  cognitive constraints on,55-68
  of linguistic competence,236-8
  of semantic and surface
    structure, 71-89
  of semantic prototypes,48-53
  of syntax, 60-8
  of WH-questions, 1-43
Augmented Transition Networks,
  172-8

Clauses
  relative, 86-9,263-87,321-32
  structure in discourse,131-2
Colour, 51
Comprehension
  in infants, 221-31
  in the pre-schooler,315-32
  of prepositions,228-9
Computer
  model of discourse production,
    125-35
  model of language acquisition,
    71-89
  programs for language
    comprehension,91-101,171-89
Computer Programming Languages
  PICO-PLANNER,181
  PLANNER, 181-3
  POP-2,181
Computer Programs
  Cedergren-Sankoff, 14-8
  PAM,100
  PST (Instance 3),71-89
  SAM, 93,96-7
Consonant clusters
  in child speech, 413-21
  in stress assignment rules,
    364,371,373-4

Constructions
  endocentric and exocentric,
    249-51
Contrast
  Given-New,178-80,289-302,
    311-4
  Marked-Unmarked,298

Dependency
  semantic,74
  syntactic,112

Embedding, 66,321

Features
  semantic, 45-8,291-302
Formants, 414
Frames, 308-14
Frequency
  conjoint,306-8,311-4

Gesture
  in speech production,347-8
    351-9
Grammar
  case,51,186,211,226
  generative,4,243,248-60
  Montague,263,270,274
  systemic,127

Kinship terms,291-302

Languages
  Black English Vernacular,39
  Classical Greek,259
  Dutch,114,116,139-60,291-302
  English,1-43,62,120,122,186-9,
    207-9,221-7,287,305-421
  Finno-Ugric, 258